# Research Techniques in Animal Ecology

**Controversies and Consequences**

LUIGI BOITANI AND TODD K. FULLER

*Editors*

COLUMBIA UNIVERSITY PRESS

NEW YORK

Columbia University Press
Publishers Since 1893
New York    Chichester, West Sussex

Copyright © 2000 by Columbia University Press
All rights reserved
Library of Congress Cataloging-in-Publication Data

Research techniques in animal ecology : controversies and consequences / Luigi Boitani
and Todd K. Fuller, editors.
    p. cm.  —  (Methods and cases in conservation science)
    Includes bibliographical references (p. ).
    ISBN 0–231–11340–4 (cloth : alk. paper)—ISBN 0–231–11341–2 (paper : alk. paper)
    1. Animal ecology—Research—Methodology.    I. Boitani, Luigi.    II. Fuller, T. K.    III.
Series.

QH541.2.R47 2000
591.7′07′2—dc21                                                            99–052230

*For*
*Stefania and Caterina*
*and*
*Susan and Mollie*
*for their patience, love, and support*

# Contents

## Chapter 3: Animal Home Ranges and Territories and Home Range Estimators
*Roger A. Powell*                                                           65

## Chapter 4: Delusions in Habitat Evaluation: Measuring Use, Selection, and Importance
*David L. Garshelis*                                                       111

## Chapter 7: Monitoring Populations
*James P. Gibbs*                                                      213

## Chapter 8: Modeling Predator–Prey Dynamics
*Mark S. Boyce*                                                       253

**Chapter 11: Modeling Species Distribution with GIS**
*Fabio Corsi, Jan de Leeuw, and Andrew K. Skidmore*                389

Luigi Boitani
Dipartimento Biologia Animale dell'Uomo
Università di Roma "La Sapienza"
Viale Università 32
00185 Rome, Italy

Mark S. Boyce
Department of Biological Sciences
University of Alberta
Edmonton, Alberta T6G 2E9, Canada

Fabio Corsi
Institute of Applied Ecology (IAE)
Via L. Spallanzani 32
00161 Rome, Italy

Joseph S. Elkinton
Department of Entomology and Graduate Program in Organismic
    and Evolutionary Biology
University of Massachusetts
Amherst, MA 01003, USA

Mark R. Fuller
USGS Forest and Rangeland Ecosystem Science Center
Snake River Field Station
    and Boise State University
970 Lusk Street
Boise, ID 83706, USA


Todd K. Fuller
Department of Natural Resources Conservation and Graduate
    Program in Organismic and Evolutionary Biology
University of Massachusetts
Amherst, MA 01003-4210, USA


David L. Garshelis
Minnesota Department of Natural Resources
1201 E. Highway 2
Grand Rapids, MN 55744, USA


James P. Gibbs
State University of New York
College of Environmental Science and Forestry
Faculty of Environmental and Forest Biology
350 Illick Hall, 1 Forestry Drive
Syracuse, NY 13210, USA


Charles J. Krebs
Department of Zoology
University of British Columbia
6270 University Blvd.
Vancouver, BC V6T 1Z4, Canada


Jan de Leeuw
Division of Agriculture, Conservation, and the Environment
International Institute for Aerospace Survey
P.O. Box 6
7500 AA Enschede, The Netherlands

John A. Litvaitis
Department of Natural Resources
University of New Hampshire
Durham, NH 03824, USA

David W. Macdonald
Wildlife Conservation Research Unit
Department of Zoology
University of Oxford
South Parks Road
Oxford OX1 3PS, UK

Dennis L. Murray
Department of Fish and Wildlife Resources
University of Idaho
Moscow, ID 83844, USA

Roger A. Powell
Department of Zoology
North Carolina State University
Raleigh, NC 27695-7617, USA

Andrew Skidmore
Division of Agriculture, Conservation, and the Environment
International Institute for Aerospace Survey
P.O. Box 6
7500 AA Enschede, The Netherlands

Paul D. Stewart
Wildlife Conservation Research Unit
Department of Zoology
University of Oxford
South Parks Road
Oxford OX1 3PS, UK

Pavel Stopka
Wildlife Conservation Research Unit
Department of Zoology
University of Oxford
South Parks Road
Oxford OX1 3PS, UK

Gary C. White
Department of Fishery and Wildlife Biology
Colorado State University
Fort Collins, CO 80523, USA

Nobuyuki Yamaguchi
Wildlife Conservation Research Unit
Department of Zoology
University of Oxford
South Parks Road
Oxford OX1 3PS, UK

# List of Illustrations

# List of Tables

As science, ecology is often accused of being weak because of its basic lack of predictive power (Peters 1991) and the many ecological concepts judged vague or tautological (Shrader-Frechette and McCoy 1993). Also, important paradigms that dominated the ecological scene for years have been discarded in favor of new concepts and theories that swamp the most recent ecological literature (e.g., the abandoning of the island biogeography theory in favor of the metapopulations theory; Hanski and Simberloff 1997). The apparent ease with which such changes seem to be accepted could be taken as an intrinsic weakness of ecological disciplines; in fact, many ecologists seem to have an inferiority complex with respect to sciences considered more rigorous, such as physics or chemistry. Thus, when ecology has to provide the basis for environmental conservation and management, this presumed weakness is easily instrumentalized by those opposing conservation. In the often sterile debates that are heard, ecology loses credibility and is easily victimized by its detractors.

It is not surprising that many ecological theories and concepts have still not been defined precisely, given the enormous complexity of ecological systems. Yet ecology is rooted in the scientific method applied to the observation and experimentation of natural facts. Rather than a discipline whose experimental practice is informed by laws and invincible paradigms, ecology is a classically bottom-up discipline in which the application of the scientific method to real facts and processes gradually builds a body of knowledge that can give rise to useful generalizations. But the complexity of ecological processes and their variability is such that any generalization conflicts with the need to account for all possible variations. It is in this light that the rigor of the results achieved in the study of real cases takes on fundamental value. Without embracing such radically critical positions as those summarized by Shrader-Frechette and McCoy (1993), we nevertheless feel that ecology, like any other discipline in the natural sciences, can only benefit from the steadily growing scientific rigor in the study of real cases.

Animal ecology, in particular, is the field in which we should strive for more scrupulous application of a scientifically rigorous methodology. Animal populations are mobile in space, they have a strong stochastic demographic component, they are involved in complex interspecific and intraspecific interactions and interactions with the abiotic environment, and they have a great environmental variance. Thus it has been more difficult to apply scientific approaches and rigorous experimental designs to them than in other scientific endeavors. Nonetheless, there is no good justification for studying animal populations without greater discipline.

These intrinsic difficulties in studying animal ecology underlie many of the weaknesses in the research methodologies available to researchers today. Certainly the quality of the research is sometimes limited by logistic and environmental adversities, by the problems of translating into practice an experimental design worked out at the drawing board, by deliberately limited samples, and by other problems that can contribute to weakening the methodological rigor of a study and therefore the validity of its results. As the methods and results of animal ecology are often applied to conservation, the practical consequences of misused techniques can mislead the implementation of conservation measures. For many species, such mistakes can have serious consequences.

This book springs from the recurring frustration we, the editors, sometimes have felt while doing our work as researchers and teachers. The scientific ecological literature (as well as a good bit of other literature) is full of publications based on false assumptions and methodological errors. Although the number of methodological errors and omissions seems to be inversely and exponentially proportional to a journal's quality, even the most scrupulous editors of the best scientific journals sometimes miss mistakes. Although the most circumspect researchers have the critical ability to recognize and respond to the errors, often they do not respond, and such critique is almost totally absent among students. Teaching students how to be critical is perhaps the most difficult and most noble objective of the teaching profession, but there has never been a text in the field of animal ecology to help us in this task. Excellent handbooks and textbooks of techniques and methods are available (e.g., Krebs 1999; Bookhout 1994) in which the techniques are well described and examples are used to illustrate when and how to apply them. Many of these techniques are well known and robust in their applications. However, several require assumptions and procedures that are not always accounted for. Conceptual limitations and methodological constraints are not often discussed in the scientific literature, and currently there are no other books from which one can learn a critical approach to use of the wide variety of methods and techniques in animal ecology.

The main purpose of this book, therefore, is to present some of the more common issues and research techniques used in animal ecology, identify their limitations and most common misuses, provide possible solutions, and address the most interesting new perspectives on how best to analyze and interpret data collected in a variety of research areas. It is not a handbook of techniques; rather, it is designed as a backup for existing handbooks, providing a critical perspective on the most common topics and techniques.

Such a critical review of methodologies is rare in animal ecology. Historically, a few individual papers have denounced misused techniques, and such papers are still cited today. Others have had to be published several times before the scientific community has taken notice. In recent years, individual papers have been discussed in some journals via a comment and reply format, and these "conversations" are among the most interesting parts of those publications. Several summarizing monographs or books have been published recently that critically address or review major topics (e.g., radiotelemetry, population estimation, survival analyses), but no single volume has presented a whole range of topics relevant to animal ecology.

In the course of the last 20 years of teaching, research, and editing, we have become increasingly convinced of the need for a book like this, with its critical look at how ecological research is conducted and interpreted, and we hope it will provide insight and reassurance for the research community. Furthermore, we hope the book, by specifically investigating the many ways in which research techniques are incorrectly applied, will contribute to increasing the consistency and reliability of the scientific method in ecology and conservation.

The book includes the topics that are most frequently reported in the scientific literature in ecology and conservation, but rarely critically reviewed in a comprehensive manner. We are aware that several other topics and extensive treatment of taxa other than vertebrates could have been included if there had been no limitations on size and readability. We prepared a priority scale of topics based on the relevance of the issue, the lack of good available critical review, the availability of outstanding contributors, and the amount of controversy and misuse found on each topic. The resulting choice is obviously subjective and can be criticized, as every scientist has his or her preferences and perspectives. However, we are confident that the book will address new topics of interest to a large proportion of researchers in animal ecology.

Each chapter explores and develops a different topic and includes an extensive review of published material and a summary of the state of knowledge on that particular topic. Techniques are usually described only briefly because the intent is to point out the underlying assumptions and constraints of the techniques and indicate ways to avoid the most common pitfalls that await us.

In the first chapter Charles Krebs presents the philosophical groundwork concerning hypotheses. He then discusses how this concept is translated in scientific studies into testable hypotheses, and then into statistical hypotheses and all of the attending problems that the simple idea of null hypotheses raises. He then explores the practical problems of hypothesis testing in ecology. Despite the fact that most ecologists and students in ecology think that good hypothesis development is self-evident to any rational person, Krebs makes a convincing case that the intellectual baggage of assumptions we all carry ought to be questioned seriously.

Marking individual animals is often a prerequisite of many research designs in animal ecology. Although most ecologists are aware that some markers may affect an animal's life history, this topic is rarely addressed in presenting research results. In chapter 2, Dennis L. Murray and Mark R. Fuller review the effects of markers on various aspects of life history, particularly on movements and energetics, and on survival and population estimation. They provide useful information on methodological or analytical modifications used to minimize the effects of markers and suggest lines of research to more fully evaluate the effects of markers on various vertebrate taxa.

The concept of home range is central to much of the animal distribution and abundance literature, and home range descriptors have received much critical attention. Nevertheless, assumptions and caveats often are ignored, especially when the most modern techniques are used. Whereas the methodological literature appears to cover extensively all critical aspects of this topic, the literature concerning the use of these methods does not reflect the same level of attention. Roger A. Powell, in chapter 3, analyzes old and recent pitfalls of home range and territory concepts and methods, and suggests the most reliable approaches for each research theme.

The evaluation of habitat use by an animal either for use, preference, and selection studies or for suitability analyses is also a theme that is found easily in any current issue of the most important journals in animal ecology. However, the topic is full of delusions, as explained by David L. Garshelis in chapter 4. There are problems in defining and measuring habitats, measuring what is really available to an animal, and assessing whether and what selection is eventually made by an individual. Adequately addressing the assumptions that form the basis of habitat selection hypotheses proves to be a formidable research design task. Equally challenging are problems with assessing habitat quality, including the basic concept of optimal habitat and the sometimes false paradigm that the best habitat always supports higher animal densities.

In chapter 5 John A. Litvaitis summarizes the current approaches and describes the most recent innovations to investigating food habits and diets. The limitations of each technique are discussed but the emphasis is on the interpretation of the results provided by these techniques. A number of fundamental assumptions are neglected far too often when extrapolating individual results to whole populations, and inadequate consideration of the spatiotemporal variance of populations is common. Litvaitis also suggests framing habitat and food use studies within an integrated approach and shows the potential of foraging theory as an aid in understanding variation in food habits.

Detection of time series of density and survival is the focus of chapter 6, by Joseph S. Elkinton. Understanding the mechanism by which population dynamics develop is of paramount importance for conservation and management, and this chapter discusses the use of density and mortality data to deduce population changes and their causes. Density dependence is an especially important parameter that is difficult to isolate from correlated factors, and Elkinton explores the statistical limitations of research design in detecting different types of density dependence.

Population monitoring is a key topic in animal ecology and in most wildlife conservation activities. However, James P. Gibbs, in chapter 7, shows that the validity of the chosen population index is rarely assessed properly and the design of a monitoring program usually is not adequate to permit a reasonable chance of detecting a trend or change. Gibbs discusses the many weaknesses and limitations of population indexes and shows how imprecise population indices often combine with inadequate study design (often imposed by logistical constraints) to severely constrain the statistical power of population-monitoring programs. After a thorough examination of the most common pitfalls of population monitoring, Gibbs points out the possible solutions. The goals set out clearly before the initiation of any monitoring program should, at a minimum, address the magnitude of change in the population index that must be detected, what probability of false detections is to be tolerated, and what frequency of failed detections is acceptable.

In chapter 8, Mark S. Boyce presents various types of predator–prey models used in ecological research and discusses the criteria by which a model is found to be good and useful. He identifies the conceptual limitations and practical constraints of old and new approaches, whether from the Lotka–Volterra model or recent structured population models. Boyce carefully analyzes the ways model are or can be validated, a necessary step in making a useful model, and he develops the need for adaptive management, where models play a role that is strictly integrated into the monitoring of model predictions.

Population viability analyses (PVAS) have become one of the most popular techniques used to assess conservation options for small populations. Several tools have been developed to carry out such analyses, but despite their great importance in conservation biology, Gary C. White, in chapter 9, discusses why the current techniques are largely unsatisfactory. He identifies the weaknesses of most estimates of population viability and points out the basic failures of most models: their inability to account for individual variation within the population and for life-long individual heterogeneity. White also explores other aspects of current PVA methods and shows that, as they stand, they are often useless for conservation purposes. White's critical approach is a powerful warning against the use of PVA results for practical conservation, but also shows the potential role of improved PVA models as research tools for understanding the dynamics of small populations.

Ethological aspects underlie many ecological studies of animals, and even though the two disciplines refer to two different theoretical and methodological frameworks, ecologists must become familiar with behavioral methods. In chapter 10, David W. Macdonald, Paul D. Stewart, Pavel Stopka, and Nobuyuki Yamaguchi provide a short guide to the main problems of measuring the dynamics of mammal societies. The greater emphasis is on social behavior, with particular attention to the many new concepts in behavioral ecology, together with the refinement of sequential statistical techniques and, very importantly, the development of many software packages to facilitate the description of social dynamics. The chapter develops the identification of the social parameters that one might choose to define the social dynamics of mammal societies, the description of the methods used to record the most important parameters, and an introduction to the style of quantitative ethological analyses currently in vogue (e.g., lag sequential analysis and multiple-matrix analysis). The chapter ends by proposing a new conceptual framework for interpreting data and asking whether parallels in the development of ecological communities and animal societies are merely analogies or evidence of similar underlying processes.

The final chapter, by Fabio Corsi, Jan de Leeuw, and Andrew Skidmore, presents state-of-the-art uses of geographic information systems (GISS) in the study of species distribution. Although the GIS is a fairly new and attractive tool that can produce a completely new set of results unavailable until few years ago, the authors warn against many conceptual limitations and potential sources of error. In particular, the chapter analyzes the growth and misuse of the concept of habitat, with its many different meanings in biological and mapping sciences; these include habitat as a multidimensional species-specific property and habitat as a Cartesian property of land. The authors discuss the accuracy of spatial wildlife habitat models, the dichotomy of inductive versus deductive modeling, and the problem of transferability of models in space and time. Finally, they warn us of the fundamental problem of scale dependency of the habitat factors and provide a set of procedures on error assessment.

This book is the result of a workshop that was held at the Ettore Majorana Centre for Scientific Culture in Erice, Sicily, from November 28 to December 3, 1996, which brought together a small number of highly qualified scientists for a 4-day discussion with a selected audience of 75 students, faculty, and scientists. Many people helped to make the workshop a success. First, we wish to thank Professor Danilo Mainardi, director of the International School of Ethology of the Ettore Majorana Centre for Scientific Culture for his insight and support in getting the project approved and funded by the Centre. We also wish to thank Marco Lambertini for his participation in organizing the workshop and the excellent staff of the center for making life in Erice a memorable event. Each manuscript was reviewed by at least two external experts in the various topic areas and we especially thank our group of 24 anonymous referees for their time and effort, which resulted in a much-improved book. We would also like to thank Ed Lugenbeel, Holly Hodder, and Roy Thomas of Columbia University Press for encouraging the publication of the book and for editorial assistance, Carol Anne Peschke for editorial skills provided throughout the editing and publication process, and Ilaria Marzetti who helped prepare the index.

Luigi Boitani
Department of Animal and Human Biology
University of Rome "La Sapienza"

Todd K. Fuller
Department of Natural Resources Conservation
University of Massachusetts, Amherst

**Literature Cited**

Bookhout, T. A., ed. 1994. *Research and management techniques for wildlife and habitats.* Bethesda, Md.: The Wildlife Society.

Hanski, I. and D. Simberloff. 1997. The metapopulation approach, its history, conceptual domain, and application to conservation. In I. Hanski and M. Gilpin, eds., *Metapopulation biology: ecology, genetics and evolution,* 5–26. New York: Academic Press.

Krebs, C. J. 1999. *Ecological methodology.* Menlo Park, California: Benjamin/Cummings (Addison Wesley Longman).

Peters, R. H. 1991. *A critique for ecology.* Cambridge, U.K.: Cambridge University Press.

Shrader-Frechette, K. S. and E. D. McCoy. 1993. *Method in ecology: Strategies for conservation.* Cambridge, U.K.: Cambridge University Press.

# Research Techniques in Animal Ecology

## Chapter 1

# Hypothesis Testing in Ecology

CHARLES J. KREBS

Ecologists apply scientific methods to solve ecological problems. This simple sentence contains more complexity than practical ecologists would like to admit. Consider the storm that greeted Robert H. Peters's (1991) book *A Critique for Ecology* (e.g., Lawton 1991; McIntosh 1992). The message is that we might profit by examining this central thesis to ask "What should ecologists do?" Like all practical people, ecologists have little patience with the philosophy of science or with questions such as this. Although I appreciate this sentiment, I would point out that if ecologists had adopted classical scientific methods from the beginning, we would have generated more light and less heat and thus made better progress in solving our problems. As a compromise to practical ecologists, I suggest that we should devote 1 percent of our time to concerns of method and leave the remaining 99 percent of our time to getting on with mouse trapping, bird netting, computer modeling, or whatever we think important. A note of warning here: None of the following discussion is original material, and all of these matters have been discussed in an extensive literature on the philosophy of science. Here I apply these thoughts to the particular problems of ecological science.

## ■ Some Definitions

Let us begin with a few definitions to avoid semantic quarrels. Scientists deal with laws, principles, theories, hypotheses, and facts. These words are often used in a confusing manner, so I offer the following definitions for the descending hierarchy of generality in science:

**Laws:** universal statements that are deterministic and so well corroborated that everyone accepts them as part of the scientific background of knowledge. There are laws in physics, chemistry, and genetics but not in ecology.

**Principles:** universal statements that we all accept because they are mostly definitions or ecological translations of physicochemical laws. For example, "no population increases without limit" is an important ecological principle that must be correct in view of the finite size of the planet Earth.

**Theories:** an integrated and hierarchical set of empirical hypotheses that together explain a significant fraction of scientific observations. The theory of island biogeography is perhaps the best known in ecology. Ecology has few good theories at present, and one can argue strongly that the theory of evolution is the only ecological theory we have.

**Hypotheses:** universal propositions that suggest explanations for some observed ecological situation. Ecology abounds with hypotheses, and this is the happy state of affairs we discuss in this chapter.

**Models:** verbal or mathematical statements of hypotheses.

**Experiments:** a test of a hypothesis. It can be mensurative (observe the system) or manipulative (perturb the system). The experimental method is the scientific method.

**Facts:** particular truths of the natural world. Philosophers endlessly discuss what a fact is. Ecologists make observations that may be faulty, and consequently every observation is not automatically a fact. But if I tell you that snowshoe hares turned white in the boreal forest of the southern Yukon in October 1996, you will probably believe me.

Ecology went through its theory stage prematurely from about 1920 to 1960, when a host of theories, now discarded, were set up as universal laws (Kingsland 1985). The theory of logistic population growth, the monoclimax theory of succession, and the theory of competitive exclusion are three examples. In each case these theories had so many exceptions that they have been discarded as universal theories for ecology. Theoretical ecology in this sense is past.

It is clear that most ecological action is at the level of the hypothesis, and I devote the rest of this chapter to a discussion of the role of hypotheses in ecological research.

## ■ What Is a Hypothesis?

Hypotheses must be universal in their application, but the meaning of *universal* in ecology is far from clear. Not all hypotheses are equal. Some are more universal than others, and we accept this as one criterion of importance. A hypothesis of population regulation that applies only to rodents in snowy environments may be useful because there are many populations of many species that live in such environments. But we should all agree that a better hypothesis would explain population regulation in all small rodents in all environments. And a hypothesis that applies to all mammals would be even better.

Hypotheses predict what we will observe in a particular ecological setting, but to move from the general hypothesis to a particular prediction we must add background assumptions and initial conditions. Hypotheses that make many predictions are better than hypotheses that make fewer predictions. Popper (1963) emphasized the importance of the falsifiability of a hypothesis, and asked us to evaluate our ecological hypotheses by asking "What does this hypothesis forbid?" Ecologists largely ignore this advice. Try to find in your favorite literature a list of predictions for any hypothesis and a list of the observations it forbids.

*Recommendation 1: Articulate a clear hypothesis and its predictions.*

If we test a hypothesis by comparing our observations with a set of predictions, what do we conclude when it fails the test? There is no topic on which ecologists disagree more. Failure to observe what was predicted may have four causes: the hypothesis is wrong, one or more of the background assumptions or initial conditions were not satisfied, we did not measure things correctly, or the hypothesis is correct but only for a limited range of conditions. All of these reasons have been invoked in past ecological arguments, and one good example is the testing of the predictions of the theory of island biogeography (MacArthur and Wilson 1967; Williamson 1989; Shrader-Frechette and McCoy 1993).

A practical illustration of this problem is found in the history of wolf control as a management tool in northern North America. The hypothesis is usually stated that wolf control will permit populations of moose and caribou to increase (Gasaway et al. 1992). The background assumptions are seldom clearly stated: that wolves are reduced to well below 50 percent of their original numbers, that the area of wolf control is large relative to wolf dispersal distances, that a sufficient time period (3–5 years) is allowed, and that the weather is not adverse. The only way to make the predictions of this hypothesis more precise is to define the background assumptions more clearly. With respect to moose, at least five tests have been made of this hypothesis (Boutin 1992). Two tests supported the hypothesis, three did not. How do we interpret these findings? Among my students I find three responses: The hypothesis is falsified by the three negative results; the hypothesis is supported in two cases, so it is probably correct; or the hypothesis is true 40 percent of the time. All of these points of view can be defended, so in this case what advice can an ecologist give to a management agency? We cannot go on forever saying that more research is needed.

I recommend that we adopt the falsificationist position more often in ecology as a way of improving our hypotheses and advancing our research agenda. In this example we would reject the original hypothesis and set up an alternative hypothesis (for example, that predation by wolves and bears together limits the increase of moose and caribou populations). Indeed, we would be better off if we started with a series of alternative hypotheses instead of just one. The method of multiple working hypotheses is not new (Chamberlin 1897; Platt 1964) but it seems to be used only rarely in ecology.

*Recommendation 2: Articulate multiple working hypotheses for anything you want to explain.*

Two cautions are in order. First, do not assume that you have an exhaustive list of alternatives. If you have alternatives A, B, C, and D, do not assume that if A, B, and C are rejected that D must be true. There are probably E and F hypotheses that you have not thought of. Second, do not generalize the method of multiple working hypotheses to the ultimate multifactorial, holistic world view, which states that all factors are involved in everything. Many factors may indeed be involved, but you will make more rapid progress in understanding if you articulate a detailed list of the factors and how they might act. We need to retain the principle of parsimony and keep our hypotheses as simple as we can. It is not scientific progress for you to articulate a hypothesis so complex that ecologists could never gather the data to test it.

## ■ Hypotheses and Models

A hypothesis implies a model, either a verbal model or a mathematical model. Analytical and simulation models have become very popular in ecology. From a series of precise assumptions you can deduce mathematically what must ensue, once you know the structure of the system under study. Whether these predictions apply to the real world is another matter altogether. Mathematical models have overwhelmed ecology with adverse consequences. The literature is now filled with unrealistic, repetitive models with simplified assumptions and no connection to variables field ecologists can measure. You can generate models more quickly than you can test their assumptions. In an ideal world there would be rapid and continuous feedback between the modeler and the empiricist so that assumptions could be tested and modified. This happens too infrequently in ecology, partly because of the time limitations of most studies. The great advantage of building a mathematical model is to enunciate clearly your assumptions. This alone is worth a modeling effort, even if you never solve the equations.

*Recommendation 3: Use a mathematical model of your hypotheses to articulate your assumptions explicitly.*

Many mathematical models, such as the Lotka–Volterra predator–prey equations, begin with very general, simple assumptions about ecological interactions. Therefore, they are useless for ecologists except as a guide of what not to do. If we have learned anything from the past 50 years it is that ecological systems do not operate on general, simple assumptions. But this *simplicity* has been the great attraction of mathematical models in ecology, along with *generality* (Levins 1966), and we need to concentrate on *precision* as a key feature of models that will bridge the gap between models and data. Precise models contain enough biological realism that they make quantitative predictions about real-world systems (DeAngelis and Gross 1992).

One unappreciated consequence for ecologists who build realistic and precise models of ecological systems is that numerical models cannot be verified or validated (Oreskes et al. 1994). A verified model is a true model and we cannot know the truth of any model in an open system, as Popper (1963) and many others have pointed out. Validation of a numerical model implies that it contains no logical or programming errors. But a numerical model may be valid but not an accurate representation of the real world. If observed data fit the model, the model may be confirmed, and at best we can obtain corroboration of our numerical models. If a numerical model fails, we learn more: that one or more of the assumptions are not correct. Mathematical models are most useful when they challenge existing ideas rather than confirm them, the exact opposite of what most ecologists seem to believe. These strictures on numerical models apply more to complex models (e.g., population viability models) than to simple models (e.g., age-based demographic models).

Numerical models in which we have reasonable confidence can be used in ecology for sensitivity analysis, a very important activity. We can explore "what-if" scenarios rapidly and the only dangers are believing the results of such simulations when the model is not yet confirmed and extrapolating beyond the bounds of the model (Walters 1993).

## ■ Hypotheses and Paradigms

Hypotheses are specified within a paradigm and the significance of the hypothesis is set by the paradigm. A paradigm is a world view, a broad approach to problems addressed in a field of science (Kuhn 1970; McIntosh 1992). The Darwinian paradigm is the best example in biology. Most ecologists do not realize the paradigms in which they operate, and there is no list of the competing paradigms of ecology. The density-dependent paradigm is one example in population ecology, and the equilibrium paradigm is an example from community ecology. Paradigms define problems that are thought to be fundamental to an area of science. Problems that loom large in one paradigm are dismissed as unimportant in an opposing paradigm, as you can attest if you read the controversies over Darwinian evolution and creationism.

Paradigms cannot be tested and they cannot be said to be true or false. They are judged more by their utility: Do they help us to understand our observations and solve our puzzles? Do they suggest connections between theories and experiments yet to be done? Hypotheses are nested within a paradigm and supporters of different paradigms often talk past each other because they use words and concepts differently and recognize different problems as significant.

The density-dependent paradigm is one that I have argued has long outlived its utility and needs replacing (Krebs 1995). The alternative view is that a few bandages will make it work well again (Sinclair and Pech 1996). My challenge for any ecological paradigm is this: Name the practical ecological problems that this paradigm has helped to solve and those it has made worse. In its preoccupation with numbers, the density-dependent paradigm neglects the quality of individuals and environmental changes, which makes the equilibrium orientation of this approach highly suspect.

Consider a simple example of a recommendation one would make from the density-dependent paradigm to a conservation biologist studying an endangered species that is declining. Because by definition density-dependent processes are alleviated at low density (figure 1.1), you should not have to do anything to save your endangered species. No ecologist would make such a poor recommendation because environmental changes in terms of habitat destruction have changed the framework of the problem. Much patchwork has been applied to camouflage the inherent bankruptcy of this approach to population problems.

Ecologists find it very difficult to discuss paradigms because they are value-laden and are part of a much broader problem of methodological value judgments (Shrader-Frechette and McCoy 1993). Scientists are unlikely to admit to value judgments, but applied areas such as conservation biology have brought this issue to a head for ecologists (Noss 1996). All scientists make value judgments as they observe nature. For example, population ecologists estimate densities of organisms, partly because they value such data more than

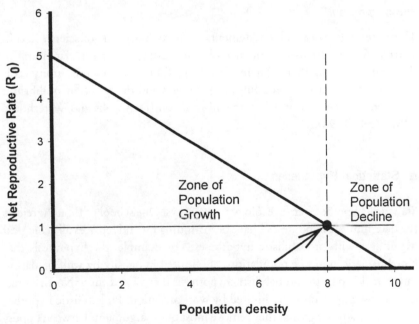

**Figure 1.1** Classic illustration of the density-dependent paradigm of population regulation. In this hypothetical example, populations above density 8 will decline and those below density 8 will increase to reach an equilibrium at density 8 (arrow). If an endangered species falls in density below 8, density-dependent processes will ensure that it recovers, without any management intervention. Of course, this is nonsense.

presence/absence data. Moreover, they prefer some estimation techniques to others because they are believed to be more accurate. Another example of methodological value judgments is the disagreement about the utility of microcosm research in ecology (Carpenter 1996).

Methodological value judgments are particularly clear in conservation biology. Why preserve biodiversity? Some ecologists answer that diversity leads to stability, and stability is a desired population and ecosystem trait. But there are two broad hypotheses about biodiversity and ecosystem function. The rivet theory, first articulated by Ehrlich and Ehrlich (1981), suggests that the loss of any species will reduce ecosystem function, whereas the redundancy theory, first suggested by Walker (1992), argues that many species in a community are replaceable and redundant, so that their loss would not affect ecosystem health. Which of these two views is closer to being correct is a value judgment at present, as is the concept of the balance of nature in conservation planning.

*Recommendation 4: Uncover and discuss the value judgments present in your research program.*

These methodological value judgments are a necessary part of science and in articulating and discussing them, ecologists advance their understanding of the problems facing them. There is a very useful tension in community ecology between the classical equilibrium paradigm and the new nonequilibrium paradigm of community structure and function (DeAngelis and Waterhouse 1987; Krebs 1994).

## ■ Statistical Hypotheses

Statistical hypotheses enter ecology in two ways. One school of thought rejects the deterministic hypotheses I have been arguing for and replaces all ecological hypotheses with probabilistic hypotheses. For example, the hypothesis that North American moose populations are limited in density by wolf predation can be replaced by the probabilistic hypothesis that 67 percent of North American moose populations are limited by wolf predation. Probabilistic hypotheses have the advantage that they remove most of the arguments between opposing schools of thought because they argue that everyone is correct part of the time. The challenge then becomes to specify more tightly the initial conditions of each hypothesis to make it deterministic. For our hypothetical example, if deer are present as alternative food, moose populations are limited by wolf predation. If deer are not present, moose are not limited by wolves. Buried in this consideration of probabilistic hypotheses are many philosophical issues and value judgments, but the major thrust is to replace ecological hypotheses with multiple-regression statistical models. Peters (1991) seemed to adopt this approach as one way of making applied ecological science predictive.

The more usual entry point for statistical hypotheses in ecology is through standard statistical tests. Ecological papers are overflowing with these statistical hypotheses and their resulting $p$-values. We spend more of our time instructing students on the mechanics of statistical hypothesis testing than we do instructing them on how to think about ecological issues. I make four points about statistical inference:

• Almost all statistical tests reported in the literature address low-level hypotheses of minor importance to the ecological issues of our day, not the major unsolved problems of ecological science. Therefore, we should not get too concerned about the resulting $p$-values.

• Achieving statistical significance is not the same as achieving ecological significance. You may have strong statistical significance but trivial ecological significance. You cannot measure ecological significance by the size of your $p$-values. What matters in ecology is what statisticians call *effect size:* How large are the differences? There is no formal guidance in what are ecologically significant effect sizes. Much depends on the structure of your ecological system. For population dynamics we can explore the impact of changes in survival and reproduction through simple life table models. Similar sensitivity analyses are not possible with questions of community dynamics.

• The null hypothesis of statistical fame, which suggests no differences between treatments or areas, is not always a good ecological model worth testing. We should apply statistics more cleverly when we expect differences between treatments and not pretend total ecological ignorance. We can often make a quantitative estimate of the differences to be expected. One-tailed tests ought to be common in ecology. Testing for differences can often be used, and specified contrasts should be the rule in ecological studies. We should use statistics as a fine scalpel, not as a machete, and we should not waste time testing hypotheses that are already firmly established.

• No important ecological issue can be answered by a statistical test. The important ecological issues, such as equilibrium and nonequilibrium paradigms, are higher-level questions that involve value judgments, not objective probability statements.

*Recommendation 5: Use statistical estimation more than statistical inference. There is more to life than p-values.*

These cautionary notes should not be misinterpreted to indicate that you do not need to learn statistics to be an ecologist. You should learn statistics well and then learn to recognize the limits of statistics as a tool for achieving knowledge. Every good study needs explicit null hypotheses and the appropriate statistical testing.

## ■ Hypotheses and Prediction

Hypotheses, once tested and confirmed, lead us to understanding but not necessarily to predictions that will be useful in applied ecology. *Prediction* is often used to mean forecasting in a temporal sense: What will happen to Lake Superior after zebra mussels are introduced? At present, applied ecologists can make only qualitative predictions in the medium term and quantitative predictions in the short term. We should focus on these strengths for the present and not berate ourselves for an inability to predict in the long term how disturbed populations and communities will change.

Short-term quantitative predictions are of enormous practical utility. If you know the number of aphids now, the numbers of their predators, and the temperature forecast for the next 2 weeks, you can predict aphid damage in the short term (Raworth et al. 1984). Ecologists should exploit the vast store of natural history data to develop these simple predictive models. This is not the route to the Nobel Prize, but it is still one of the most important contributions ecologists can make to society.

Medium-term predictions are more difficult, and ecologists often have to settle for qualitative predictions. A good example is provided by the search for habitat models that can be used in conservation planning. Not all habitat patches are occupied by all species, and metapopulation theory builds on this observation. But a habitat can be declared suitable only if it has the food and shelter a species requires and if the species can disperse there. Suitable habitats may have all the structural features needed but become unsuitable if a predator takes up residence (Doncaster et al. 1996). The scale of the difficulty in achieving medium-term predictions can be seen by work on the spotted owl in

Oregon and Washington (Bart and Forsman 1992; Carey et al. 1992; Lande 1988; Taylor and Gerrodette 1993). Attempts to predict what habitat configuration will permit the owl to survive are ecologically sophisticated because of the extensive background of descriptive studies on this owl. But even with maximum effort, the medium-term predictions are more uncertain than a conservation biologist would like, particularly in the mixed logging-partial preservation strategies.

If ecologists cannot at present achieve long-term predictions, we do have an extensive storehouse of knowledge about what management policies will *not* work. The catalog of disasters is now large enough that, without additional hypothesis testing, we can provide management agencies with sound advice about many ecological problems. For example, designating no-fishing zones or refuges for marine fisheries is an important conservation measure that we can recommend without detailed studies of the mechanisms of dispersal and community organization in the marine community affected by overfishing.

Because ecological communities are open systems and are subject to a changing climate, it is unlikely that we will ever be able to provide broad ecological laws that apply universally in time and space. We should concentrate on understanding and developing predictions for short-term changes in communities and populations. This understanding will be local and specific, and we should not worry that our spotted owl understanding cannot be applied universally to all owls or all birds on all continents.

*Recommendation 6: Concentrate on short-term predictions to solve local problems. Learn to walk before running.*

This recommendation to focus on the local and the particular is the complete antithesis of what Brown (1995) recommends as a macroecological future for ecology. There is a sense of frustration among ecologists that their chosen subject does not advance as rapidly as genetics or nuclear chemistry. Why is it so difficult to design theory in ecology? Is it because we are not studying the right questions? Not using the right methods? Do the textbooks we are using teach us to focus on unsolvable problems, as Peters (1991) suggests? Lawton (1996) gives an example of what he considers a critical question in biodiversity: Why are there 2 species of a taxonomic group in one ecosystem, 20 in a second system, and 200 in a third? I suggest that this is an unanswerable question, the ecologist's analog of angels-on-the-pinhead, and you could waste your scientific life trying to find an answer to it. But you will find in the literature almost no discussion of which types of questions in ecology have proven to be unsolvable and which have been fruitful, which have contributed to solving practical problems and which have been interesting but of limited utility.

*Recommendation 7: Address significant problems. Do not waste your thesis research or your career on trivial issues.*

What is trivial to one ecologist is the major problem of ecology to another. What can we do about this unsatisfactory state of affairs? In the long run, history sorts out these issues, but for ecologists facing biodiversity issues now, history will take too long. We cannot escape these judgments and more discussion ought to be devoted to them in ecological journals. If medical research councils devoted equal amounts of money to acupuncture and schizophrenia research, we would be alarmed at the poor judgment. We should not hesitate to make similar value judgments for ecological research. No person or group is infallible in their judgments, and this call for discussion of the relative importance of ecological questions must not be misinterpreted as a call for the regimentation of research ideas.

In this chapter I have concentrated on the role of hypothesis testing in ecology, and one may ask whether any of this applies to ethology as well. I am not a professional ethologist, so my judgment on this matter can be questioned. In my experience the problems I have outlined do indeed apply to ethology as well as ecology. I suspect that much of organismal biology could profit from a more rigorous approach to hypothesis testing.

In our haste to become scientists (with a capital *S*), we should be careful to focus on what we desire to achieve as ethologists and as ecologists. This debate, more about values than about scientific facts, is important for you to join. By your decisions you will affect the future developments of these sciences.

### Acknowledgments

I thank Alice Kenney, Rudy Boonstra, and Dennis Chitty for their comments on the manuscript, and the Canada Council for a Killam Fellowship that provided time to write. Joe Elkinton helped me at the Erice meeting by summarizing questions and comments on this chapter.

### Literature Cited

Bart, J. and E. D. Forsman. 1992. Dependence of northern spotted owls *Strix occidentalis caurina* on old-growth forests in the western USA. *Biological Conservation* 62: 95–100.

Boutin, S. 1992. Predation and moose population dynamics: A critique. *Journal of Wildlife Management* 56: 116–127.

Brown, J. H. 1995. *Macroecology.* Chicago: University of Chicago Press.

Carey, A. B., S. P. Horton, and B. L. Biswell. 1992. Northern spotted owls: Influence of prey base and landscape character. *Ecological Monographs* 62: 223–250.

Carpenter, S. R. 1996. Microcosm experiments have limited relevance for community and ecosystem ecology. *Ecology* 77: 677–680.

Chamberlin, T. C. 1897, reprinted 1965. The method of multiple working hypotheses. *Science* 148: 754–759.

DeAngelis, D. L. and L. J. Gross, eds. 1992. *Individual-based models and approaches in ecology: Populations, communities, and ecosystems.* New York: Chapman & Hall.

DeAngelis, D. L. and J. C. Waterhouse 1987. Equilibrium and nonequilibrium concepts in ecological models. *Ecological Monographs* 57: 1–21.

Doncaster, C. P., T. Micol, and S. P. Jensen. 1996. Determining minimum habitat requirements in theory and practice. *Oikos* 75: 335–339.

Ehrlich, P. R. and A. H. Ehrlich. 1981. *Extinction: The causes and consequences of the disappearance of species.* New York: Random House.

Gasaway, W. C., R. D. Boertje, D. V. Grangaard, D. G. Kelleyhouse, R. O. Stephenson, and D. G. Larsen. 1992. The role of predation in limiting moose at low densities in Alaska and Yukon and implications for conservation. *Wildlife Monographs* 120: 1–59.

Kingsland, S. E. 1985. *Modeling nature.* Chicago: University of Chicago Press.

Krebs, C. J. 1994 (4th ed.). *Ecology: The experimental analysis of distribution and abundance.* New York: HarperCollins.

Krebs, C. J. 1995. Two paradigms of population regulation. *Wildlife Research* 22: 1–10.

Kuhn, T. 1970. *The structure of scientific revolutions.* Chicago: University of Chicago Press.

Lande, R. 1988. Demographic models of the northern spotted owl (*Strix occidentalis caurina*). *Oecologia* 75: 601–607.

Lawton, J. 1991. Predictable plots. *Nature* 354: 444.

Lawton, J. 1996. Patterns in ecology. *Oikos* 75: 145–147.

Levins, R. 1966. The strategy of model building in population biology. *American Scientist* 54: 421–431.

MacArthur, R. and E. O. Wilson. 1967. *The theory of island biogeography.* Princeton, N.J.: Princeton University Press.

McIntosh, R. P. 1992. Whither ecology? *Quarterly Review of Biology* 67: 495–498.

Noss, R. F. 1996. Conservation biology, values, and advocacy. *Conservation Biology* 10: 904.

Oreskes, N., K. Shrader-Frechette, and K. Belitz. 1994. Verification, validation, and confirmation of numerical models in the earth sciences. *Science* 263: 641–646.

Peters, R. H. 1991. *A critique for ecology.* Cambridge, U.K.: Cambridge University Press.

Platt, J. R. 1964. Strong inference. *Science* 146: 347–353.

Popper, K. R. 1963. *Conjectures and refutations: The growth of scientific knowledge.* London: Routledge & Kegan Paul.

Raworth, D. A., S. McFarlane, N. Gilbert, and B. D. Frazer. 1984. Population dynamics of the cabbage aphid, *Brevicoryne brassicae* (Homoptera: Aphididae) at Vancouver, B.C. III. Development, fecundity, and morph determination vs. aphid density and plant quality. *Canadian Entomologist* 116: 879–888.

Shrader-Frechette, K. S. and E. D. McCoy. 1993. *Method in ecology: Strategies for conservation*. Cambridge, U.K.: Cambridge University Press.

Sinclair, A. R. E. and R. P. Pech. 1996. Density dependence, stochasticity, compensation and predator regulation. *Oikos* 75: 164–173.

Taylor, B. L. and T. Gerrodette. 1993. The uses of statistical power in conservation biology: The vaquita and northern spotted owl. *Conservation Biology* 7: 489–500.

Walker, B. H. 1992. Biodiversity and ecological redundancy. *Conservation Biology* 6: 18–23.

Walters, C. J. 1993. Dynamic models and large scale field experiments in environmental impact assessment and management. *Australian Journal of Ecology* 18: 53–62.

Williamson, M. 1989. The MacArthur and Wilson theory today: True but trivial. *Journal of Biogeography* 16: 3–4.

## Chapter 2

# A Critical Review of the Effects of Marking on the Biology of Vertebrates

Dennis L. Murray and Mark R. Fuller

Vertebrates often are marked to facilitate identification of free-ranging individual animals or groups for studies of behavior, population biology, and physiology. Marked animals provided data for many of the topics discussed in this volume, including home range use, resource selection, social behavior, and population estimation. Markers can be classified into three general categories: mutilations, tags and bands, and radiotransmitters. The appropriate marking technique for a study depends on several considerations, including study objectives, target species, marker cost, marker efficacy, and marker effects on the animals (Day et al. 1980; Nietfeld et al. 1994).

Studies using marked animals are characterized by the assumption that marking does not affect animals or that negative effects are not important (Ricker 1956; Day et al. 1980; Nietfeld et al. 1994). The assumption of no significant marking effects is critical because it is the basis for generalizing data collected from marked individuals to unmarked animals and populations. However, the assumption has not been tested rigorously for most marker types or animal species, despite the often necessary use of seemingly invasive marking techniques. The general paucity of marker evaluation studies apparently is related to the difficulties associated with conducting such tests in the field, as well as the belief that marker evaluation is tangential to most study objectives and therefore of minor importance to the researcher. In addition, studies that evaluate marker effects often suffer from small samples, thus leading to qualitative conclusions or weak statistical inference (White and Garrott 1990). As a result, researchers tend to choose markers that intuitively seem least likely to induce abnormal behavior or survival, even though data supporting that assertion usually are weak or lacking. However, if the assumption of no marking effects is violated and the effect is not evaluated, then data collected from marked animals will be biased. It follows that if significant marker effects remain undetected or unaddressed, conservation and management actions based on those results might not be appropriate. In addition, recent guidelines established by institutional animal care and use committees require that marking protocols minimize pain and stress to study animals (Friend et al. 1994). If researchers collectively ignore the development, evaluation, and application of animal markers acceptable to such committees, and fail to publish results of studies not finding significant effects, then some research might be needlessly jeopardized or precluded.

The purpose of this chapter is to present examples of the effects markers can have on animals and to examine critically the treatment of potential marking effects by ecologists. We use the word *effect* to mean unusual or abnormal behavior, an abnormal function, or abnormal reproduction or survival. We use *significant* to indicate statistical results and *important* to indicate an observed effect and implication for studies. We emphasize the shortcomings of various marking techniques to animal biology. Our discussion is restricted to effects of markers, and thus does not include a specific review of handling effects. Furthermore, we do not present results specific to causes of pain or stress because essentially no data exist from wildlife. First, we present the variety of marking techniques that are available for, and explore possible implications of markers on, various taxonomic groups. Next, we review recently published articles to examine how researchers consider potential marking effects. Finally, we discuss how potential marking effects can be minimized and evaluated in future studies. Consistent with the theme of this volume, the approach we have taken is often critical of existing information and protocols. However, such an approach is necessary if researchers are to improve the overall quality of data being generated from ecological studies (Peters 1991).

## ■ Review of the Literature

Nietfeld et al. (1994) described available marking techniques (excluding marking with radiotransmitters) and generally reviewed marking techniques for vertebrates (excluding fish). Samuel and Fuller (1994) provided similar information about radiotransmitters. Stonehouse (1978) edited a book about animal marking, and other overviews dealing with selected vertebrate groups include Stasko and Pincock (1977), Wydowsky and Emery (1983), and Parker et al. (1990) for fish; Ferner (1979) for amphibians and reptiles; and Marion and Shamis (1977), Calvo and Furness (1992), and Bub and Oelke (1980) for birds. These sources will lead the reader to the literature dealing with many species, many marking methods, and various considerations associated with different techniques, different species, and study objectives.

## WHICH MARKERS TO USE?

It is worthwhile to reiterate some important factors that Nietfeld et al. (1994) and others noted as important when deciding which markers to use for a study. Expense can be an important consideration because marking materials can range widely in cost (e.g., tags versus radiotelemetry via satellites). The procedures required to initially capture and mark animals and to obtain results from intensive field observations or recapture efforts also are important. Markers should be easily assembled and attached, recognized in the field, and durable enough to remain functional throughout the study. Additionally, all marking techniques should result in minimum pain or stress to the animal during application and use. Finally, markers should not cause abnormal behavior or affect survival. Clearly, it is difficult to address all these criteria satisfactorily before the initiation of a study, so some marking has undesirable effects on animals and research results. The adverse effects of marking often are species-specific and might occur only in conjunction with certain behavior (e.g., courtship) or environmental conditions (e.g., extreme temperature). Also, the magnitude and importance of such effects are highly variable among marker types. We present examples of marking techniques and their effects on vertebrate species. This material will help address questions about adverse effects that were raised by Young and Kochert (1987) and Nietfeld et al. (1994): Does the information obtained from the study justify marking of animals? Can the effects of marking be identified during data analysis? If marking effects are accounted for in the analysis, can the study objectives still be achieved? Such questions should be posed at the outset of any study involving the marking of animals. If one or more answers to these questions is negative or unknown, an alternative marker should be sought or the effects of the marker under consideration should be evaluated thoroughly.

## EFFECTS OF MARKERS AMONG TAXA

We reviewed a sample of articles that had as a primary objective the evaluation of marker effects. The articles consisted of qualitative or quantitative assessments of the effect of specific marker types on study animals. We acknowledge that marker evaluation studies probably are biased toward those showing effects because results indicating no effects might be published less often. This implies that our sample of the literature overestimates the occurrence of marker effects in evaluation studies. However, the objective of our review is not to determine how often marker effects occur, but rather to provide examples of the range and diversity of negative effects among marker types, species, and sex, and thus encourage biologists to consider seriously the effects of marking animals. Our review begins with these examples, presented by taxonomic group in the following sections and associated tables.

*Fish*

TAGGING    Marking has been used widely in fish population estimation; accordingly, the earliest tests evaluating marker effects in vertebrates occurred in fish. Historically, most evaluations of marking effects were anecdotal (Mellas and Haynes 1985), but by the 1940s researchers were suspicious of the potential effects of markers and thus began evaluating their merit in the field. Early fish research often involved the use of commercially made plastic or metal tags, and fish tagging was considered an effective marking system because tags were inexpensive, easily applied and seen, and rarely lost by tagged fish. However, studies evaluating potential effects of tags often found that tags altered aspects of fish biology (table 2.1). For example, several field studies used mark–recapture techniques and concluded that tags reduced survival and growth of fish. In some situations (DeRoche 1963), negative effects persisted throughout the life of a fish, whereas in others (Carline and Brynildson 1972), the effects seemed to be short-lived. Tagged fish were found to experience reduced swimming ability because of increasing drag (Clancy 1963), but not all effects of tagging can be attributed directly to the tags themselves. For instance, choice of tag placement on the fish's body can elicit marker effects (Bardach and LeCren 1948; Stroud 1953; Kelly and Barker 1963; Rawstron 1973; Rawstron and Pelzam 1978), and it is generally considered that tags placed in and around the mouth may interfere with feeding. It is notable that not all tag evaluation studies have shown negative effects of tagging (table 2.1), and with additional study some tags will be shown to be more appropriate than others.

The recent development of passive integrated transponder (PIT) tags has allowed researchers to mark fish and other vertebrates with smaller tags than those used previously. PIT tags are electromagnetically charged microchips implanted either subcutaneously or intraabdominally, and are read remotely via a portable scanner (Nietfeld et al. 1994). Insertion of PIT tags usually is performed using a syringe, thus eliminating the need for extensive invasive surgery. So far, no negative effect of PIT tags has been found in fish (Prentice et al. 1990; Jenkins and Smith 1990), suggesting that this technique can become an important tool for marking fish and other vertebrates. However, one drawback of PIT tagging is that at present tags can be read only when near a scanner.

MUTILATION    Marking by mutilation, usually by fin removal or partial removal, is a permanent marking technique often used by fish researchers. However, fin removal often affects fish growth and survival (table 2.2). For example, evaluations using mark–recapture methods (Shetter 1951; Mears and Hatch 1976) show that fin removal causes lower probability of recapture and, by inference, lower survival. Excision of multiple fins generally appears to be more harmful than single-fin excision, and removal of the adipose fin usually is less harmful than removal of other fins (Nicola and Cordone 1973; Mears and Hatch 1976). Removal of dorsal or anal fins can be particularly damaging (Coble 1967), partly because under certain conditions such excisions may predispose some species to bacterial or fungal infections (Stott 1968) or predation (Coble 1971). However, as with tagging, not all studies evaluating fin removal have detected significant effects, suggesting that for certain species or age classes, or under specific conditions, this marking technique could be acceptable. Clearly, evaluation of the effects of fin clipping on fish biology requires more attention, particularly under controlled laboratory conditions.

RADIOTRANSMITTERS    Radiotelemetry has become an important technique in fishery research, allowing biologists to accurately monitor long-term movements and survival of many species that would otherwise be difficult to study. Transmitter sizes and types available for fish are variable, and they have been attached to animals either externally or internally (see review by Stasko and Pincock 1977). Laboratory studies have shown that externally mounted transmitters increase drag and reduce or prevent swimming, particularly in high-speed currents (Mellas and Haynes 1985). It has been suggested that fusiform, lotic fishes are more influenced by external mounts than non-fusiform, lentic, or pelagic species (McCleave and Stred 1975). Internal implantation can be achieved by force-feeding stomach transmitters, or by surgery to attach the transmitter either in the peritoneal cavity or intramuscularly. Implants are more commonly used than external transmitters and have the

**Table 2.1  Survey of Marker Evaluation Studies in Fish**

| Marking Technique | Species | Parameters Tested | Lab (L) or Field (F) | Effect of mark | Reference |
|---|---|---|---|---|---|
| Tagging | Alosa aestivalis | Su, Be | F | n.s. | Bulak 1983 |
| | Ameiurus nebulosus | Gr, Su | F | n.s. | Stroud 1953 |
| | Cynoscion nebulosus | Ph | L | Ph | Vogelbein and Overstreet 1987 |
| | Esox lucius | Co | F | Gr | Scheirer and Coble 1991 |
| | Esox niger | Gr, Su | F | n.s. | Stroud 1953 |
| | Gadus morhua | Gr | F | n.s. | Jensen 1967 |
| | Leiostomus xanthurus | Ph | L | Ph | Vogelbein and Overstreet 1987 |
| | Morone americana | Gr, Su | F | n.s. | Stroud 1953 |
| | Micropterus salmoides | Gr, Su | F | n.s | Stroud 1953 |
| | | Su | F | Su | Rawstron and Pelzam 1978 |
| | | Gr, Ma, Su | F | n.s. | Tranquilli and Childers 1982 |
| | Oncorhynchus gorbuscha | Ph, Ma, Su | L | Ph, Ma, Su | Saddler and Caldwell 1971 |
| | Oncorhynchus tshawytscha | Gr, Su | F | n.s. | Eames and Hino 1983 |
| | Perca flavescens | Gr, Su | F | n.s. | Stroud 1953 |
| | | Ph | F | Ph | Stobo 1972 |
| | Salmo gairdneri | Gr, Su | F | Gr, Su | Shetter 1967 |
| | Salmo salar | Gr, Su, Ma | F | Gr, Su | Saunders and Allen 1967 |
| | | Ph | L | Ph | Roberts et al. 1973 |
| | Salmo trutta | Ma | F | Ma | Schuck 1942 |
| | Salvelinus fontinalis | Gr, Su | F | n.s. | Stroud 1953 |
| | | Gr, Su | F | n.s | Stroud 1953 |
| | Salvelinus namaycush | Gr, Su | F | Gr | Carline and Brynildson 1972 |
| | | Gr | F | Gr | DeRoche 1963 |
| Tetracycline | Onchorhynchus nerka | Su | F | n.s. | Weber and Wahle 1969 |

| Fluorescent pigment | Onchorhynchus spp. | Su | F | n.s. | Phinney et al. 1967 |
|---|---|---|---|---|---|
|  | Salmo gairdneri | Su | F | n.s. | Phinney et al. 1967 |
| PIT tagging | Morone saxatilis | Be, Ph, Gr, Su | L | n.s. | Jenkins and Smith 1990 |
|  | Onchorynchus spp. | Gr, Su, Ph | L | n.s. | Prentice et al. 1990 |
|  | Salmo salar | Gr, Su, Ph | L | n.s. | Prentice et al. 1990 |
|  | Sciaenops ocellatus | Be, Ph, Gr, Su | L | n.s. | Jenkins and Smith 1990 |
| Fin removal | Cristivomer namaycush | Gr, Su | F | n.s. | Armstrong 1949 |
|  |  | Pr, Gr | F | n.s. | Shetter 1952 |
|  |  | Gr, Su | F | n.s. | Shetter 1951 |
|  | Esox masquinongy | Gr, Su | F | n.s. | Patrick and Haas 1971 |
|  | Esox lucius | Gr, Co | F | Gr | Scheirer and Coble 1991 |
|  | Huro salmoides | Gr, Su | F | Gr, Su | Ricker 1949 |
|  | Lepomis macrochirus | Gr, Su | F | n.s. | Ricker 1949 |
|  |  | Be, Gr, Su | F | Be | Crawford 1958 |
|  |  | Pr | L | Pr | Coble 1972 |
|  | Micropterus dolomieui | Gr, Su | F | Su | Coble 1971 |
|  | Perca flavescens | Gr, Su | F | n.s. | Ricker 1949 |
|  |  | Gr, Su | F | Su | Coble 1967 |
|  | Salmo gairdneri | Mo | L | Mo | Clancy 1963 |
|  |  | Gr, Su | F | Su | Shetter 1967 |
|  |  | Mo | L | n.s. | Horak 1969 |
|  | Salmo salar | Gr, Su | F | Gr, Su | Nicola and Cordone 1973 |
|  | Salmo trutta | Gr, Su, Ma | F | Gr, Su | Saunders and Allen 1967 |
|  | Salvelinus fontinalis | Gr | F | n.s. | Brynildson and Brynildson 1967 |
|  |  | Su | F | Su | Mears and Hatch 1976 |
|  | Sebastes marinus | Gr | F | Gr | Kelly and Barker 1963 |
|  | Onchorhynchus nerka | Su | F | Su | Weber and Wahle 1969 |

*(continued)*

**Table 2.1** Continued

| Marking Technique | Species | Parameters Tested | Lab (L) or Field (F) | Effect of mark | Reference |
|---|---|---|---|---|---|
| Radiotelemetry | *Ictalurus punctatus* | Gr, Su | L | n.s. | Summerfelt and Mosier 1984 |
| | | Ph | L | Ph | Marty and Summerfelt 1986 |
| | *Morone americana* | Mo, Be, Pa | L | Pa | Mellas and Haynes 1985 |
| | *Pylodictis olivaris* | Gr, Su, Be | F | Be | Hart and Summerfelt 1975 |
| | *Roccus chrysops* | Su | F | S | Henderson et al. 1966 |
| | *Salmo gairdneri* | Mo | L | Mo | Lewis and Muntz 1984 |
| | | Mo, Be | L | Mo | Mellas and Haynes 1985 |
| | | Su, Gr | F | n.s. | Lucas 1989 |
| | *Salmo salar* | Mo | L | Mo | McCleave and Stred 1975 |

Articles surveyed were published in the peer-reviewed literature and consist of qualitative or quantitative evaluations of marking effects. We report the effect of markers as being important/significant or not (n.s.), as interpreted by the authors in the article.
Gr = growth, Su = survival, Be = behavior, Mo = movements, Ma = mass, Co = condition, Pa = parasitism/disease, Pr = predation, Ph = physiology.

advantages of lying near a fish's center of gravity, not being lost or entangled in the environment, and not creating drag forces. However, these advantages can be offset by reduction in swimming performance, increased handling time, and stress associated with surgery, as well as the higher chance of infection following release. Also, implanted transmitters occasionally can be passively expelled from the body, although sometimes without causing mortality or morbidity (Lucas 1989). Some species appear more predisposed than others to postoperative complications and transmitter expulsion (Mellas and Haynes 1985; Marty and Summerfelt 1986), meaning that it may be necessary to tailor surgical technique and specific implantation site to the target species. However, in some species, stomach implants seem to have fewer effects than either external mounts or surgically implanted transmitters (Henderson et al. 1966).

In all telemetry studies, transmitter size is an important consideration, and smaller transmitters are always more desirable than larger ones from the standpoint of effects on the animal (Stasko and Pincock 1977; Marty and Summerfelt 1986). However, the general question regarding the effects of transmitter mass on fish still must be addressed in controlled studies (Stasko and Pincock 1977).

## Reptiles and amphibians

TAGGING    The use of marking in reptile and amphibian research is fairly new, so fewer studies have evaluated marker effects in these taxonomic groups. Many species of reptiles and amphibians have proven difficult to mark because of their epidermal sensitivity, small size, and potential for tissue regeneration. Tagging of reptiles and amphibians has included various types of branding and the use of polymers, pigments, dyes, and radioactive substances (Ferner 1979; Ashton 1994; Donnelly et al. 1994; table 2.2). Many of these markers are of limited utility because they were not tested adequately for marking effects (Donnelly et al. 1994); such limitations are particularly important for amphibians, given the sensitivity of their skin. A field test of marking by dye injection did not find any effects on larval amphibians (Seale and Boraas 1974), but a controlled laboratory study did identify stunting in dyed tadpoles (Travis 1981). Although these studies used different dyes, the results call into question previous suggestions that some dyes are largely benign (Guttman and Creasey 1973) and suggest that laboratory studies might be more sensitive to detection of marking effects. Other color markers, such as fluorescent paint, often are used to monitor amphibians in the field (Taylor and Deegan 1982; Nishikawa and Service 1988; Ireland 1991), despite the fact that such paint

**Table 2.2  Survey of Marker Evaluation Studies in Reptiles and Amphibians**

| Marking Technique | Species | Parameters Tested | Lab (L) or Field (F) | Effect of Mark | Reference |
|---|---|---|---|---|---|
| Paint marking | *Sceloporus jarrovi* | Su | F | n.s. | Simon and Bissinger 1983 |
| | *Sceloporus undulatus* | Su | F | n.s. | Jones and Ferguson 1980 |
| Staining | *Ambystoma tigrinum* | Gr, Pa, Mo | F | n.s. | Seale and Boraas 1974 |
| | *Hyla gratiosa* | Gr | L | Gr | Travis 1981 |
| | *Rana catesbiana* | Gr, Pa, Mo | F | n.s. | Seale and Boraas 1974 |
| | *Rana clamitans* | Su, Mo | L | Su, Mo | Guttman and Creasey 1973 |
| | *Rana pipiens* | Gr, Pa, Mo | F | n.s. | Seale and Boraas 1974 |
| Tagging | *Rana catesbiana* | Be, Mo, Ma | F | n.s. | Emlen 1968 |
| | *Thamnophis siritalis* | Ph | F | n.s. | Pough 1970 |
| PIT tagging | *Sistrurus miliarius* | Gr, Mo, Su | F | n.s. | Jemison et al. 1995 |
| | *Thamnophis proximus* | Ma | L | n.s. | Keck 1994 |
| Clipping | *Alligator mississippiensis* | Gr, Su | L | n.s. | Jennings et al. 1991 |
| | *Coluber constrictor* | Be | F | n.s | Brown 1976 |
| Toe clipping | *Bufo woodhousei* | Su | F | Su | Clarke 1972 |
| | *Cnemidophorus sexlineatus* | Mo | L | n.s. | Dodd 1993 |
| | *Sceloporus merriami* | Mo | L | n.s. | Huey et al. 1990 |
| Branding | *Alligator mississippiensis* | Gr, Su | L | n.s. | Jennings et al. 1991 |
| | *Ascaphus truei* | Ph, Ca | F | n.s. | Daugherty 1976 |
| Radiotransmitter | *Nerodia sipedon* | Be | L | Be | Lutterschmidt and Reinert 1990 |
| | *Thamnophis elegans* | Ph | L | n.s. | Charland 1991 |
| | *Thamnophis marcianus* | Mo | L | n.s. | Lutterschmidt 1994 |
| | *Thamnophis siritalis* | Ph | L | n.s. | Charland 1991 |

Articles surveyed were published in the peer-reviewed literature and consist of qualitative or quantitative evaluations of marking effects. We report the effect of markers as being important/significant or not (n.s.), as interpreted by the authors in the article.
Gr = growth, Su = survival, Be = behavior, Mo = movements, Ma = mass, Ca = capture probability, Pa = parasitism/disease, Ph = physiology.

apparently has not been evaluated for negative effects on behavior, physiology, or vulnerability to predation.

Most metal or plastic tags used on reptiles and amphibians were originally designed for attachment to fish or birds. Such tags tend to be large and cumbersome, and their effect on study animals remains largely untested despite early suspicions that they affected behavior and physical condition (Raney 1940; Woodbury 1956). It is promising that both studies evaluating the effect of PIT tags on reptiles failed to detect effects (table 2.2), and with additional study, they might become the standard for tagging many species of reptiles and amphibians. However, in addition to expense and distance requirements for reading (Germano and Williams 1993), PIT tags have the disadvantage of being lost at a high rate by some free-ranging reptiles (Parmenter 1993; Rossi and Rossi 1993).

MUTILATION    Until recently, most studies of reptiles and amphibians used mutilation marking to identify individuals. One of the most common forms of mutilation, toe clipping, has been widely used on lizards, frogs, and salamanders because it provides an inexpensive method of identifying individuals. High frequency of natural toe loss in some populations of free-ranging lizards has been used to justify its use as an acceptable marking tool (Middleburg and Strijbosch 1988; Hudson 1996), but the natural occurrence of missing toes does not indicate that toe loss is not traumatic. Although toe clipping apparently does not affect the sprint performance of some lizards (Guttman and Creasey 1973), another study (Clarke 1972) inferred from the low rate of recapture of toe-clipped toads that the marking technique reduced survival. Clarke (1972) also noted that recapture rates were inversely related to the number of toes removed, and that toe-clipped toads experienced reduced dexterity when handling large prey. In other species, regeneration of clipped toes can occur, thus causing problems associated with misidentification of marked animals. However, despite the potential negative effects of toe clipping on reptiles and amphibians, it has remained a widely used form of marking. Clearly, the effect of this technique on reptiles and amphibians requires additional study (American Society of Ichthyologists and Herpetologists, the Herpetologists' League, and the Society for the Study of Amphibians and Reptiles 1987).

RADIOTRANSMITTERS    Movements of reptiles and amphibians occasionally have been monitored using a thread-loaded bobbin that unrolls a trail of thread as the animal travels (Scott and Dobie 1980). More often, however, radiotelemetry is used to monitor movements, behavior, and physiology of reptiles (Larsen 1987) and occasionally amphibians (Bradford 1984; Smits 1984). For amphibians, the main constraint appears to be related to transmitter size, and as a general rule it is recommended that packages not exceed 10 percent of the body mass of the study animal (Richards et al. 1994). Radiotelemetry is problematic with many reptiles and amphibians, and snakes in particular offer challenges because external mounting is not feasible. Considerable effort has been invested in developing an effective method for implanting transmitters in snakes (Weatherhead and Anderka 1984). However, the value of stomach implants is questioned on the grounds that they may affect aspects of snake behavior (Fitch and Shirer 1971; Jacob and Painter 1980; Reinert and Cundall 1982). For instance, stomach-implanted snakes seem to behave similarly to nonimplanted snakes that have recently ingested food (Lutterschmidt and Reinert 1990), suggesting that activity patterns of implanted snakes are not representative of those of nonimplanted animals. Stomach transmitters also can affect other behaviors or physiological processes, and it might be that such markers simply are not acceptable in snakes because of effects on the animal. Alternatively, transmitters can be implanted in snakes either intraperitoneally or subcutaneously, and these modes of attachment generally appear to be effective (Weatherhead and Anderka 1984).

## Birds

There is more literature about the effects of marking on birds than for other taxa (table 2.3). Therefore, we provide a sample of recent (i.e., largely post-1989) references for numerous avian marking techniques, and refer the reader to recent reviews by Nietfeld et al. (1994) and Calvo and Furness (1992) for earlier references.

BANDS AND COLLARS    Selecting correct band size and material is a very important step in the marking process because different band materials and configurations can have different effects on birds. For example, aluminum butt-end and lock-on bands can cause more injury and reduce the probability of recovery or recapture of some birds, compared to stainless steel bands (Meyers 1994). Some authors (Hatch and Nisbet 1983a, 1983b; Nisbet and Hatch 1985) recommend use of incoloy (a metal alloy) bands as a substitute for aluminum bands because aluminum bands can cause abrasion to legs of some bird species. Young birds whose legs are still growing can be the most subject to harmful effects of improperly fitting bands, but one method alleviating such effects is by using plasticine to fill the space between standard size bands and the leg (Blums et al. 1994). With this innovation, young birds can be banded without the risk of the large band injuring the bird or slipping off the leg.

Band color can influence bird behavior. Burley (1985) noted that for captive zebra finches (*Poephila guttata*), interactions among the sexes and mortality were influenced by legband color. Among wild zebra finches, Zann (1994) found no differences in survival or body condition associated with legband color, but in one colony females that paired with red-banded males laid more eggs than females paired with males not banded with red. These results and others (table 2.3) reveal that the effects of color banding can be complex because they vary by species and experimental and environmental conditions. The conspicuousness of color bands has been enhanced by attaching streamers of the same color to the band. However, the durability of streamers and fading as well as birds attempting to remove the streamers (Platt 1980) are problems. Color marking also has been accomplished by placing colored tape that contrasts with the plumage on several adjacent flight or tail feathers (Ritchison 1984). Also, feathers can be dyed or painted to enhance detectability, or a portion of colored feather can be used to replace a natural feather (Young and Kochert 1987; Handel and Gill 1983). However, like color banding, color marking of feathers can cause either significant (Goforth and Baskett 1965) or negligible (Wendeln et al. 1996) effects, depending on various conditions.

Neckbands, which are similar to legbands, have been used on long-necked bird species because they are more easily seen and read. In some cases, neckbands have been found to affect bird survival (Castelli and Trost 1996). Patagial tags, also known as wing tags, are used to enable identification of individual waterfowl. However, in a study of American coots (*Fulica americana*), patagial tags were associated with loss of body mass when compared with neckbanded controls (Bartelt and Rusch 1980). In other species, the use of wing tags may result in wounding, changes in migration times, and reduced reproductive success (Sallaberry and Valencia 1985; Southern and Southern 1985).

There are other types of bird markers, many having been shown to have effects on birds (table 2.3). For example, titanium dioxide, which was a useful marker on some species, was found to be deleterious to several raptor species (Barton and Houston 1991). Fluorescent bead markers can be spread in water, from which they attach to waterfowl, and apparently cause no irritation or detectable physical change in the birds (Godfrey et al. 1993). However, the overall efficacy of the marker is subject to exposure time, bird activity, marker transfer among birds, and equipment required for recognition of marked animals.

**Table 2.3  Survey of Marker Evaluation Studies in Birds**

| Marking Technique | Species | Parameters Tested | Lab (L) or Field (F) | Effect of Mark | Reference |
|---|---|---|---|---|---|
| Legband | *Sterna fuscata* | Lo | F | Lo | Spear 1980 |
| | *Larus occidentalis* | Lo | F | Lo | Bailey et al. 1987 |
| | *Actitic macularia* | Ph | F | n.s. | Reed and Oring 1993 |
| | *Brenta berricula* | Lo | F | Lo | Lensink 1988 |
| | *Carpodocus mexicanus* | Be, Lo | F | Be, Lo | Stedman 1990 |
| | *Pica pica* | Be | F | Be | Reese 1980 |
| | *Amazona* spp. | Lo | F | Lo | Meyers 1994 |
| | *Strix occidentalis* | Lo | F | Lo | Forsman et al. 1996 |
| | *Fulmorus glaciatis* | Lo | F | Lo | Anderson 1980a |
| | *Poephila guttata* | Be | F | Be | Burley 1988 |
| | *Poephila guttata* | Be, Re | L | Be, Re | Burley et al. 1982 |
| | *Poephila guttata* | Be, Re | L | Be, Re | Ratcliffe and Boag 1987 |
| | *Agelaius phoeniceus* | Be | F | Be | Metz and Weatherhead 1993 |
| | *Junco byemalis* | Be | L | n.s. | Cristol et al. 1992 |
| | *Picoides borealis* | Re, Su | F | Re, Su | Hagan and Reed 1988 |
| | *Tachycineta bicolor* | Re | F | Re | Burtt and Tuttle 1983 |
| Wing tag | Raptors, raven | Lo | F | Lo | Kochert et al. 1983 |
| | *Columba fasciata* | Lo | F | Lo | Curtis et al. 1983 |
| Dye marking | *Larus argentatus* | Re, Lo | F | Re, Lo | Belant and Seamans 1993 |
| | *Sterna hirundo* | Re, Lo | F | n.s. | Wendeln et al. 1996 |
| | *Larus* spp. | Lo | F | Lo | Cavanagh et al. 1992 |

| | | | | | |
|---|---|---|---|---|---|
| Neckband | *Branta canadensis* | Su | F | Su | Zicus et al. 1983 |
| | | Lo | F | Lo | Campbell and Becker 1991 |
| | | Lo | F | Lo | Samuel et al. 1990 |
| | | Lo | F | n.s. | Zicus and Pace 1986 |
| | *Anser albifrons* | Be | F | Be | Ely 1990 |
| | *Cygnus columbianus* | Be | F | Be | Hawkins and Simpson 1985 |
| Egg marker | *Larus dalawrensis* | Su | F | n.s. | Hayward 1982 |
| Radio marking | *Anas platyrhynchos* | Re | F | Re | Rotella et al. 1993 |
| Suture to back | | Re | Fe | Re | Bergmann et al. 1994 |
| Neck collar | *Phasianus colchicus* | Be | F | Be | Kenward et al. 1993 |
| Abdomen implant | *Anas discors* | Lo | F | Lo | Garrettson and Rohwer 1996 |
| Subcutaneous | *Aythya valisinevia* | Be, Ph | F | n.s. | Korschgen et al. 1996a |
| | *Falco mexicanus* | Be, Re, Fo | C | n.s. | Vekasy et al. 1996 |
| Back pack | *Alectoris chukar* | Re | F | n.s. | Slaugh et al. 1990 |
| | *Anas platyrhynchos* | Lo | F | Lo | Rotella et al. 1993 |
| | *Pygoscelis antarctica* | Re | F | Re | Croll et al. 1991, 1996 |
| Glue back | *Passerines* | Su, Be, Lo | F | Lo | Johnson et al. 1991 |
| | *Phasianus colchicus* | Su | F | n.s. | Kenward et al. 1993 |
| Tail mount | *Rissa tridactyla* | Be, Fo | F | n.s. | Wanless 1992 |
| | *Falco naumanni* | Re, Su | F | n.s. | Hiraldo et al. 1994 |

Articles surveyed were published in the peer-reviewed literature and consist of qualitative or quantitative evaluations of marking effects. We report the effect of markers as being important/significant or not (n.s.), as interpreted by the authors in the article. Su = survival, Be = behavior, Ph = physiology, Lo = loss of marker, Re = reproduction, Fo = foraging.

RADIOTRANSMITTERS    Radiotelemetry is an important tool for the study of avian biology (Kenward 1993; Kenward and Walls 1994; Custer et al. 1996). Radiomarking and use of recording devices (e.g., for flight velocity, diving depth) on birds has generated considerable study of effects of these markers because they are large compared to most other bird markers. The effect of transmitter or device size on birds can be influenced by where the package is placed or how it is attached. Packages have been placed on legs, necks, wings, backs, retrices and other feathers, under the skin, in the body, by banding, collaring, wing tagging, harnessing, gluing, tying, suturing, clamping, and implanting (Kenward 1987; Samuel and Fuller 1994). Since the earliest uses of radiomarking in birds, it has been recognized that the transmitter attachment method can affect a variety of aspects of behavior and survival. For instance, neck collars were shown to be effective in some cases (Marcstrom et al. 1989; Meyers 1996), but in other cases their use was accompanied by negative effects (Sorenson 1989). In one series of studies, tail-mounted transmitters did not affect mass or survival of northern goshawks (*Accipiter gentilis;* Kenward 1978, 1985), but it is understood that such transmitters must be light (2 percent or less of bird body mass), thereby limiting battery size and transmitter longevity. Reid et al. (1996) describe a method for replacing a tail-mounted transmitter when individuals can be recaptured. As with legbands and dyes, transmitter color also must be considered: Wilson and Wilson (1989) and Wilson et al. (1990) found that penguins pecked significantly less at black recorders (attached to the dorsal feathers by tape) than at other colors.

Radiotransmitters are similar to other markers in having variable effects that are influenced by attachment method, and the bird's species, age, and sex are necessary considerations. In an effort to securely attach large packages to birds, a number of researchers have experimented with various harness designs to hold transmitters on the bird's back. Several authors (Houston and Greenwood 1993; Kenward and Walls 1994; Kenward et al. 1996; Neudorf and Pitcher 1997) failed to show differences in survival or behavior of birds carrying various harness transmitter packages, but other examples show that such transmitters can have negative effects on bird behavior or survival (Hooge 1991; Klaasen et al. 1992; Pietz et al. 1993; Gammoneley and Kelly 1994; Ward and Flint 1995; table 2.3).

In an attempt to minimize the deleterious effects of harnesses on birds, researchers have experimented with different types of implants. Partial-implant (Mauser and Jarvis 1991; Pietz et al. 1995) as well as full-implant transmitters (Dzus and Clark 1996) have been used with some success, but such implants can result in short-term preening over the incision site and cause low rates of seroma and infection (Harms et al. 1997). One constraint with transmitter implants is that transmission distance is reduced when the antenna is implanted in the abdominal cavity, although some researchers (Korschgen et al. 1995; Petersen et al. 1995; Korschgen et al. 1996a, 1996b) have developed a technique to exit the antenna from the body, thereby augmenting transmission distance.

Since the mid-1980s several investigators have considered the effects of radiomarking on bird energetics. This issue is important for birds because radiomarking compounds the increase in energy required to carry additional mass by adding aerodynamic or hydrodynamic drag (Pennycuick 1975; Wilson et al. 1986; Culik and Wilson 1991). Streamlining of transmitters reduces aerodynamic drag and therefore minimizes their negative effects (Obrecht et al. 1988). Although some researchers failed to detect effects of transmitters weighing less than 4 percent of body mass (Sedinger et al. 1990; Gessaman et al. 1991b; Bakken et al. 1996), others (Pennycuick and Fuller 1987; Gessaman and Nagy 1988; Pennycuick et al. 1988, 1990, 1994; Gessaman et al. 1991a; Wilson and Culik 1992) indicate that, either for different species or larger transmitters, radiomarking can affect bird metabolism. This may be particularly important for large birds because they have proportionally less surplus power than smaller birds (Caccamise and Hedin 1985).

## Mammals

TAGGING   Although studies of mammals often involve marking, marker effects have been evaluated in few instances (Leuze 1980; Kenward 1982; White and Garrott 1990). This appears to be particularly true for externally mounted metal or plastic tags, with few studies evaluating effects of such markers (table 2.4) despite their widespread use. Internal PIT tags have been used in several species of mammals and evaluation tests (Fagerstone and Johns 1987; Schooley et al. 1993) so far have failed to detect significant negative effects.

MUTILATION   Toe clipping is a widely used tool for marking small mammals, and many studies have evaluated the effects of this technique on survival and body condition (table 2.4). Several studies have detected effects of toe clipping, but similar numbers of studies have failed to observe negative effects. In one case, different studies on the same species provided conflicting results (Ambrose 1972; Pavone and Boonstra 1985), suggesting that study methodology can influence outcome of marker evaluation studies. In two cases (Pavone and Boonstra 1985; Wood and Slade 1990), significant effects of toe clipping

**Table 2.4  Survey of Marker Evaluation Studies in Mammals**

| Marking Technique | Species | Parameters Tested | Lab (L) or Field (F) | Effect of Mark | Reference |
|---|---|---|---|---|---|
| Tagging | *Enhydra lutris* | Ph | F | Ph | Hatfield and Rathbun 1996 |
| | *Monachus schauinslandi* | Ac, Su | F | Ac | Henderson and Johanos 1988 |
| | *Odocoileus hemionus* | Pa, Ph | F | n.s. | Queal and Hlavachick 1968 |
| PIT tagging | *Mustela nigripes* | Ph | F | n.s. | Fagerstone and Johns 1987 |
| | *Mustela putorius* | Ph | L | n.s. | Fagerstone and Johns 1987 |
| | *Spermophilus townsendii* | Su | F | n.s | Schooley et al. 1993 |
| Fluorescent powder or paste | *Callorhinus ursinus* | Ph, Be | F | n.s. | Griben et al. 1984 |
| | *Mus domesticus* | Ac | L | Ac | Mikesic and Drickhamer 1992 |
| | *Peromyscus maniculatus* | Ph | F | n.s. | Stapp et al. 1994 |
| | | Mo | F | n.s. | Mullican 1988 |
| Toe clipping | *Apodemus sylvaticus* | Be | F | Be | Fairley 1982 |
| | | Ma | F | n.s. | Korn 1987 |
| | | Ma | F | n.s. | Fullagar and Jewell 1965 |
| | *Canis latrans* | Su, Be | F | n.s. | Andelt and Gipson 1980 |
| | *Clethrionomys glareolus* | Be | F | n.s. | Trojan and Wojciechowska 1964 |
| | | Ma | F | n.s. | Korn 1987 |
| | *Lepus americanus* | Be, Pa | F | n.s. | Baumgartner 1940 |
| | *Microtus ochrogaster* | Ma, Mo, Be, Su | F | Be, Mo | Wood and Slade 1990 |
| | *Microtus pennsylvanicus* | Su | F | Su | Pavone and Boonstra 1985 |
| | | Pr | L | n.s. | Ambrose 1972 |
| | *Sciurus niger* | Mo | F | n.s. | Dell 1957 |
| | *Tatera brantsii* | Ma | F | n.s. | Korn 1987 |
| | *Tatera leucogaster* | Ma | F | n.s. | Korn 1987 |

| | | | | | |
|---|---|---|---|---|---|
| Fur clipping | *Meles meles* | Ma, Co | F | n.s. | Stewart and Macdonald 1998 |
| Freeze branding | *Mus musculus* | Su, Gr | L | n.s. | Hadow 1972 |
| | *Rattus norvegicus* | Su | L | n.s. | Hadow 1972 |
| | *Sciurus niger* | Su | L | n.s. | Hadow 1972 |
| Radiotransmitters | *Apodemus sylvaticus* | Su, Ma, Mo, Be | L, F | Su | Wolton and Trowbridge 1985 |
| | *Arvicola terrestris* | Ac, Be, So, Re, Ma | F | Ac | Leuze 1980 |
| | *Castor canadensis* | Ph, Su, Be | F | Su | Davis et al. 1984 |
| | | Ph, Ma | F | n.s. | Guynn et al 1987 |
| | *Dipodomys merriami* | Pr, Mo | F | Pr, Mo | Daly et al. 1992 |
| | *Enhydra lutris* | Ph, Su | F | Ph, Su | Garshelis and Siniff 1983 |
| | *Lasiurus cinereus* | Fo | F | n.s. | Hickey 1992 |
| | *Lepus americanus* | Ma, Pr | F | n.s. | Brand et al. 1975 |
| | *Lutra canadensis* | Re | F | n.s. | Reid et al. 1986 |
| | *Lycaon pictus* | St | F | n.s. | Creel et al. 1997 |
| | *Marmota flaviventris* | Su, Gr, Re | F | n.s. | Van Vuren 1989 |
| | *Microtus pennsylvanicus* | En | L, F | n.s. | Berteaux et al. 1996 |
| | | So, Ma, Ac | L | So, Ma, Ac | Berteaux et al. 1994 |
| | | Ac | L, F | Ac | Hamley and Falls 1975 |
| | | Fe, Ac, Ma, Su | L, F | Fe, Ac, Ma, Su | Webster and Brooks 1980 |
| | *Microtus pinetorum* | Mo | F | n.s. | Madison et al. 1985 |
| | *Mus domesticus* | Ac, So, Su | L, F | Ac | Pouliquen et al. 1990 |
| | | Ac | L | Ac | Mikesic and Drickhamer 1992 |
| | *Myotis yumanensis* | Mo | L | Mo | Aldridge and Brigham 1988 |
| | | Ac, Ma, Fo | F | n.s. | Birks and Linn 1982 |
| | *Mustela vison* | Su, Gr, Re | F | Su | Eagle et al. 1984 |

*(continued)*

**Table 2.4** Continued

| Marking Technique | Species | Parameters Tested | Lab (L) or Field (F) | Effect of Mark | Reference |
|---|---|---|---|---|---|
| Radiotransmitters (continued) | Odocoileus hemionus | Re | F | n.s. | Hamlin et al. 1982 |
| | | Pr | F | n.s. | Garrott et al. 1985 |
| | Odocoileus virginianus | Su | F | n.s. | Ozoga and Clute 1988 |
| | Peromyscus leucopus | Gr, Re, Su | L, F | n.s. | Smith 1980 |
| | | Su, Re, Ma, Ac | F | n.s. | Ormiston 1985 |
| | | Pa | F | Pa | Ostfeld et al. 1993 |
| | Peromyscus maniculatus | Pr | F | n.s. | Douglass 1992 |
| | | Mo | F | n.s. | Mullican 1988 |
| | Plecopus auritus | Mo | L | Mo | Hughes and Rayner 1991 |
| | Rhinolophus ferrumequinum | Be, Su | F | n.s. | Stebbins 1982 |
| | Sciurus carolinensis | Ma, Su, Re | F | n.s. | Kenward 1982 |
| | Spermophilus franklinii | Su, Gr, Re | F | Su | Eagle et al. 1984 |
| | Vulpes macrotis | Su, Gr | F | Su, Gr | Cypher 1997 |

Effect of mark section corresponds to author's interpretation of effects. Articles surveyed were published in the peer-reviewed literature and consist of qualitative or quantitative evaluations of marking effects. We report the effect of markers as being significant or not, as interpreted by the authors in the article.
Gr = growth, Su = survival, Be = behavior, Mo = movements, Ac = activity, Ma = mass, Co = condition, Pa = parasitism/disease, Pr = predation, Ph = physiology, Sr = stress, Fo = foraging, Re = reproduction, En = energetics, So = social behavior.

were restricted to a particular sex, implying that not all animals are equally vulnerable to negative effects. However, the prevalence of toe clipping as a tool for marking small (and some larger, e.g., Andelt and Gipson 1980) mammals necessitates that additional studies address its potential effects. Other forms of mutilation include freeze branding, tattooing, and fur clipping (Hadow 1972; Cheeseman and Harris 1982; Fullagar and Jewel 1965; Stewart and Macdonald 1997), but in most cases effects of these markers have not been evaluated. However, Stewart and Macdonald (1997) did show that European badgers (*Meles meles*) could be effectively marked via fur clipping and that the mark has no effect on badger body condition. However, the applicability of this technique in colder climates, where loss of guard hairs may affect thermoregulation, requires further study.

RADIOTRANSMITTERS    Radiotelemetry of mammals can involve either external or internal attachment of packages. We found that small mammals had received more attention than other groups in marker evaluation studies and that only recently (Cypher 1997; Creel et al. 1997) had terrestrial carnivores received consideration for potential transmitter effects. Most studies we surveyed failed to find significant effects of transmitters (table 2.4), but discrepancies among studies performed on the same species were noted. For instance, the meadow vole (*Microtus pennsylvanicus*) was subjected to four evaluation tests: Bertaux et al. (1996) did not find a negative effect of transmitters (6.7 to 9.0 percent of body weight) on vole energetics, but other studies (Hamley and Falls 1975; Webster and Brooks 1980; Berteaux et al. 1994) showed that transmitters affected vole activity patterns. The fact that differential activity of radiomarked voles was not detected as higher energy expenditure (Berteaux et al. 1996) highlights the difficulty associated with attempting to generalize study results. It also has been shown that effects of radiotransmitters on small mammal behavior and movements often are either short-term (Wolton and Trowbridge 1985; Henderson and Johanos 1988; Mikesic and Drickhamer 1992) or specific to a particular sex (Daly et al. 1992), but the potential demographic implications of such marker effects have not been assessed.

## CRITIQUE OF MARKER EVALUATION STUDIES

A primary shortcoming of many marker evaluation studies is experimental design. Sometimes this is manifested as a lack of appropriate controls (i.e., unmarked animals) to which marked animals can be readily compared. The lack of control animals was common during the initial period of marker evaluation studies (i.e., 1940–1960), but more recent evaluation studies also possess this flaw (Davis et al. 1984; Eagle et al. 1984; Reid et al. 1986; Koehler et al. 1987; Mullican 1988). In other cases, authors include in the design a cohort of previously marked (Fairley 1982) or alternatively marked (Garrott et al. 1985; Wood and Slade 1990) animals as controls. However, this approach assumes that "control" animals are representative of the unmarked population even though alternative markers might cause important effects on their own. In this case, any comparison of marked versus "control" animals could result in an underestimation of effects of the targeted marker.

In some studies, control animals are not subjected to the same handling procedure as marked animals, thereby making marking effects indistinguishable from those of handling (Mears and Hatch 1976; Scheirer and Coble 1991). This can be particularly problematic in situations where handling causes significant stress or long-term effects, and as a result researchers may find it difficult to identify which procedure (marking or handling) requires modification. However, some studies (Lucas 1989) have correctly subjected controls to all the same handling procedures as the marked sample, thus allowing a more rigorous evaluation of the effects of the marker itself.

Marker evaluation studies often have sample sizes that are simply too small to detect a reasonable difference between marked and unmarked samples (White and Garrott 1990; Daly et al. 1992). Inadequate statistical power increases the likelihood of committing a type II error (Sokal and Rohlf 1981), thereby increasing the chance of failing to reject a null hypothesis of no significant marking effects when effects actually occur. Marker effects tend to be more readily detected in the laboratory because field studies often have smaller sample sizes and larger within-sample variance. Many field studies for which marker evaluation is apparently an offshoot (Guynn et al. 1987; Douglass 1992), or those that evaluate marker effects on large mammals (Hamlin et al. 1982), lack statistical power. Thus determining the detectable effect size and statistical power associated with a given marker evaluation study should be a necessary precursor to implementation of that study. Also, whenever possible, studies probably should be initiated under controlled laboratory conditions to reduce confounding effects of the environment. However, laboratory studies should be followed by evaluations in the field.

A common characteristic of marker evaluation studies is the use of subjective or qualitative measures of marking effects (Seale and Boraas 1974; Goldberg and Haas 1978; Andelt and Gipson 1980, 1981; Garshelis and Siniff 1983; Griben et al. 1984; Reid et al. 1986; Van Vuren 1989). Without rigorous statistical treatment of measured effects, results of such studies are of limited utility. Also, indices are sometimes used to infer direct effects (e.g., calculating capture–recapture rates to infer marking effects on survival), but if the index also measures other aspects of species biology (e.g., dispersal), such inferences might be spurious. Other evaluation studies are too short to derive meaningful conclusions regarding long-term effects, even though the latter effects may very well be the most demographically significant (see discussions by Daly et al. 1992 and Berteaux et al. 1996). Finally, sometimes statistically significant results are not considered to be biologically important because they are too small or uncommon (Korn 1987). Each of these approaches reduces the likelihood of identifying marker effects that may adversely affect the animal or the study results.

Our review of the marker evaluation literature reveals that a marker can affect a variety of aspects of animal biology, and that different types of evaluations provide different results. When biologists plan an evaluation of marker effects or when they interpret and apply results from previous evaluations, they must make decisions about which methods of evaluation are most appropriate for their objectives and subject species. Also, they must decide which results are most relevant to assessing the importance of an effect on the animals and their study objectives. For example, a biomechanical analysis of the effect of a marker provides an estimate of how much extra energy is needed to carry the marker, but it does not determine whether that increase in energy expenditure has other biological implications for the animal, such as reduced food delivery to young. A metabolic measurement might not indicate a significant change in $O_2$ or $CO_2$ between the marked and unmarked animals. Should we thus conclude there is no effect? Observations of behavior of the same animal might reveal that the marked animals spend more time resting than unmarked animals. All these evaluations could produce "significant" results, and yet contradict each other or provide different types of information. It is the responsibility of the researcher to conduct or consider the most appropriate marker evaluations relevant to the study objectives and the well-being of the study animals. The biologist must decide which effects are important.

## REVIEW OF CURRENT GUIDANCE AVAILABLE FOR CHOOSING MARKERS

Numerous criteria must be considered when selecting markers to be used in a given ecological study, including potential effects of markers on animals (Marion and Shamis 1977; Day et al. 1980; Friend et al. 1994; Nietfeld et al. 1994; Samuel and Fuller 1994). Researchers can review the literature, refer to colleagues, or consult with marker manufacturers or merchants. Limited guidance also is available through guidelines published by scientific journals (Animal Behavior Society/Animal Society for Animal Behavior 1986) and government agencies (Canadian Council on Animal Care 1980; Canadian Wildlife Service and U.S. Fish and Wildlife Service 1996) and in general references (Day et al. 1980; Friend et al. 1994; Heyer et al. 1994; Nietfeld et al. 1994; Wilson et al. 1996). In addition, several professional zoological societies (American Society of Ichthyologists and Herpetologists, American Fisheries Society, American Institute of Fisheries Biologists, Herpetologists' League, Society for the Study of Amphibians and Reptiles, American Ornithologists' Union, and American Society of Mammalogists) have published guidelines for the use of animals in field research. In general, these societies suggest that markers incur as little pain as possible and not restrict excessively behavior, physiology, and survival of study animals. We provide a brief overview of some current recommendations provided by zoological societies (see Animal Behavior Society 1986; American Society of Ichthyologists and Herpetologists, et al. 1987a, 1987b; Ad Hoc Committee 1988; Animal Care and Use Committee 1998).

## Tagging and mutilation

Tags used on fish, reptiles and amphibians should be of appropriate size and shape, but the use of tags that protrude from the body or are brightly colored is discouraged. It is recommended that for birds all bands be of appropriate size, but the use of nasal disks, saddles, patagial markers, dyes, and ultraviolet markers is discouraged.

Fin clipping is suggested as having minimal impact on survival and social structure of fish, and it is recommended as an appropriate technique if the specific fins to be clipped are expendable by the target species. Free-ranging reptiles and amphibians should not be toe-clipped unless the technique has been shown not to impair normal activity in the target species or a close relative, whereas for mammals it is recommended that all types of mutilations be avoided. Although birds are occasionally marked via mutilation (i.e., nail clipping, web-punching, feather clipping), no guidelines for use of this technique have been provided.

Marking fish, reptiles, and amphibians using techniques such as tissue removal, branding, freeze branding, and electrocauterization is generally acceptable, but the use of tattoos and paint is less desirable because of problems associated with dye visibility and legibility. Although fish can be marked with paint, the technique is discouraged for use on amphibian skin, for which nontoxic stains and dyes should be used. In cases where toxicity is unknown, laboratory trials should be undertaken before any field use. Few guidelines are provided for the use of brands, dyes, or paints with birds and mammals.

## Radiotransmitters

Most professional zoological societies address the issue of radiotransmitters specifically. Many fish, reptile, and amphibian species are not suitable for radiotelemetry because of their small size. However, for species that are amenable to telemetry, stomach implants and internally mounted transmitters should be small and coated with a biologically inert coating, and not interfere with physiology and behavior. Externally mounted transmitters should be shaped and attached to reduce chances of entanglement, irritation, or constriction. In the case of large birds, it is suggested that radiotransmitters weigh less than 1 percent of body mass to reduce negative effects on biomechanical performance. For smaller birds transmitters should not exceed 5 or 10 percent of body mass. Before use in the field, biologists should observe individuals in captivity to evaluate effects of radiomarking on behavior. For reptiles, amphibians, and most mammals, it is recommended that transmitters not exceed 10 percent of body mass.

## CRITIQUE OF GUIDELINES AVAILABLE FOR CHOOSING MARKERS

In general, the guidelines provided by zoological societies are too general for choosing a specific marker for a given study objective or species. Some recommendations made by professional societies even appear to ignore the findings of previous marker evaluation studies. For instance, fin removal is recommended by several fish societies as an appropriate method of marking many species, despite numerous instances in which the technique has been shown to affect fish biology (table 2.1). Also, it is recommended that mass of transmitters never exceed 10 percent of body mass of vertebrates, even though some transmitters weighing less than this have been shown to produce negative effects. Given that, at least for birds, the effect on flight power increases with body mass, the 10 percent threshold is clearly an arbitrary construct that does not apply to all species. Furthermore, a given mass or drag has different effects depending on the type of flight (e.g., soaring, flapping, sprint; Pennycuick and Fuller 1987; Pennycuick 1989; Pennycuick et al. 1989). Therefore, general guidelines (e.g., 10 percent, 5 percent, or 3 percent) can be misleading.

Although for terrestrial animals costs associated with carrying a load increase linearly with load mass (Taylor et al. 1980), intra- and interspecific differences in biology should preclude the establishment of a threshold that transcends taxonomic groups (Bertaux et al. 1994). Societies should assume a more comprehensive responsibility to indicate to their members possible shortcomings associated with various marking procedures, although it remains the responsibility of individual researchers to address the potential effects of markers on their study species.

Where information is lacking on a particular marker or its effect, zoological societies should encourage additional evaluation studies. It is imperative that markers used in research be acceptable to animal care and use committees before project initiation, and zoological societies should strive to harmonize their standards with those set by such committees and disseminate information regarding acceptable markers and protocols to committees. Otherwise, research using newly tested markers might be needlessly precluded because of an uncertainty regarding marker effects.

## ■ Survey of Recent Ecological Studies

In light of the measurable and obvious effects of markers to numerous vertebrate species (tables 2.1–2.4), we examined the treatment of potential marking effects in a sample of recent peer-reviewed literature. Specifically, we assessed the frequency with which researchers addressed potential marker effects, either by using methodologies that reduced potential effects or by testing for such effects qualitatively or statistically. We surveyed nine journals (table 2.5) that publish studies on a broad range of taxonomic groups for articles in which vertebrates had been marked; our survey included only articles in which the primary objective was to address general issues related to animal biology rather than to study marking effects.

We found that in most instances (90 percent, $n = 238$), authors did not address potential effects of marking, or at least did not report any such consideration in the article (table 2.5). Undoubtedly, some authors attempted to minimize or evaluate marker effects but chose not to report these efforts in the article; however, we suspect that such cases were uncommon. We found no indication that the failure to evaluate marker effects was weighted toward a given taxonomic group or journal. In 3 percent of cases authors assumed explicitly that markers did not affect animals. Such assumptions were usually based on previous reports showing no significant marking effects in the target (or related) species (Artiss and Martin 1995; Ralls et al. 1995) or on

**Table 2.5    Review of Treatment of Potential Marking Effects in the Ecological Literature**

| Journal | Number of Papers Surveyed | No Marking Effects (implicit) | No Marking Effects (explicit) | Marking Tests or Modifications |
|---|---|---|---|---|
| *American Naturalist* | 4 | 4 | 0 | 0 |
| *Animal Behaviour* | 62 | 59 | 2 | 1 |
| *Canadian Journal of Zoology* | 50 | 46 | 0 | 4 |
| *Conservation Biology* | 10 | 8 | 0 | 2 |
| *Ecology* | 31 | 28 | 2 | 1 |
| *Journal of Animal Ecology* | 15 | 15 | 0 | 0 |
| *Journal of Wildlife Management* | 37 | 30 | 3 | 4 |
| *Oecologia* | 12 | 10 | 0 | 2 |
| *Oikos* | 17 | 15 | 0 | 2 |
| Total | 238 | 215 | 7 | 16 |
| Percentage | | 90 | 3 | 7 |

Journals, numbers of papers surveyed in which free-ranging vertebrates were marked, and percentage of papers that assumed (explicitly or implicitly) that marking had no effect on measurements. All papers reviewed were published in 1995.

requirements of the statistical method chosen to analyze the data (Burger et al. 1995).

Only 7 percent of the articles surveyed included information directly pertaining to potential marking effects. Some authors attempted to minimize marker effects by allowing postmarking recovery to take place in captivity (Baupre 1995; Forrester 1995; Nelson 1995; Shine and Fitzgerald 1995). Although this approach might be effective when handling and marking are stressful or invasive (i.e., internally mounted radiotransmitters), it might also bias study results if captivity affects behavior or survival after release. For species that are highly vulnerable to stress from captivity, prolonged recovery under controlled situations could be less desirable than immediate release (Hart and Summerfelt 1975).

Other attempts to minimize marker effects included adjusting markers to fit individual animals (Powell and Bjork 1995) or using expandable and breakaway markers (Adams et al. 1995). Some authors justified the use of a marker by describing qualitatively how animals appeared to behave normally after marking (Baupre 1995; Brawn et al. 1995; Christian and Bedford 1995; Riley et al. 1995). Subjective evaluation of potential marking effects is a good first step, but whenever possible quantitative measurements of the biology of marked versus unmarked animals (Vekasy et al. 1996) or comparisons of pre- and postmarking behavior of individuals should be included.

When markers were suspected of exerting short-term effects on animals, researchers excluded data obtained during an arbitrary period (2–14 days after marking; Baupre 1995; Bloomer et al. 1995; Roberts et al. 1995; Migoya and Baldassarre 1995; Miller et al. 1995). However, this practice can result in data bias if marking causes subsequent increases in mortality of young or frail animals; the result would be an inflation of survival estimates. Two studies (Cotter and Gratto 1995; Cucco and Malacarne 1995) evaluated the effect of one marker type using an alternatively marked cohort as control, without validating that negative effects of the alternative marker did not occur.

We found no indication from our survey that authors addressed potential long-term effects of markers, although one study (Booth 1995) inferred from the lack of external damage in study animals that long-term tagging mortality was absent or negligible. In another case (Höglung et al. 1995) marked and unmarked animals were monitored and data from both cohorts were apparently pooled, but no mention was made of trying to evaluate possible marker effects. It is also notable that almost all (more than 90 percent, $n = 16$) instances in which authors addressed marking effects the studies involved animals that were marked with radiotransmitters. This implies that various forms of mutilation and tagging or banding are either assumed by researchers to have little or no effect or else not afforded the same scrutiny in the editorial process.

To summarize, our review of ecological studies found that in general authors tended to overlook or not acknowledge potential marking effects. In few cases potential marking effects were addressed, but usually only in a superficial manner or one that could produce bias in the resulting data.

## ■ Future Approaches

This review clearly indicates that there are a large number of reported marking effects from commonly used markers, there is a lack of thorough guidance from zoological societies in choosing a marker, and there is a common failure to address potential marking effects in the ecological literature. Thus, we provide the following recommendations for using markers and for additional research into the effects of marking wildlife.

## STUDY PROTOCOLS AND TECHNOLOGICAL ADVANCES

Depending on study objectives, target species, and possible problems associated with the handling and marking process, it is not always necessary or desirable to apply marks to animals. For example, in some studies genetic or mineral markers in animal tissues, excreta, or blood, may allow for identification of individuals or populations. However, reliable identification using genetics is, at present, labor intensive and costly, can require recapture, and is largely untested under field conditions. As another alternative, it might be possible to use naturally occurring variable color markings or unique morphological features for recognition of individuals. For example, some amphibian species are amenable to visual recognition of individuals, and photographs or sketches can be used to record characteristics of individuals in a population (Forester 1977; Tilley 1977; Andreone 1986; Loafman 1991). Individual recognition via natural markings also has been used occasionally for fish (Nakano 1995) and some bird species (Scott 1978), but has probably found its greatest utility with large mammals (Pennycuick and Rudnai 1970; Clutton-Brock and Guiness 1975; Ingram 1978). For mammals, researchers usually rely on unique facial features to identify individuals; in most cases the method appears reliable although validation is desirable and there are limits to the number of animals in a population that are individually recognizable. Pennycuick (1978) reviewed limitations and difficulties associated with the use of natural markings in free-ranging animals; where the use of this technique meets study objectives, it is more desirable than artificial marking.

It is essential that field personnel be adequately trained in the proper handling and marking of animals (Ad Hoc Committee on the Use of Wild Birds in Research 1988; Livezey 1990; Friend et al. 1994; Samuel and Fuller 1994). This may involve practice with captive individuals of the targeted or surrogate species, or animal models. Improper marker application by poorly trained field personnel can result in serious problems for study animals (Perry and Beckett 1966), and well-trained personnel not only minimize the potential for negative marker effects but also are able to process and release study animals more rapidly (Korschgen et al. 1996a, 1996b), thereby reducing potential negative effects of handling and marking.

Although significant progress has been made in the development of less harmful marks, additional steps are necessary. Few technological developments can contribute to minimizing effects of mutilations, although routine use of surgical instruments and sterile conditions lessens occurrence of disease and mortality. Development of smaller tags and bands can contribute to less invasive marking. In the past, one constraint of reducing tag size was the ability to read numbers or color markings identifying individual animals. However, carefully designed protocols can increase effectiveness (Howitz 1981; Ottaway et al. 1984). The use of PIT tags eliminates size constraints for many animals. However, other tags also hold promise, such as passively applied (i.e., usually ingested) group recognition markers (Crier 1970; Lindsey 1983; Follmann et al. 1987; Johnston et al. 1998). These may prove particularly useful when animal handling is undesirable and individual recognition is not necessary.

Radiotransmitters probably will remain an important and widely used form of marking, and technological developments should include aerodynamically or hydrodynamically shaped, smaller, and lighter packages (Obrecht et al. 1988; Bannasch et al. 1994) and development of new attachment methods. Because most of the bulk caused by transmitters is attributable to batteries, packages could be significantly lightened if smaller high-capacity batteries were developed. However, in some cases significant reductions in transmitter size or mass are not possible without forgoing battery longevity.

Thorough testing and reporting of animal marking procedures (Korschgen et al. 1996a; Meyers 1996) provides other investigators information useful for choosing among various attachments and trying them with a new species or under new circumstances. Detailed descriptions of procedures (Snyder et al. 1989) and modifications to procedures (Keister et al. 1988; Reid et al. 1996; Adams and Campbell 1996; Nagendran et al. 1994) provide potential users with additional options. Finally, the results of applying various attachments in field situations are very important and should include information about effects of the capture and handling that must accompany marking (Hill and Talent 1990). Sometimes the procedures associated with capture (Vehrencamp and Halpenny 1981) and handling (Rotella and Ratti 1990; Caccamise and Stauffer 1994; McGowan and Caffrey 1994) can be as important a component in the marking process as the marker itself. Our review reveals such a variety of marker effects (depending on marker type, species, age, sex, attachment method, environment, and season) that careful development, testing, documentation, and reporting of experience with markers will continue to be very useful in identifying and dealing with marker effects. Zoological societies should assume the responsibility for setting reasonable standards for the marking of animals by providing detailed guidelines.

## MARKER EVALUATION STUDIES

We recommend increased efforts to test for effects of markers, changes in existing protocols and recommendations, and more publication of the results of marker evaluation studies. Evaluation studies are a critical step in the marker validation process, but many commonly used markers have been evaluated for only a few species. Our review highlighted several specific deficiencies in our understanding of marker effects, such as effects of radiotransmitters in fish, efficacy of internally versus externally mounted transmitters (particularly stomach implants), influence of external tags on amphibians and mammals, and effects of transmitters on large mammals, particularly carnivores. It will not be possible to evaluate the effects of all markers in all species, but results of evaluation studies will be more useful if researchers focus their initial evaluation studies on broadly applicable markers and species that share several biological features with other species of interest. We think that PIT tags will become very important for marking animals in the future, and thus urge researchers to undertake the necessary evaluation studies without delay. All evaluation studies should be designed to measure marker effects across ages and sexes.

Evaluation studies must include comparison to unmarked (or alternatively marked) controls and adequate sample sizes. Statistical power analysis can be used to calculate the minimum sample size required to detect significant effects. Captive or laboratory conditions should be used whenever possible, although researchers should ensure that biological parameters being measured are not significantly affected by captivity (Berteaux et al. 1996). Ultimately, results of captive studies should be validated in the field (Wilson and Culik 1992; Wallace et al. 1994; Cohen 1994). Also, evaluation studies should take advantage of sensitive techniques for estimating subtle negative effects, such as biomechanical modeling (Pennycuick 1989) and doubly labeled water (Gessaman and Nagy 1988). Marker evaluation studies should attempt to address indirect effects of markers on survival (e.g., predation, starvation, disease) and behavior (e.g., altered foraging or parental abandonment of juveniles to compensate for higher energy demands), and focus not only on short-term but also on long-term effects of markers (Reed and Oring 1993; Buehler et al. 1995; Meyers et al. 1996).

Marker evaluation studies should be published in the peer-reviewed literature. It was not possible for us to determine whether a publication bias existed in our sample of marker evaluation studies, but we suspect a tendency exists for publishing studies that show significant negative effects over those failing to show effects. Such a bias can misrepresent the effect of markers, and it inhibits dissemination of information about the best marking methods. Clearly, some journals must encourage authors to perform and publish marker evaluation studies. A similar suggestion was proposed to overcome the apparent publication bias in biological studies showing nonsignificant results (Csada et al. 1996). Publication of all marker evaluation studies would provide researchers and animal care and use committees with current knowledge regarding acceptable animal markers.

It remains the responsibility of reviewers and editors to ensure that appropriate marker evaluation studies are cited or conducted in the published ecological and wildlife literature. In some cases, it may be sufficient to justify the use of a given marker by citing previous studies that have failed to detect effects, but when researchers are applying a marker to a new (or unrelated) species, or using a marker that has not been evaluated previously, appropriate tests must be undertaken. Again, experimental design and sample size are paramount considerations for tests. Publication of technique validation studies is common in other scientific fields (see Notice to Contributors, *Journal of Reproduction and Fertility,* 1991 93[1]:ii–iv) and should be encouraged in animal ecology. Because current funding sources often do not support technique evaluation studies, researchers must find other sources of funding for such work, such as private animal welfare organizations, and they must encourage traditional sources of funding to support evaluation of marking methods. Only through additional studies and published results will it be possible for colleagues to thoroughly evaluate potential marker effects and choose the most appropriate markers.

### Acknowledgments

We are grateful to Cynthia Kapke and Kirk Bates for assisting with our search of the literature. Wanda Manning of the library at the Patuxent Wildlife Research Center and Carolyn Fritschle at the R. R. Olendorf Memorial Library at the Snake River Field Station also assisted in obtaining information. Georgia Garling, Kirk Bates, and Peggy Zumwalt assisted with manuscript preparation. We thank the participants of the conference in Erice, Italy, for their helpful comments, and R. M. DeGraaf, J. A. Litvaitis, and P. R. Sievert for their reviews.

### Literature Cited

Ad Hoc Committee on the Use of Wild Birds in Research. 1988. Guidelines for the use of wild birds in research. *Auk* 105 (1, suppl.).

Adams, I. T. and G. D. Campbell. 1996. Improved radio-collaring for southern flying squirrels. *Wildlife Society Bulletin* 24: 4–7.

Adams, L. G., F. J. Singer, and B. W. Dale. 1995. Caribou calf mortality in Denali National Park, Alaska. *Journal of Wildlife Management* 59: 584–594.

Aldridge, H. D. J. N. and R. M. Brigham. 1988. Load carrying and maneuverability in an insectivorous bat: A test of the 5% "rule" of radio-telemetry. *Journal of Mammalogy* 69: 379–382.

Ambrose, H. W. 1972. Effect of habitat familiarity and toe-clipping on rate of owl predation in *Microtus pennsylvanicus*. *Journal of Mammalogy* 53: 909–912.

American Society of Ichthyologists and Herpetologists, American Fisheries Society, and the American Institute of Fisheries Research Biologists. 1987a. Guidelines for the use of fishes in field research. *Copeia* 1987(suppl.).

American Society of Ichthyologists and Herpetologists, the Herpetologists' League, and the Society for the Study of Amphibians and Reptiles. 1987b. Guidelines for the use of live amphibians and reptiles in field research. *Journal of Herpetology* 4(suppl.): 1–14.

Andelt, W. F. and P. S. Gipson. 1980. Toe-clipping coyotes for individual identification. *Journal of Wildlife Management* 43: 944–951.

Andelt, W. F. and P. S. Gipson. 1981. Toe-clipping coyotes: A reply. *Journal of Wildlife Management* 45: 1007–1009.

Anderson, A. 1980a. Band wear in the fulmar. *Journal of Field Ornithology* 51: 101–109.

Anderson, A. 1980b. The effects of age and wear on color bands. *Journal of Field Ornithology* 51: 213–219.

Andreone, F. 1986. Considerations on marking methods in newts, with particular reference to a variation of the "belly pattern" marking technique. *Bulletin of the British Herpetological Society* 16: 36–37.

Animal Behavior Society/Animal Society for Animal Behavior. 1986. ABS/ASAB guidelines for the use of animals in research. *Animal Behavior Society Newsletter* 31: 7–8.

Animal Care and Use Committee. 1998. Guidelines for the capture, handling, and care of mammals as approved by the American Society of Mammalogists. *Journal of Mammalogy* 79: 1416–1431.

Armstrong, G. C. 1949. Mortality, rate of growth, and fin regeneration of marked and unmarked lake trout fingerlings at the provincial fish hatchery, Port Arthur, Ontario. *American Fisheries Society* 78: 129–131.

Artiss, T. and K. Martin. 1995. Male vigilance in white-tailed ptarmigan, *Lagopus lagopus:* Mate guarding or predator detection? *Animal Behaviour* 49: 1249–1258.

Ashton, R. E. 1994. Tracking with radioactive tags. In R. H. Heyer, M. H. Donnelly, R. W. McDiarmid, L. A. C. Hayek, and M. S. Foster, eds., *Measuring and monitoring biological diversity: Standard methods for amphibians,* 159–163. Washington, D.C.: Smithsonian Press.

Bailey, E. E., G. E. Woolfenden, and W. B. Robertson, Jr. 1987. Abrasion and loss of bands from Dry Tortugas sooty terns. *Journal of Field Ornithology* 58: 413–424.

Bakken, G. S., P. S. Reynolds, K. P. Kenow, C. E. Korschgen, and A. F. Boysen. 1996. Thermoregulatory effects of radiotelemetry transmitters on mallard ducklings. *Journal of Wildlife Management* 60: 669–678.

Bannasch, R., R. P. Wilson, and B. Culik. 1994. Hydrodynamic aspects of design and attachment of a back-mounted device in penguins. *Journal of Experimental Biology* 194: 83–96.

Bardach, J. E. and E. D. LeCren. 1948. A preopercular tag for perch. *Copeia* 1948: 222–224.

Bartelt, G. A. and D. H. Rusch. 1980. Comparison of neck bands and patagial tags for marking American coots. *Journal of Wildlife Management* 44: 236–241.

Barton, N. W. H. and D. C. Houston. 1991. The use of titanium dioxide as an inert marker for digestion studies in raptors. *Comparative Biochemistry and Physiology* 100A: 1025–1029.

Baumgartner, L. L. 1940. Trapping, handling, and marking fox squirrels. *Journal of Wildlife Management* 4: 444–450.

Baupre, S. J. 1995. Effects of geographically variable thermal environment on the bioenergetics of mottled rock rattlesnakes. *Ecology* 76: 1655–1665.

Belant, J. L. and T. W. Seamans. 1993. Evaluation of dyes and techniques to color-mark incubating herring gulls. *Journal of Field Ornithology* 64: 440–451.

Bergmann, P. J., L. D. Flake, and W. L. Tucker. 1994. Influence of brood rearing on female mallard survival and effects of harness-type transmitters. *Journal of Field Ornithology* 65: 151–159.

Berteaux, D., R. Duhamel, and J. M. Bergeron. 1994. Can radio-collars affect dominance relationships in *Microtus? Canadian Journal of Zoology* 72: 785–789.

Berteaux, D., F. Masseboeuf, J.-M. Bonzom, J.-M. Bergeron, D. W. Thomas, and H. Lapierre. 1996. Effect of carrying a radiocollar on expenditure of energy by meadow voles. *Journal of Mammalogy* 72: 359–363.

Birks, J. D .S. and I. J. Linn. 1982. Studies of home range of the feral mink, *Mustela vison. Symposium of the Zoological Society of London* 49: 231–257.

Bloomer, S. E. M., T. Willebrand, I. M. Keith, and L. B. Keith. 1995. Impact of helminth parasitism on a snowshoe hare population in central Wisconsin: A field experiment. *Canadian Journal of Zoology* 73: 1891–1898.

Blums, P., A. Mednis, and J. D. Nichols. 1994. Retention of web tags and plasticine-filled leg bands applied to day-old ducklings. *Journal of Wildlife Management* 58: 76–81.

Booth, D. J. 1995. Juvenile groups in a coral-reef damselfish: Density-dependent effects on individual fitness and population demography. *Ecology* 76: 91–106.

Bradford, D. F. 1984. Temperature modulation in a high-elevation amphibian, *Rana muscosa. Copeia* 1984: 966–976.

Brand, C. J., R. H. Vowles, and L. B. Keith. 1975. Snowshoe hare mortality monitored by telemetry. *Journal of Wildlife Management* 39: 741–747.

Brawn, J. D., J. R. Karr, and J. D. Nichols. 1995. Demography of birds in a neotropical forest: Effects of allometry, taxonomy, and ecology. *Ecology* 76: 41–51.

Brown, W. S. 1976. A ventral scale clipping system for permanently marking snakes (Reptilia, Serpentes). *Journal of Herpetology* 10: 247–249.

Brynildson, O. M. and C. L. Brynildson. 1967. The effect of pectoral and ventral fin removal on survival and growth of wild brown trout in a Wisconsin stream. *Transactions of the American Fisheries Society* 96: 353–355.

Bub, H. and H. Oelke. 1980. *Markierungsmethoden für Vögel.* Wittenberg, Germany: Neue Brehm Bücherei. A. Ziemsen Verlag.

Buehler, D. A., J. D. Fraser, M. R. Fuller, L. S. McAllister, and J. K. D. Seegar. 1995. Captive and field-tested radio transmitter attachments for bald eagles. *Journal of Field Ornithology* 66: 173–180.

Bulak, J. S. 1983. Evaluation of Floy Anchor tags for short-term mark–recapture studies with blueback herring. *North American Journal of Fisheries Management* 3: 91–94.

Burger, L. W., T. V. Dailey, E. W. Kurzejeski, and M. R. Ryan. 1995. Survival and cause-specific mortality of northern bobwhite in Missouri. *Journal of Wildlife Management* 59: 401–410.

Burley, N. 1985. Leg-band color and mortality patterns in captive breeding populations of zebra finches. *Auk* 102: 647–651.

Burley, N. T. 1988. Wild zebra finches have band-colour preferences. *Animal Behaviour* 36: 1235–1237.

Burley, N., G. Krantzberg, and P. Radman. 1982. Influence of colour-banding on the conspecific preferences of zebra finches. *Animal Behaviour* 30: 444–455.

Burtt, E. H. Jr. and R. M. Tuttle. 1983. Effect of timing of banding on reproductive success of tree swallows. *Journal of Field Ornithology* 54: 319–323.

Caccamise, D. F. and R. S. Hedin. 1985. An aerodynamic basis for selecting transmitter loads in birds. *Wilson Bulletin* 97: 306–318.

Caccamise, D. F. and P. C. Stouffer. 1994. Risks of using alpha-chloralose to capture crows. *Journal of Field Ornithology* 65: 458–460.

Calvo, B. and R. W. Furness. 1992. A review of the use and the effects of marks and devices on birds. *Ringing and Migration* 13: 129–151.

Campbell, B. H. and E. F. Becker. 1991. Neck collar retention in dusky Canada geese. *Journal of Field Ornithology* 62: 521–527.

Canadian Council on Animal Care. 1980. *Guide to the care and use of experimental animals,* Vol. 1. Ottawa: Canadian Council on Animal Care.

Canadian Wildlife Service and U.S. Fish and Wildlife Service. 1996. *North American bird banding,* Vol. 1. Ottawa/Washington, D.C.: Canadian Wildlife Service and U.S. Fish and Wildlife Service.

Carline, R. F. and O. M. Brynildson. 1972. Effects of the Floy anchor tag on the growth and survival of brook trout (*Salvelinus fontinalis*). *Journal of the Fisheries Research Board of Canada* 29: 458–460.

Castelli, P. M. and R. E. Trost. 1996. Neck bands reduce survival of Canada geese in New Jersey. *Journal of Wildlife Management* 60: 891–898.

Cavanagh, P. M., C. R. Griffin, and E. M. Hoopes. 1992. A technique to color-mark incubating gulls. *Journal of Field Ornithology* 63: 263–267.

Charland, M. B. 1991. Anesthesia and transmitter implantation effects on gravid garter snakes (*Thamnophis siritalis* and *T. elegans*). *Herpetological Review* 22: 46–47.

Cheeseman, C. L. and S. Harris. 1982. Methods of marking badgers (*Meles meles*). *Journal of Zoology* 197: 289–292.

Christian, K. A. and G. S. Bedford. 1995. Seasonal changes in thermoregulation by the frillneck lizard, *Chlamydosaurus kingii,* in tropical Australia. *Ecology* 76: 124–132.

Clancy, D. W. 1963. The effect of tagging with Petersen disc tags on the swimming ability of fingerling steelhead trout (*Salmo gairdneri*). *Journal of the Fisheries Research Board of Canada* 20: 969–981.

Clarke, R. D. 1972. The effect of toe clipping on survival in Fowler's toad (*Bufo woodhousei fowleri*). *Copeia* 1972: 182–185.

Clutton-Brock, T. H. and F. Guiness. 1975. Behaviour of red deer (*Cervus elaphus* L.) at calving time. *Behaviour* 55: 287–300.

Coble, D. W. 1967. Effects of fin-clipping on mortality and growth of yellow perch with a review of similar investigations. *Journal of Wildlife Management* 31: 173–180.

Coble, D. W. 1971. Effects of fin clipping and other factors on survival and growth of smallmouth bass. *Transactions of the American Fisheries Society* 100: 460–473.

Coble, D. W. 1972. Vulnerability of fin clipped bluegill to largemouth bass predation in tanks. *Transactions of the American Fisheries Society* 101: 563–565.

Cohen, R. R. 1994. Tarsal widths and band sizes for tree swallows and violet-green swallows. *North American Bird Bander* 19: 137–139.

Cotter, R. C. and C. J. Gratto. 1995. Effects of nest and brood visits and radio transmitters on rock ptarmigan. *Journal of Wildlife Management* 59: 93–98.

Crawford, R. W. 1958. Behaviour, growth and mortality in the bluegill, *Lepomis macrochirus* Rafinesque, following fin clipping. *Copeia* 1958: 330–331.

Creel, S., N. M. Creel, and S. Monfort. 1997. Radiocollaring and stress hormones in African wild dogs. *Conservation Biology* 11: 544–548.

Crier, J. K. 1970. Tetracyclines as a fluorescent marker in bones and teeth of rodents. *Journal of Wildlife Management* 34: 829–834.

Cristol, D. A., C. S. Chiu, S. M. Peckham, and J. F. Stoll. 1992. Color bands do not affect dominance status in captive flocks of wintering dark-eyed juncos. *Condor* 94: 537–539.

Croll, D. A., J. K. Jansen, M. E. Goebel, P. L. Boveng, and J. L. Bengtson. 1996. Foraging behavior and reproductive success in chinstrap penguins: The effects of transmitter attachment. *Journal of Field Ornithology* 67: 1–9.

Croll, D. A., S. D. Osmek, and J. L. Bengtson. 1991. An effect of instrument attachment on foraging trip duration in chinstrap penguins. *Condor* 93: 777–779.

Csada, R. D., P. C. James, and R. H. M. Espie. 1996. The "file drawer problem" of non-significant results: Does it apply to biological research? *Oikos* 76: 591–593.

Cucco, M. and R. Malacarne. 1995. Increase of parental effort in experimentally enlarged broods of pallid swifts. *Canadian Journal of Zoology* 73: 1387–1395.

Culik, B. and R. P. Wilson. 1991. Swimming energetics and performance of instrumented Adélie penguins *Pygoscelis adeliae*. *Journal of Experimental Biology* 158: 355–368.

Curtis, P. D., C. E. Braun, and R. A. Ryder. 1983. Wing markers: Visibility, wear, and effects on survival of band-tailed pigeons. *Journal of Field Ornithology* 54: 381–386.

Custer, C. M., T. W. Custer, and D. W. Sparks. 1996. Radio telemetry documents 24-hour feeding activity of wintering lesser scaup. *Wilson Bulletin* 108: 556–566.

Cypher, B. L. 1997. Effects of radiocollars on San Joaquin kit foxes. *Journal of Wildlife Management* 61: 1412–1423.

Daly, M., M. I. Wilson, P. R. Behrends, and L. F. Jacobs. 1992. Sexually differentiated effects of radiotransmitters on predation risk and behaviour in kangaroo rats *Dipodomys merriami*. *Canadian Journal of Zoology* 70: 1851–1855.

Daugherty, C. H. 1976. Freeze-branding as a technique for marking anurans. *Copeia* 1976: 836–838.

Davis, J. R., A. F. Von Recum, and D. C. Guynn. 1984. Implantable telemetry in beaver. *Wildlife Society Bulletin* 12: 322–324.

Day, R. D., S. D. Schemnitz, and R. D. Taber. 1980. Capturing and marking wild animals. In S. D. Shemnitz, ed., *Wildlife management techniques manual,* 61–98. Bethesda, Md.: Wildlife Society.

Dell, J. 1957. Toe clipping varying hares for track identification. *New York Fish and Game Journal* 4: 61–68.

DeRoche, S. E. 1963. Slowed growth of lake trout following tagging. *Transactions of the American Fisheries Society* 92: 185–186.

Dodd, C. K. 1993. The effect of toeclipping on sprint performance of the lizard *Cnemidophorus sexlineatus. Journal of Herpetology* 27: 209–213.

Donnelly, M. A., C. Guyer, J. E. Juterbock, and R. A. Alford. 1994. Techniques for marking amphibians. In R. H. Heyer, M. H. Donnelly, R. W. McDiarmid, L. A. C. Hayek, and M. S. Foster, eds., *Measuring and monitoring biological diversity: Standard methods for amphibians,* 277–284. Washington, D.C.: Smithsonian Institution Press.

Douglass, R. J. 1992. Effects of radio-collaring on deer mouse survival and vulnerability to ermine predation. *American Midland Naturalist* 127: 198–199.

Dzus, E. H. and R. G. Clark. 1996. Effects of harness-style and abdominally implanted transmitters on survival and return rates of mallards. *Journal of Field Ornithology* 67: 549–557.

Eagle, T. C., J. Choromanski-Norris, and V. B. Kuechle. 1984. Implanting radio transmitters in mink and ground squirrels. *Wildlife Society Bulletin* 12: 180–184.

Eames, M. J. and M. K. Hino. 1983. An evaluation of four tags suitable for marking juvenile chinook salmon. *Transactions of the American Fisheries Society* 112: 464–468.

Ely, C. R. 1990. Effects of neck bands on the behavior of wintering greater white-fronted geese. *Journal of Field Ornithology* 61: 249–253.

Emlen, S. T. 1968. A technique for marking anuran amphibians for behavioral studies. *Herpetology* 24: 172–173.

Fagerstone, K. A. and B. E. Johns. 1987. Transponders as a permanent identification markers for domestic ferrets, black-footed ferrets, and other wildlife. Journal of Wildlife Management 51: 294–297.

Fairley, J. S. 1982. Short-term effects of ringing and toe-clipping on the recapture of wood mice (*Apodemus sylvaticus*). *Journal of Zoology* 197: 295–297.

Ferner, J. W. 1979. A review of marking techniques for amphibians and reptiles. *Society for the Study of Amphibians and Reptiles* 9, 42 pp.

Fitch, H. S. and H. W. Shirer. 1971. A radiotelemetric study of spatial relationships in some common snakes. *Copeia* 1971: 118–128.

Follmann, E. H., P. J. Savarie, D. G. Ritter, and G. M. Baer. 1987. Plasma marking of Arctic foxes with iophenoxic acid. *Journal of Wildlife Diseases* 23: 709–712.

Forester, D. C. 1977. Comments on the female reproductive cycle and philopatry by *Desmognathus ochrophaeus* (Amphibia, Urodela, Plethodontidae). *Journal of Herpetology* 11: 311–316.

Forrester, G. E. 1995. Strong density-dependent survival and recruitment regulate the abundance of a coral reef fish. *Oecologia* 103: 275–282.

Forsman, E. D., A. B. Franklin, F. M. Oliver, and J. P. Ward. 1996. A color band for spotted owls. *Journal of Field Ornithology* 67: 507–510.

Friend, M., D. E. Toweill, R. L. Brownell, V. F. Nettles, D. S. Davis, and W. J. Foreyt. 1994. Guidelines for proper care and use of wildlife in field research. In T. A. Bookhout, ed., *Research and management techniques for wildlife and habitats,* 96–124. Bethesda, Md.: Wildlife Society.

Fullagar, P. J. and P. A. Jewell. 1965. Marking small rodents and the difficulties of using leg rings. *Journal of Zoology* 197: 224–228.

Gammonley, J. H. and J. R. Kelley, Jr. 1994. Effects of back-mounted radio packages on breeding wood ducks. *Journal of Field Ornithology* 65: 530–533.

Garrettson, P. R. and F. C. Rohwer. 1996. Loss of an abdominally implanted radio transmitter by a wild blue-winged teal. *Journal of Field Ornithology* 67: 355–357.

Garrott, R. A., R. M. Bartmann, and G. C. White. 1985. Comparison of radio-transmitter packages relative to deer fawn mortality. *Journal of Wildlife Management* 49: 758–759.

Garshelis, D. L. and D. B. Siniff. 1983. Evaluation of radio-transmitter attachments for sea otters. *Wildlife Society Bulletin* 11: 378–383.

Germano, D. J. and D. F. Williams. 1993. Field evaluation of using passive integrated transponders (PIT) tags to permanently mark lizards. *Herpetological Review* 24: 54–56.

Gessaman, J. A., M. R. Fuller, P. J. Pekins, and G. E. Duke. 1991a. Resting metabolic rate of golden eagles, bald eagles, and barred owls with a tracking transmitter or an equivalent load. *Wilson Bulletin* 103: 261–265.

Gessaman, J. A. and K. A. Nagy. 1988. Transmitter loads affect the flight speed and metabolism of homing pigeons. *Condor* 90: 662–668.

Gessaman, J. A., G. W. Workman, and M. R. Fuller. 1991b. Flight performance, energetics and water turnover of tippler pigeons with a harness and dorsal load. *Condor* 93: 546–554.

Godfrey, R. D. Jr., A. M. Fedynich, and E. G. Bolen. 1993. Fluorescent particles for marking waterfowl without capture. *Wildlife Society Bulletin* 21: 283–288.

Goforth, W. R. and T. S. Baskett. 1965. Effects of experimental color marking on pairing of captive mourning doves. *Journal of Wildlife Management* 29: 543–553.

Goldberg, J. S. and W. Haas. 1978. Interactions between mule deer dams and their radio-collared and unmarked fawns. *Journal of Wildlife Management* 42: 422–425.

Griben, M. R., H. R. Johnson, B. B. Gallucci, and V. F. Gallucci. 1984. A new method to mark pinnipeds as applied to the northern fur seal. *Journal of Wildlife Management* 48: 945–949.

Guttman, S. I. and W. Creasey. 1973. Staining as a technique for marking tadpoles. *Journal of Herpetology* 7: 388.

Guynn, D. C., J. R. Davis, and A. F. Von Recum. 1987. Pathological potential of intraperitoneal transmitter implants in beavers. *Journal of Wildlife Management* 51: 605–606.

Hadow, H. H. 1972. Freeze-branding: A permanent marking technique for pigmented mammals. *Journal of Wildlife Management* 36: 645–649.

Hagan, J. M. and J. M. Reed. 1988. Red color bands reduce fledging success in red-cockaded woodpeckers. *Auk* 105: 498–503.

Hamley, J. M. and J. B. Falls. 1975. Reduced activity in transmitter-carrying voles. *Canadian Journal of Zoology* 53: 1476–1478.

Hamlin, K. L., R. J. Mackie, J. G. Mundinger, and D. F. Pac. 1982. Effect of capture and marking on fawn production in deer. *Journal of Wildlife Management* 46: 1086–1089.

Handel, C. M. and R. E. Gill, Jr. 1983. Yellow birds stand out in a crowd. *North American Bird Bander* 8: 6–9.

Harms, C. A., W. J. Fleming, and M. K. Stoskopf. 1997. A technique for dorsal subcutaneous implantation of heart rate biotelemetry transmitters in black ducks: Application in an aircraft noise response study. *Condor* 99: 231–237.

Hart, L. G. and R. C. Summerfelt. 1975. Surgical procedures for implanting ultrasonic transmitters into catfish (*Pylodictis olivaris*). *Transactions of the American Fisheries Society* 104: 56–59.

Hatch, J. J. and I. C. T. Nisbet. 1983a. Band wear and band loss in common terns. *Journal of Field Ornithology* 54: 1–16.

Hatch, J. J. and I. C. T. Nisbet. 1983b. Band wear in arctic terns. *Journal of Field Ornithology* 54: 91.

Hatfield, B. B. and G. B. Rathbun. 1996. Evaluation of a flipper-mounted transmitter on sea otters. *Wildlife Society Bulletin* 24: 551–554.

Hawkins, L. L. and S. G. Simpson. 1985. Neckband a handicap in an aggressive encounter between tundra swans. *Journal of Field Ornithology* 56: 182–184.

Hayward, J. L. Jr. 1982. A simple egg-marking technique. *Journal of Field Ornithology* 53: 173.

Henderson, H. F., A. D. Hasler, and G. G. Chipman. 1966. An ultrasonic transmitter for use in studies of movements of fishes. *Transactions of the American Fisheries Society* 95: 350–356.

Henderson, J. B. and T. C. Johanos. 1988. Effects of tagging on weaned Hawaiian monk seals. *Wildlife Society Bulletin* 16: 312–317.

Heyer, W. R., M. A. Donnelly, R. W. McDiarmid, L. A. C. Hayek, and M. S. Foster. 1994. *Measuring and monitoring biological diversity: Standard methods for amphibians.* Washington, D.C.: Smithsonian Institution Press.

Hickey, M. B. C. 1992. Effect of radiotransmitters on the attack success of hoary bats, *Lasiurus cinerus*. *Journal of Mammalogy* 73: 344–346.

Hill, L. A., and L. G. Talent. 1990. Effects of capture, handling, banding, and radio-marking on breeding least terns and snowy plovers. *Journal of Field Ornithology* 61: 310–319.

Hiraldo, F., J. A. Donazar, and J. J. Negro. 1994. Effects of tail-mounted radio-tags on adult lesser kestrels. *Journal of Field Ornithology* 65: 466–471.

Höglung, J., R. V. Alatalo, R. M. Gibson, and A. Lundberg. 1995. Mate-choice copying in black grouse. *Animal Behaviour* 49: 1627–1633.

Hooge, P. N. 1991. The effect of radio weight and harnesses on time budgets and movements of acorn woodpeckers. *Journal of Field Ornithology* 62: 230–238.

Horak, D. L. 1969. The effect of fin removal on stamina of hatchery-reared rainbow trout. *Progressive Fish-Culturist* 31: 217–220.

Houston, R. A. and R. J. Greenwood. 1993. Effects of radio transmitters on nesting captive mallards. *Journal of Wildlife Management* 57: 703–709.

Howitz, J. L. 1981. Determination of total color band combinations. *Journal of Field Ornithology* 52: 317–324.

Hudson, S. 1996. Natural toe loss in southeastern Australian skinks: Implications for marking lizards by toe-clipping. *Journal of Herpetology* 30: 106–110.

Huey, R. B., A. E. Dunham, K. L. Overall, and R. A. Newman. 1990. Variation in locomotor performance in demographically known populations of the lizard *Scleoporus merriami. Physiological Zoology* 63: 845–872.

Hughes, P. M. and J. M. V. Rayner. 1991. Addition of artificial loads to long-eared bats *Plecotus auritus:* Handicapping flight performance. *Journal of Experimental Biology* 161: 285–298.

Ingram, J. C. 1978. Primate markings. In B. Stonehouse, ed., *Animal marking: Recognition marking of animals in research,* 169–174. Baltimore, Md.: University Park Press.

Ireland, P. H. 1991. A simplified fluorescent marking technique for identification of terrestrial salamanders. *Herpetological Review* 22: 21–22.

Jacob, J. S. and C. W. Painter. 1980. Overwinter thermal ecology of *Crotalus viridis* in the north central plains of New Mexico. *Copeia* 1980: 799–805.

Jemison, S. C., L. A. Bishop, P. G. May, and T. M. Farrell. 1995. The impact of PIT-tags on growth and movement of the rattlesnake, *Sistrurus miliarius. Journal of Herpetology* 29: 129–132.

Jenkins, W. E. and T. I. J. Smith. 1990. Use of PIT tags to individually identify striped bass and red drum brood stocks. *American Fisheries Society Symposium* 7: 341–345.

Jennings, M. L., D. N. David, and K. M. Portier. 1991. Effect of marking techniques on growth and survivorship of hatchling alligators. *Wildlife Society Bulletin* 19: 204–207.

Jensen, A. C. 1967. Effects of tagging on the growth of cod. *Transactions of the American Fisheries Society* 96: 7–41.

Johnson, G. D., J. L. Pebworth, and H. O. Krueger. 1991. Retention of transmitters attached to passerines using a glue-on technique. *Journal of Field Ornithology* 62: 486–491.

Johnston, J. J., L. A. Windberg, C. A. Furcolow, R. M. Engeman, and M. Roetto. 1998.

Chlorinated benzenes as physiological markers for coyotes. *Journal of Wildlife Management* 62: 410–421.

Jones, S. M. and G. W. Ferguson. 1980. The effect of paint marking on mortality in a Texas population of *Scleoporus undulatus*. *Copeia* 1980: 850–854.

Keck, M. B. 1994. Test for detrimental effects of pit tags on neonatal snakes. *Copeia* 1994: 226–228.

Keister, G. P., C. E. Trainer, and M. J. Willis. 1988. A self-adjusting collar for young ungulates. *Wildlife Society Bulletin* 16: 321–323.

Kelly, G. F. and A. M. Barker. 1963. Effect of tagging on redfish growth rate at Eastport, Maine. International Committee of the Northwest Atlantic Fishery special publication no. 4, pp. 210–213.

Kenward, R. E. 1978. Radio transmitters tail-mounted on hawks. *Ornis Scandinavia* 9: 220–223.

Kenward, R. E. 1982. Techniques for monitoring the behaviour of grey squirrels by radio. *Symposium of the Zoological Society of London* 49: 175–196.

Kenward, R. E. 1985. Raptor radio tracking and radio telemetry. In I. Newton and R. D. Chancellor, eds., *Conservation studies of raptors*, 409–420. ICBP technical bulletin no. 5.

Kenward, R. E. 1987. *Wildlife radio tagging: Equipment, field techniques and data analysis.* London: Academic Press.

Kenward, R. E. 1993. Modelling raptor populations: To ring or to radio-tag? In J. D. Lebreton and P. M. North, eds., *Marked individuals in the study of bird population*, 157–167. Basel, Switzerland: Birkhauser Verlag.

Kenward, R. E., R. H. Pfeffer, E. A. Bragin, A. Levin, and A. F. Kovshar. 1996. The status of saker falcons in Kazakhstan. In J. Samour, ed., *Proceedings of the Specialist Workshop Middle East Falcon Research Group*, 131–142. Abu Dhabi, United Arab Emirates.

Kenward, R. E., P. A. Robertson, A. S. Coates, S. V. Marcstrom, and M. Karlbom. 1993. Techniques for radio-tagging pheasant chicks. *Bird Study* 40: 51–54.

Kenward, R. E. and S. S. Walls. 1994. The systematic study of radio-tagged raptors: I. Survival, home-range and habitat-use. In B. U. Meyburg and R. D. Chancellor, eds., *Raptor conservation today*, 303–315. East Sussex, U.K.: WWGBP/Pica Press.

Klaassen, M., P. H. Becker, and M. Wagener. 1992. Transmitter loads do not affect the daily energy expenditure of nesting common terns. *Journal of Field Ornithology* 63: 181–185.

Kochert, M. N., K. Steenhof, and M. Q. Moritsch. 1983. Evaluation of patagial markers for raptors and ravens. *Wildlife Society Bulletin* 11: 271–281.

Koehler, D. K., T. D. Reynolds, and S. H. Anderson. 1987. Radio-transmitter implants in 4 species of small mammals. *Journal of Wildlife Management* 51: 105–108.

Korn, H. 1987. Effects of live-trapping and toe-clipping on body weight of European and African rodent species. *Oecologia* 71: 597–600.

Korschgen, C. E., K. P. Kenow, J. E. Austin, C. O. Kochanny, W. L. Green, C. H. Simmons, and M. Janda. 1995. An automated telemetry system for studies of migrating diving ducks. In M. R. Neuman, C. J. Amalaner, Jr., and C. Cristalli, eds., *Biotelemetry XIII.* Terre Haute: Indiana State University Press.

Korschgen, C. E., K. P. Kenow, A. Gendron-Fitzpatrick, W. L. Green, and F. J. Dein. 1996b. Implanting intra-abdominal radiotransmitters with external whip antennas in ducks. *Journal of Wildlife Management* 60: 132–137.

Korschgen, C. E., K. P. Kenow, W. L. Green, M. D. Samuel, and L. Sileo. 1996a. Techniques for implanting radio transmitters subcutaneously in day-old ducklings. *Journal of Field Ornithology* 67: 392–397.

Larsen, K. W. 1987. Movements and behavior of migratory garter snakes, *Thamnophis siritalis. Canadian Journal of Zoology* 65: 2241–2247.

Lensink, C. J. 1988. Survival of aluminum and monel bands on black brant. *North American Bird Bander* 13: 33–35.

Leuze, C. C. K. 1980. The application of radio tracking and its effect on the behavioral ecology of the water vole, *Arvicola terrestris.* In C. J. Almaner and D. W. Macdonald, eds., *A handbook on biotelemetry and radio tracking,* 361–366. Oxford, U.K.: Pergamon.

Lewis, A. E. and W. R. A. Muntz. 1984. The effects of external ultrasonic tagging on the swimming performance of rainbow trout, *Salmo gairdneri* Richardson. *Journal of Fisheries Biology* 25: 577–585.

Lindsey, G. D. 1983. Rhodamine B: A systemic fluorescent marker for studying mountain beavers (*Aplodontia rufa*) and other animals. *Northwest Science* 57: 16–21.

Lindsey, G. D., K. A. Wilson, and C. Herrmann. 1995. Color change in Hughes's celluloid leg bands. *Journal of Field Ornithology* 66: 289–295.

Livezey, K. B. 1990. Toward the reduction of marking-induced abandonment of newborn ungulates. *Wildlife Society Bulletin* 18: 193–203.

Loafman, P. 1991. Identifying individual spotted salamanders by spot pattern. *Herpetological Review* 22: 91–92.

Lucas, M. C. 1989. Effects of implanted dummy transmitters on mortality, growth and tissue reaction in rainbow trout, *Salmo gairdneri* Richardson. *Journal of Fisheries Biology* 35: 577–587.

Lutterschmidt, W. I. 1994. The effect of surgically implanted transmitters upon locomotory performance of the checkered garter snake, *Thamnophis m. marcianus. Herpetology Journal* 4: 11–14.

Lutterschmidt, W. I. and H. K. Reinert. 1990. The effect of ingested transmitters upon the temperature preference of the northern water snake, *Nerodia s. sipedon. Herpetology* 46: 39–42.

Madison, D. M., R. W. Fitzgerald, and W. J. McShea. 1985. A user's guide to the successful radiotracking of small mammals in the field. In R. W. Weeks and F. M. Long, eds., *Proceedings Fifth International Conference on Wildlife Biotelemetry,* 28–39. International Conference on Wildlife Biotelemetry, Laramie, Wyoming.

Marcstrom, V., R. E. Kenward, and M. Karlbom. 1989. Survival of ring-necked pheasants with backpacks, necklaces, and leg bands. *Journal of Wildlife Management* 53: 808–810.

Marion, W. R. and J. D. Shamis. 1977. An annotated bibliography of bird marking techniques. *Bird-Banding* 48: 42–61.

Marty, G. D. and R. C. Summerfelt. 1986. Pathways and mechanisms for expulsion of surgically implanted dummy transmitters from channel catfish. *Transactions of the American Fisheries Society* 115: 577–589.

Mauser, D. M. and R. L. Jarvis. 1991. Attaching radio transmitters to 1-day-old mallard ducklings. *Journal of Wildlife Management* 55: 488–491.

McCleave, J. D. and K. A. Stred. 1975. Effect of dummy telemetry transmitters on stamina of Atlantic salmon (*Salmo salar*) smolts. *Journal of the Fisheries Research Board of Canada* 32: 559–563.

McGowan, K. J. and C. Caffrey. 1994. Does drugging crows for capture cause abnormally high mortality? *Journal of Field Ornithology* 65: 453–457.

Mears, H. C. and R. W. Hatch. 1976. Overwinter survival of fingerling brook trout with single and multiple fin clips. *Transactions of the American Fisheries Society* 6: 669–674.

Mellas, E. J. and J. M. Haynes. 1985. Swimming performance and behavior of rainbow trout (*Salmo gairdneri*) and white perch (*Morone americana*): Effects of attaching telemetry transmitters. *Canadian Journal of Fisheries and Aquatic Sciences* 42: 488–493.

Metz, K. J. and P. J. Weatherhead. 1993. An experimental test of the contrasting-color hypothesis of red-band effects in red-winged blackbirds. *Condor* 95: 395–400.

Meyers, J. M. 1994. Leg bands cause injuries to parakeets and parrots. *North American Bird Bander* 19: 133–136.

Meyers, J. M. 1996. Evaluation of 3 radio transmitters and collar designs for Amazona. *Wildlife Society Bulletin* 24: 15–20.

Meyers, J. M., W. J. Arendt, and G. D. Lindsey. 1996. Survival of radio-collared nestling Puerto Rican parrots. *Wilson Bulletin* 108: 159–163.

Middelburg, J. J. M. and H. Strijbosch. 1988. The reliability of the toe-clipping method with the common lizard (*Lacerta vivipara*). *Herpetology Journal* 1: 291–293.

Migoya, R. and G. A. Baldassarre. 1995. Winter survival of female northern pintails in Sinaloa, Mexico. *Journal of Wildlife Management* 59: 16–22.

Mikesic, D. G. and L. C. Drickhamer. 1992. Effects of radiotransmitters and fluorescent powders on activity of wild house mice (*Mus musculus*). *Journal of Mammalogy* 73: 663–667.

Miller, M. S., D. J. Buford, and R. S. Lutz. 1995. Survival of female Rio Grande turkeys during the reproductive season. *Journal of Wildlife Management* 59: 766–771.

Mullican, T. R. 1988. Radio telemetry and fluorescent pigments: A comparison of techniques. *Journal of Wildlife Management* 52: 627–631.

Nagendran, M., H. Higuchi, and A. Sorokin. 1994. A harnessing technique to deploy transmitters on cranes. In H. Higuchi and I. Minton, eds., *The Future of Cranes and Wetlands, Proceedings of the International Symposium,* Tokyo, pp. 57–60.

Nakano, S. 1995. Individual differences in resource use, growth and emigration and the influence of a dominance hierarchy in fluvial red-spotted masu salmon in a natural habitat. *Journal of Animal Ecology* 64: 75–84.

Nelson, J. 1995. Intrasexual competition and spacing behaviour in male field voles, *Microtus agrestis,* under constant female density and spatial distribution. *Oikos* 73: 9–14.

Neudorf, D. L. and T. E. Pitcher. 1997. Radio transmitters do not affect nestling feeding rates by female hooded warblers. *Journal of Field Ornithology* 68: 64–68.

Nicola, S. J. and A. J. Cordone. 1973. Effects of fin removal on survival and growth of rainbow trout (*Salmo gairdneri*) in a natural environment. *Transactions of the American Fisheries Society* 102: 753–758.

Nietfeld, M. T., M. W. Barrett, and N. Silvy. 1994. Wildlife marking techniques. In T. A. Bookhout, ed., *Research and management techniques for wildlife and habitats,* 140–168. Bethesda, Md.: Wildlife Society.

Nisbet, I. C. T. and J. J. Hatch. 1985. Influence of band size on rates of band loss by common terns. *Journal of Field Ornithology* 56: 178–181.

Nishikawa, K. C. and P. M. Service. 1988. A fluorescent marking technique for individual recognition of terrestrial salamanders. *Journal of Herpetology* 22: 351–353.

Obrecht, H. H., C. J. Pennycuick, and M. R. Fuller. 1988. Wind tunnel experiments to assess the effect of back-mounted radio transmitters on bird body drag. *Journal of Experimental Biology* 135: 265–273.

Ormiston, B. G. 1985. Effects of a subminiature radio collar on activity of free-living white-footed mice. *Canadian Journal of Zoology* 63: 733–735.

Ostfeld, R. S., M. C. Miller, and J. Schnurr. 1993. Ear tagging increases tick (*Ixodes dammini*) infestation rates of white-footed mice (*Peromyscus leucopus*). *Journal of Mammalogy* 74: 651–655.

Ottaway, J. R., R. Carrick, and M. D. Murray. 1984. Evaluation of leg bands for visual identification of free-living silver gulls. *Journal of Field Ornithology* 55: 287–308.

Ozoga, J. J. and R. M. Clute. 1988. Mortality rates of marked and unmarked fawns. *Journal of Wildlife Management* 52: 549–551.

Parker, N. C., A. E. Giorgi, R. C. Heidinger, D. B. Jester, E. D. Prince, and G. A. Winans. 1990. *Fish-marking techniques.* Bethesda, Md.: American Fisheries Society.

Parmenter, C. J. 1993. A preliminary evaluation of the performance of passive integrated transponders and metal tags in a population study of the flatback sea turtle (*Nanator depressus*). *Wildlife Research* 20: 375–381.

Patrick, B. and R. Haas. 1971. Fin pulling as a technique for marking muskellunge fingerlings. *Progressive Fish-Culturist* 33: 116–119.

Pavone, L. V. and R. Boonstra. 1985. The effects of toe clipping on the survival of the meadow vole (*Microtus pennsylvanicus*). *Canadian Journal of Zoology* 63: 499–501.

Pennycuick, C. J. 1975. Mechanics in flight. In D. S. Farner and J. R. King, eds., *Avian biology,* Vol. 5, 1–75. New York: Academic Press.

Pennycuick, C. J. 1978. Identification using natural markings. In B. Stonehouse, ed., *Animal marking: Recognition marking of animals in research,* 147–159. Baltimore, Md.: University Park Press.

Pennycuick, C. J. 1989. *Bird flight performance: A practical calculation manual.* New York: Oxford University Press.

Pennycuick, C. J., and M. R. Fuller. 1987. Considerations of effects of radio-transmitters on bird flight. In H. P. Kimmich and M. R. Neuman, eds., *Biotelemetry IX*, 327–330. Zurich: International Symposium on Biotelemetry.

Pennycuick, C. J., M. R. Fuller, and L. McAllister. 1989. Climbing performance of Harris' hawks *Parabuteo unicinctus* with added load: Implications for muscle mechanics and for radiotracking. *Journal of Experimental Biology* 142: 17–29.

Pennycuick, C. J., M. R. Fuller, J. J. Oar, and S. J. Kirkpatrick. 1994. Falcon versus grouse: Flight adaptations of a predator and its prey. *Journal of Avian Biology* 25: 39–49.

Pennycuick, C. J., H. H. Obrecht, and M. R. Fuller. 1988. Empirical estimates of body drag of large waterfowl and raptors. *Journal of Experimental Biology* 135: 253–264.

Pennycuick, C. J. and J. Rudnai. 1970. A method of identifying individual lions *Panthera leo* with an analysis of the reliability of identification. *Journal of Zoology* 160: 497–508.

Pennycuick, C. J., F. C. Schaffner, M. R. Fuller, H. H. Obrecht, and L. Sternberg. 1990. Foraging flights of the white-tailed tropicbird *Phaethon lepturus:* Radiotracking and doubly-labelled water. *Colonial Waterbirds* 13: 96–102.

Perry, A. E. and G. Beckett. 1966. Skeletal damage as a result of band injury in bats. *Journal of Mammalogy* 47: 131–132.

Peters, R. H. 1991. *A critique for ecology.* Cambridge, U.K.: Cambridge University Press.

Petersen, M. R., D. C. Douglas, and D. M. Mulcahy. 1995. Use of implanted satellite transmitters to locate spectacled eiders at sea. *Condor* 97: 276–278.

Phinney, D. E., D. M. Miller, and M. L. Dahlberg. 1967. Mass-marking young salmonids with fluorescent pigment. *Transactions of the American Fisheries Society* 96: 157–162.

Pietz, P. J., D. A. Brandt, G. L. Krapu, and D. A. Buhl. 1995. Modified transmitter attachment method for adult ducks. *Journal of Field Ornithology* 66: 408–417.

Pietz, P. J., G. L. Krapu, R. J. Greenwood, and J. T. Lokemoen. 1993. Effects of harness transmitters on behavior and reproduction of wild mallards. *Journal of Wildlife Management* 57: 696–703.

Platt, S. W. 1980. Longevity of herculite leg jess color markers on the prairie falcon *Falco mexicanus. Journal of Field Ornithology* 51: 281–282.

Pough, F. H. 1970. A quick method for permanently marking snakes and turtles. *Herpetology* 26: 428–430.

Pouliquen, O., M. Leishman, and T. D. Redhead. 1990. Effects of radio collars on wild mice, *Mus domesticus. Canadian Journal of Zoology* 68: 1607–1609.

Powell, G. V. N. and R. Bjork. 1995. Implications of intratropical migration on reserve design: A case study using *Pharomacrus mocino. Conservation Biology* 9: 354–362.

Prentice, E. F., T. A. Flagg, and C. S. McCutcheon. 1990. Feasibility of using implantable passive integrated transponder (PIT) tags in salmonids. *American Fisheries Society Symposium* 7: 317–322.

Queal, L. M. and B. D. Hlavachick. 1968. A modified marking technique for young ungulates. *Journal of Wildlife Management* 32: 628–629.

Ralls, K. A., B. B. Hatfield, and D. B. Siniff. 1995. Foraging patterns of California sea otters as indicated by telemetry. *Canadian Journal of Zoology* 73: 1387–1395.

Raney, E. C. 1940. Summer movements of the bullfrog, *Rana catesbiana* Shaw, as determined by the jaw-tag method. *American Midland Naturalist* 23: 733–745.

Ratcliffe, L. M. and P. T. Boag. 1987. Effects of colour bands on male competition and sexual attractiveness in zebra finches *Poephila guttata*. *Canadian Journal of Zoology* 65: 333–338.

Rawstron, R. R. 1973. Comparisons of disk dangler, trailer, and internal anchor tags on three species of salmonids. *California Fish and Game* 59: 266–280.

Rawstron, R. R. and R. J. Pelzam. 1978. Comparison of Floy internal anchor and Disk Dangler tags on largemouth bass (*Micropterus salmoides*) at Merle Collins Reservoir. *California Fish and Game* 64: 121–122.

Reed, J. M. and L. W. Oring. 1993. Banding is infrequently associated with foot loss in spotted sandpipers. *Journal of Field Ornithology* 64: 145–148.

Reese, K. 1980. The retention of colored plastic leg bands by black-billed magpies. *North American Bird Bander* 5: 136–137.

Reid, D. G., W. E. Melquist, J. D. Woolington, and J. M. Noll. 1986. Reproductive effects of intraperitoneal transmitter implants in river otters. *Journal of Wildlife Management* 50: 92–94.

Reid, J. A., R. B. Horn, and E. D. Forsman. 1996. A method for replacing tail-mounted radio transmitters on birds. *Journal of Field Ornithology* 67: 177–180.

Reinert, H. K. and D. Cundall. 1982. An improved surgical implantation method for radio-tracking snakes. *Copeia* 1982: 702–705.

Richards, S. J., U. Sinsch, and R. A. Alford. 1994. Radio tracking. In R. H. Heyer, M. H. Donnelly, R. W. McDiarmid, L. A. C. Hayek, and M. S. Foster, eds., *Measuring and monitoring biological diversity: Standard methods for amphibians,* 155–158, Washington, D.C.: Smithsonian Institution Press.

Ricker, W. E. 1949. Effects of removal of fins upon the growth and survival of spiny-rayed fishes. *Journal of Wildlife Management* 13: 29–40.

Ricker, W. E. 1956. Uses of marking animals in ecological studies: The marking of fish. *Ecology* 37: 665–670.

Riley, H. T., D. M. Bryant, R. E. Carter, and D. T. Parkin. 1995. Extra-pair fertilizations and paternity defence in house martins, *Delichon urbica. Animal Behaviour* 49: 495–509.

Ritchison, G. 1984. A new marking technique for birds. *North American Bird Bander* 9:8.

Roberts, R. J., A. MacQueen, W. M. Shearer, and H. Young. 1973. The histopathology of salmon tagging. I. The tagging lesion in newly tagged parr. *Journal of Fisheries Biology* 5: 497–503.

Roberts, S. D., J. M. Coffey, and W. F. Porter. 1995. Survival and reproduction of female wild turkeys in New York. *Journal of Wildlife Management* 59: 437–447.

Rossi, J. V. and R. Rossi. 1993. Field evaluation of using passive integrated transponder (PIT) tags to permanently mark lizards. *Herpetological Review* 24: 54–56.

Rotella, J. J., D. W. Howerter, T. P. Sankowski, and J. H. Devries. 1993. Nesting effort by wild mallards with 3 types of radio transmitters. *Journal of Wildlife Management* 57: 690–695.

Rotella, J. J. and J. T. Ratti. 1990. Use of methoxyflurane to reduce nest abandonment of mallards. *Journal of Wildlife Management* 54: 627–628.

Saddler, J. B. and R. Caldwell. 1971. The effect of tagging upon the fatty acid metabolism of juvenile pink salmon. *Compendium of Biochemistry and Physiology* 39A: 709–721.

Sallaberry, M. and J. Valencia. 1985. Wounds due to flipper bands on penguins. *Journal of Field Ornithology* 56: 275–277.

Samuel, M. D. and M. R. Fuller. 1994. Wildlife radiotelemetry. In T. A. Bookout, ed., *Research and management techniques for wildlife and habitats,* 370–418. Bethesda, Md.: Wildlife Society.

Samuel, M. D., N. T. Weiss, D. H. Rusch, S. R. Craven, R. E. Trost, and F. D. Caswell. 1990. Neck-band retention for Canada geese in the Mississippi flyway. *Journal of Wildlife Management* 54: 612–621.

Saunders, R. L. and K. R. Allen. 1967. Effects of tagging and fin-clipping on the survival and growth of Atlantic salmon between smolt and adult stages. *Journal of the Fisheries Research Board of Canada* 24: 2595–2611.

Scheirer, J. W. and D. W. Coble. 1991. Effects of Floy FD-67 Anchor tags on growth and condition of northern pike. *North American Journal of Fisheries Management* 11: 369–373.

Schooley, R. L., B. Van Horne, and K. P. Burnham. 1993. Passive integrated transponders for marking free-ranging Townsend's ground squirrels. *Journal of Mammalogy* 74: 480–484.

Schuck, H. A. 1942. The effect of jaw-tagging upon the condition of trout. *Copeia* 1942: 33–39.

Scott, A. F. and J. L. Dobie. 1980. An improved technique for a thread-trailing device used to study movements of terrestrial turtles. *Herpetological Review* 11: 106–107.

Scott, D. K. 1978. Identification of individual Bewick's swans by bill patterns. In B. Stonehouse, ed., *Animal marking: Recognition marking of animals in research,* 160—168. Baltimore, Md.: University Park Press.

Seale, D. and M. Boraas. 1974. A permanent mark for amphibian larvae. *Herpetology* 30: 160–162.

Sedinger, J. S., R. G. White, and W. E. Hauer. 1990. Effects of carrying radio transmitters on energy expenditure of Pacific black brant. *Journal of Wildlife Management* 54: 42–45.

Shetter, D. S. 1951. The effect of fin removal on fingerling lake trout (*Cristivomer namaycush*). *Transactions of the American Fisheries Society* 80: 260–277.

Shetter, D. S. 1952. The mortality and growth of marked and unmarked lake trout fingerlings in the presence of predators. *Transactions of the American Fisheries Society* 81: 17–34.

Shetter, D. S. 1967. Effects of jaw tags and fin excision upon the growth, survival, and exploitation of hatchery rainbow trout fingerlings in Michigan. *Transactions of the American Fisheries Society* 96: 394–399.

Shine, R. and M. Fitzgerald. 1995. Variation in mating systems and sexual size dimorphism between populations of the Australian python *Morelia spilota* (Serpentes: Pythonidae). *Oecologia* 103: 490–498.

Simon, C. A. and B. E. Bissinger. 1983. Paint-marking lizards: Does the color affect survivorship? *Journal of Herpetology* 17: 184–186.

Slaugh, B. T., J. T. Flinders, J. A. Roberson, and N. P. Johnston. 1990. Effect of backpack radio transmitter attachment on chukar mating. *Great Basin Naturalist* 50: 379–380.

Smith, H. R. 1980. Growth, reproduction and survival in *Peromyscus leucopus* carrying intraperitoneally implanted transmitters. In C. J. Almaner and D. W. Macdonald, eds., *A handbook on biotelemetry and radio tracking*, 367–374. Oxford, U.K.: Pergamon.

Smits, A. W. 1984. Activity patterns and thermal biology of the toad *Bufo boreas halophilus*. *Copeia* 1984: 689–696.

Snyder, N. F. R., S. R. Beissinger, and M. R. Fuller. 1989. Solar radio-transmitters on snail kites in Florida. *Journal of Field Ornithology* 60: 171–177.

Sokal, R. R. and F. J. Rohlf. 1981. *Biometry.* New York: W.H. Freeman.

Sorenson, M. D. 1989. Effects of neck collar radios on female redheads. *Journal of Field Ornithology* 60: 523–528.

Southern, L. K. and W. E. Southern. 1985. Some effects of wing tags on breeding ring-billed gulls. *Auk* 102: 38–42.

Spear, L. 1980. Band loss from the western gull on Southeast Farallon Island. *Journal of Field Ornithology* 51: 319–328.

Stapp, P., J. K. Young, S. VandeWoude, and B. Van Horne. 1994. An evaluation of the pathological effects of fluorescent powder on deer mice (*Peromyscus maniculatus*). *Journal of Mammalogy* 75: 704–709.

Stasko, A. B. and D. G. Pincock. 1977. Review of underwater biotelemetry, with emphasis on ultrasonic techniques. *Journal of the Fisheries Research Board of Canada* 34: 1262–1285.

Stebbins, R. E. 1982. Radio tracking greater horseshoe bats with preliminary observations on flight patterns. *Symposium of the Zoological Society of London* 49: 161–173.

Stedman, S. 1990. Band opening and removal by house finches. *North American Bird Bander* 15: 136–138.

Stewart, P. D. and D. W. Macdonald. 1997. Age, sex, and condition as predictors of moult and the efficacy of a novel fur clip technique for individual marking of the European badger (*Meles meles*). *Journal of Zoology* 241: 543–550.

Stobo, W. T. 1972. The effects of dart tags on yellow perch. *Transactions of the American Fisheries Society* 101: 365–366.

Stonehouse, B. 1978. *Animal marking: Recognition marking of animals in research.* Baltimore, Md.: University Park Press.

Stott, B. 1968. Marking and tagging. In W. E. Ricker, ed., *Methods for assessment of fish production in fresh waters.* IBP handbook no. 3, 78–93. Oxford, U.K.: Blackwell Scientific.

Stroud, R. H. 1953. Notes on the reliability of some fish tags used in Massachusetts. *Journal of Wildlife Management* 17: 268–275.

Summerfelt, R. C. and D. Mosier. 1984. Transintestinal expulsion of surgically implanted dummy transmitters by channel catfish. *Transactions of the American Fisheries Society* 115: 760–766.

Taylor, C. R., N. C. Heglund, T. A. McMahon, and T. R. Looney. 1980. Energetic cost of generating muscular force during running. *Journal of Experimental Biology* 86: 9–18.

Taylor, J. and L. Deegan. 1982. A rapid method for mass marking of amphibians. *Journal of Herpetology* 16: 172–173.

Tilley, S. G. 1977. Studies of life histories and reproduction in North American plethodontid salamanders. In S. Guttman and D. Taylor, eds., *Reproductive biology of Plethodontid salamanders,* 1–39. New York: Plenum.

Tranquilli, J. A. and W. F. Childers. 1982. Growth and survival of largemouth bass tagged with Floy anchor tags. *North American Journal of Fisheries Management* 2: 184–187.

Travis, J. 1981. The effect of staining on the growth of *Hyla gratiosa* tadpoles. *Copeia* 1972: 193–196.

Trojan, P. and B. Wojciechowska. 1964. Studies on the residency of small forest rodents. *Ecologia Polska* Ser. A 3: 33–50.

Van Vuren, D. 1989. Effects of intraperitoneal transmitter implants on yellow-bellied marmots. *Journal of Wildlife Management* 53: 320–323.

Vehrencamp, S. L. and L. Halpenny. 1981. Capture and radio-transmitter attachment techniques for roadrunners. *North American Bird Bander* 6: 128–132.

Vekasy, M. S., J. M. Marzluff, M. N. Kochert, R. N. Lehman, and K. Steenhof. 1996. Influence of radio transmitters on prairie falcons. *Journal of Field Ornithology* 67: 680–690.

Vogelbein, W. K. and R. M. Overstreet. 1987. Histopathology of the internal anchor tag in spot and spotted seatrout. *Transactions of the American Fisheries Society* 116: 745–756.

Wallace, M. P., M. R. Fuller, and J. Wiley. 1994. Patagial transmitters for large vultures and condors. In B. U. Meyburg and R. D. Chancellor, eds., *Raptor conservation today,* 381–387. East Sussex, U.K.: WWGBP/Pica Press.

Wanless, S. 1992. Effects of tail-mounted devices on the attendance behavior of kittiwakes during chick rearing. *Journal of Field Ornithology* 63: 169–176.

Ward, D. H. and P. L. Flint. 1995. Effects of harness-attached transmitters on premigration and reproduction of brant. *Journal of Wildlife Management* 59: 39–46.

Weatherhead, P. J. and F. W. Anderka. 1984. An improved radio transmitter and implantation technique for snakes. *Journal of Herpetology* 18: 264–269.

Weber, D. and R. J. Wahle. 1969. Effect of finclipping on survival of sockeye salmon (*Oncorhynchus nerka*). *Journal of the Fisheries Research Board of Canada* 26: 1263–1271.

Webster, A. B. and R. J. Brooks. 1980. Effects of radiotransmitters on the meadow vole, *Microtus pennsylvanicus. Canadian Journal of Zoology* 58: 997–1001.

Wendeln, H., R. Nagel, and P. H. Becker. 1996. A technique to spray dyes on birds. *Journal of Field Ornithology* 67: 442–446.

White, G. C. and R. A. Garrott. 1990. *Analysis of radio-tracking data.* New York: Academic Press.

Wilson, D. E., F. R. Cole, J. D. Nichols, R. Rudran, and M. S. Foster, eds. 1996. *Measuring and monitoring biological diversity: Standard methods for mammals.* Washington, D.C.: Smithsonian Institution Press.

Wilson, R. P. and B. M. Culik. 1992. Packages on penguins and device-induced data. In G. I. Priede and S. M. Swift, eds., *Wildlife telemetry: Remote monitoring and tracking of animals,* 573–580. New York: Ellis Horwood.

Wilson, R. P., W. S. Grant, and D. C. Duffy. 1986. Recording devices on free-ranging marine animals: Does measurement affect performance? *Ecology* 67: 1091–1093.

Wilson, R. P., H. J. Spariani, N. R. Coria, B. M. Culik, and D. Adelung. 1990. Packages for attachment to seabirds: What color do Adélie penguins dislike least? *Journal of Wildlife Management* 54: 447–451.

Wilson, R. P. and M. P. T. Wilson. 1989. A peck activity record for birds fitted with devices. *Journal of Field Ornithology* 60: 104–108.

Wolton, R. J. and B. J. Trowbridge. 1985. The effects of radio-collars on wood mice, *Apodemus sylvaticus. Journal of Zoology* Ser. A 206: 222–224.

Wood, M. D. and N. A. Slade. 1990. Comparison of ear-tagging and toe clipping in prairie voles, *Microtus ochrogaster. Journal of Mammalogy* 71: 252–255.

Woodbury, A. M. 1956. Uses of marking animals in ecological studies: Marking amphibians and reptiles. *Ecology* 63: 583–585.

Wydowski, R. and L. Emery. 1983. Tagging and marking. In L. A. Nielsen, D. L. Johnson, and S. S. Lampton, eds., *Fisheries techniques,* 215–237. Bethesda, Md.: American Fisheries Society.

Young, L. S. and M. N. Kochert. 1987. Marking techniques. In B. A. Giron Pendelton, B. A. Millsap, K. W. Cline, and D. M. Bird, eds., *Raptor management techniques manual,* 125–156. Washington, D.C.: National Wildlife Federation.

Zann, R. 1994. Effects of band color on survivorship, body condition and reproductive effort of free-living Australian zebra finches. *Auk* 111: 131–142.

Zicus, M. C. and R. M. Pace, III. 1986. Neckband retention in Canada geese. *Wildlife Society Bulletin* 14: 388–391.

Zicus, M. C., D. F. Schultz, and J. A. Cooper. 1983. Canada goose mortality from neckband icing. *Wildlife Society Bulletin* 11: 286–290.

## Chapter 3

# Animal Home Ranges and Territories and Home Range Estimators

Roger A. Powell

## ■ Definition of Home Range

Most animals are not nomadic but live in fairly confined areas where they enact their day-to-day activities. Such areas are called home ranges.

Burt (1943:351) provided the verbal definition of a mammal's home range that is the foundation of the general concept used today: "that area traversed by the individual in its normal activities of food gathering, mating, and caring for young. Occasional sallies outside the area, perhaps exploratory in nature, should not be considered part of the home range." This definition is clear conceptually, but it is vague on points that are important to quantifying animals' home ranges. Burt gave no guidance concerning how to quantify occasional sallies or how to define the area from which the sallies are made. The vague wording implicitly and correctly allows a home range to include areas used in diverse ways for diverse behaviors. Members of two different species may use their home ranges very differently with very different behaviors, but for both the home ranges are recognizable as home ranges, not something different for each species.

How does an animal view its home range? Obviously, with our present knowledge we cannot know, but to be able to know would provide tremendous insight into animals' lives. Aldo Leopold (1949:78) wrote, "The wild things that live on my farm are reluctant to tell me, in so many words, how much of my township is included within their daily or nightly beats. I am curious about this, for it gives me the ratio between the size of their universe and mine, and it conveniently begs the much more important question, who is the more thoroughly acquainted with the world in which he lives?" Leopold continued, "Like people, my animals frequently disclose by their actions what they decline to divulge in words."

We do know that members of some species, probably many species, have cognitive maps of where they live (Peters 1978) or concepts of where different resources and features are located within their home ranges and of how to travel between them. Such cognitive maps may be sensitive to where an animal finds itself within its home range or to its nutritional state; for example, resources that the animal perceives to be close at hand or resources far away that balance the diet may be more valuable than others. From extensive research on optimal foraging (Ellner and Real 1989; Pyke 1984; Pyke et al. 1977), we know that animals often rank resources in some manner. Consequently, *we* might envision an animal's cognitive map of its home range as an integration of contour maps, one (or more) for food resources, one for escape cover, one for travel routes, one for known home ranges of members of the other sex, and so forth.

Why do animals have home ranges? Stamps (1995:41) argued that animals have home ranges because individuals learn "site-specific serial motor programs," which might be envisioned as near reflex movements that take an animal along well traveled routes to safety. These movements should enhance the animal's ability to maneuver through its environment and thereby to avoid or escape predators. Stamps argued that the willingness of an animal to incur costs to remain in a familiar area implies that being familiar with that area provides a fitness benefit greater than the costs. For animals with small home ranges that live their lives as potential prey, Stamps's hypothesis makes sense. However, many animals, especially predatory mammals and birds, have home ranges too large and use specific places too seldom for site-specific serial motor programs to have an important benefit. Site-specific serial motor patterns of greatest use to a predator would have to match the escape routes of each prey individual, but each of these might be used only once after it is learned. The reason that animals maintain home ranges must be broader than Stamps's hypothesis.

Nonetheless, Stamps has undoubtedly identified the key reason that animals establish and maintain home ranges: The benefits of maintaining a home range exceed the costs. Let $C_D$ be the daily costs for an animal, excluding the costs, $C_R$, of monitoring, maintaining, defending, developing, and remembering the critical resources on which it based its decision to establish a home range. In the long term, $C_D$ plus $C_R$ must be equal to or less than the benefits, $B$, gained from the home range, or

$$C_D + C_R \leq B$$

Costs and benefits must ultimately be calculated in terms of an animal's fitness, but if the critical resources are food, then costs and benefits might be indexed by energy. If the benefits are nest sites or escape routes, energy is not an adequate index. If $C_D$ plus $C_R$ exceeds $B$ for an animal in the short term, then the animal might be able to live on a negative balance until conditions change. If $C_D$ plus $C_R$ exceeds $B$ in the long term, then the animal must reduce $C_D$, or $C_R$, both of which have lower limits. $C_D$ generally cannot be reduced below basic maintenance costs, or basal metabolism; however, hibernation and estivation are methods some animals can use to reduce $C_D$ below basal metabolism. Reducing $C_R$ might reduce $B$ because benefits can be experienced only through attending to critical, local resources, which is $C_R$. If $C_R$ can be reduced through increased efficiency, $B$ need not be reduced when $C_R$ is reduced or need not be reduced as much as $C_R$ is reduced. Ultimately, in the long term, if $C_D + C_R > B$ then the animal cannot survive using local resources. If the animal cannot survive using local resources, it must go to another locale where benefits exceed costs, or it must be nomadic and not exhibit site fidelity.

Because maintaining a home range requires site fidelity, site fidelity can be used as an indicator of whether an animal has established a home range. Operational definitions of home ranges exist using statistical definitions of site fidelity (Spencer et al. 1990). The goals of such definitions are good but the methods sometimes fail to define home ranges for animals that exhibit true and localized site fidelity. For example, Swihart and Slade (1985a, 1985b) used data for a female black bear (*Ursus americanus*) that I studied in 1983–1985 and determined that she did not have a home range because the sequence of her locations did not show site fidelity as defined by their statistical model. However, the bear's locations were strictly confined for 3 years to a distinct, well-defined area (figure 3.1). Consequently, researchers must sometimes use subjective measures of site fidelity, such as figure 3.1, to augment objective measures that sometimes fail, probably because statistical models have assumptions that are not appropriate for animal movements. Nonetheless, tests of site fidelity should be disregarded only when other objective approaches to site fidelity exist.

An animal's cognitive map must change as the animal learns new things about its environment and, hence, the map changes with time. As new resources develop or are discovered and as old ones disappear, appropriate changes must be made on the map. Such changes may occur quickly because an animal has an instantaneous concept of its cognitive map. A researcher, in contrast, can learn of the changed cognitive map only by studying the changes in the locations that the animal visits over time. An animal's home range usually cannot

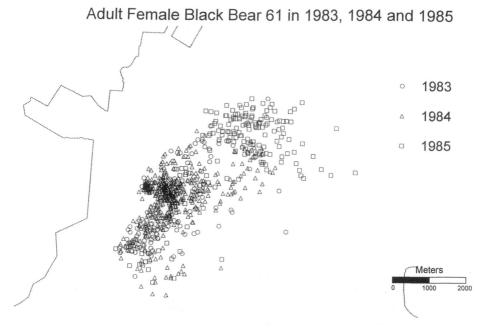

**Figure 3.1** Location estimates for adult female bear 61 in studied in 1983, 1984, and 1985 in the Pisgah Bear Sanctuary, North Carolina, U.S.A. Note that in each year, bear 61's locations were confined to a distinct area and that the area did not change much over the course of 3 years. This bear showed site fidelity, even though her location data did not conform to the rules of site fidelity for Swihart and Slade's (1985a, 1985b) model. The lightly dotted black line marks the study area border.

be quantified, practically, as an instantaneous concept because the home range can only be deduced from locations of an animal within its home range and the locations occur sequentially (but see Doncaster and Macdonald 1991). Thus, for most approaches, a home range must be defined for a specific time interval (e.g., a season, a year, or possibly a lifetime). The longer the interval, the more data can be used to quantify the home range, but the more likely that the animal has changed its cognitive map since the first data were collected.

In addition, no standard exists as to whether one should include in an animal's home range areas that the animal seldom visits or never visits after initial exploration. Many researchers define home ranges operationally to include only areas of use. Nonetheless, animals may be familiar with areas that they do not use. An arctic fox (*Alopex lagopus*) may be familiar with areas larger than 100 km2, yet use only a small portion (ca. 25 km2) where food is concentrated (Frafjord and Prestrud 1992). Areas with no food are not visited often, if ever, despite *and because of* the animal's familiarity with them. Should such areas be included in the fox's home range? Other areas with food might not have been visited in a given year simply by chance. Should those areas be included in the fox's home range? Pulliainen (1984) asserted that any area larger than 4 ha (an arbitrary size) not traversed by the Eurasian martens (*Martes martes*) he and his coworkers followed should not be included in the martens' home ranges. Through a winter, a marten crosses and recrosses old travel routes, leaving progressively smaller and smaller areas of irregular shape surrounded by tracks. Pulliainen presumed that a marten's radius of familiarity, or radius of perception, would cover an area of 4 ha or less. But how wide might an animal's radius of perception be? Some mammals can smell over a kilometer, see a few hundred meters, but feel only what touches them. Which radius should be used, or should a multiscale radius be used? In addition, areas not traversed may have been avoided by choice. Hence, should no radius of familiarity be considered? If we do not allow some radius of familiarity, or perception, around an animal, we are reduced, *reductio absurdum,* to counting as an animal's home range only the places where it actually placed its feet. Clearly, this is not satisfactory.

Related to this final problem is how to define the edges of an animal's home range. For many animals, the edges are areas an animal uses little but knows; the animal may actually care little about the precision of the boundaries of its home range because it spends the vast majority of its time elsewhere. Except for some territorial animals, the interior of an animal's home range is often more important both to the animal and to understanding how the animal lives and why the animal lives in that place. Gautestad and Mysterud (1993, 1995) and others have noted that the boundaries of home ranges are diffuse and general, making the area of a home range difficult to measure. That the boundary and area of a home range are difficult to measure does not reduce in any way the importance of the home range to the animal and to our understanding of the animal, however. Even crudely estimated areas for home ranges have led to insights into animal behavior and ecology (see the review by Powell 1994 of home ranges of *Martes* species), suggesting that home range areas should be quantified. However, we must keep in mind that home range boundaries and areas are imprecise, at least in part, because the boundaries are probably imprecise to the animals themselves.

## ■ Territories

A territory is an area within an animal's home range over which the animal has exclusive use, or perhaps priority use. A territory may be the animal's entire home range or it may be only part of the animal's home range (its core, for example). Territories may be defended with tooth and claw (or beaks, talons, or mandibles) but generally are defended through scent marking, calls, or displays (Kruuk 1972, 1989; Peters and Mech 1975; Price et al. 1990; Smith 1968), which are safer, more economical, and evolutionarily stable (Lewis and Murray 1993; Maynard Smith 1976). Members of many species, such as red squirrels (*Tamiasciurus hudsonicus;* Smith 1968), defend individual territories against all conspecifics, but tremendous variation in territorial behavior exists. In some species, individuals defend territories only against members of the same sex. In other species, mated pairs defend territories. In still other species, extended family groups, sometimes containing non–family members, defend territories. Whether territories are defended by an individual, mated pair, or family appears to depend on the productivity, predictability, and fine-grained versus coarse-grained patchiness of the limiting resources (Bekoff and Wells 1981; Doncaster and Macdonald 1992; Kruuk and Parish 1982; Macdonald 1981, 1983; Macdonald and Carr 1989; Powell 1989).

Members of many species in the Carnivora exhibit intrasexual territoriality and maintain territories only with regard to members of their own sex (Powell 1979, 1994; Rogers 1977, 1987). These species exhibit large sexual dimorphism in body size and males of these species are polygynous (and females undoubtedly selectively polyandrous). Females raise young without help from males and the large body sizes of males may be considered a cost of reproduction (Seaman 1993). For species that affect food supplies mostly through resource depression (i.e., have rapidly renewing food resources such as ripening berries and nuts or prey on animals that become wary when they perceive a predator and later relax), intrasexual territoriality appears to have a minor cost compared to intersexual territoriality because the limiting resource renews. This cost may be imposed on females by males (Powell 1993a, 1994).

Males of many songbird species defend territories. In migratory species, the males usually establish their territories on the breeding range before the females arrive and a male will continue to defend his territory if his mate is lost early in the breeding season. For these territories, the limiting resource may be a complex mix of the food and other resources that females need for successful reproduction and the females themselves. In red-cockaded woodpeckers (*Picoides borealis*) and scrub jays (*Aphelocoma coerulescens*), however, extended family groups defend territories. Male offspring, or occasionally female offspring, remain in their parents' (fathers') territories (Walters et al. 1988, 1992). Wolves (*Canis lupus*), beavers (*Castor canadensis*), and dwarf mongooses (*Helogale parvula*) also defend territories as extended families (Jenkins and Busher 1979; Mech 1970; Rood 1986).

Although territorial behavior might intuitively appear to help clarify the problem of identifying home range boundaries, this is not always the case. The territorial behavior of wolves actually highlights the imprecise nature of the boundaries of their territories. Peters and Mech (1975) documented that territorial wolves scent marked at high rates in response to the scent marks of neighboring wolf packs. In addition, the alpha male of a pack often ventured up to a couple hundred meters into a neighboring pack's territory to leave a scent mark. Such behavior changes a territory boundary into a space a few hundred meters wide, not a distinct, linear boundary. Hence, distinct boundaries of territories are little easier to identify than are boundaries of undefended home ranges.

Animals are territorial only when they have a limiting resource, that is, a critical resource that is in short supply and limits population growth (Brown 1969). The ultimate regulator of a population of territorial animals is the limiting resource that stimulates territorial behavior. Although population regulation through territoriality has received extensive theoretical attention (Brown 1969; Fretwell and Lucas 1970; Maynard Smith 1976; Watson and Moss 1970), the general conclusion of such theory is that territoriality can regulate populations only proximally. The most common limiting resource is food and, for territorial individuals, territory size tends to vary inversely with food availability (Ebersole 1980; Hixon 1980; Powers and McKee 1994; Saitoh 1991; Schoener 1981) For red-cockaded woodpeckers, however, the limiting resource is nest holes (Walters et al. 1988, 1992). For coral reef fish, the limiting resource is usually space (Ehrlich 1975). For pine voles (*Microtus pinetorum*), the limiting resource appears to be tunnel systems (Powell and Fried 1992). And for beavers, the limiting resource may be dams and lodges. Wolff (1989, 1993) warned that the limiting resource may not be food even if it appears superficially to be food.

Territorial behavior is not a species characteristic. In some species, individuals defend territories in certain parts of the species' range but not in other parts. This is the case for black bears (Garshelis and Pelton 1980, 1981; Powell et al. 1997; Rogers 1977, 1987). Similarly, many nectarivorous birds defend territories only when nectar production is at certain levels (Carpenter and

MacMillen 1976; Hixon 1980; Hixon et al. 1983). To understand why members of these species display flexibility in their territorial behavior, one must start with the concept that a territory must be economically defensible (Brown 1969). Carpenter and MacMillen (1976) showed theoretically that an animal should be territorial only when the productivity of its food (or whatever its limiting resource is) is between certain limits. When productivity is low, the costs of defending a territory are not returned through exclusive access to the limiting resource. When productivity is high, requirements can be met without exclusive access. The model developed by Carpenter and MacMillen (1976) is broadly applicable because it expresses clearly the limiting conditions required for territoriality to exist and it incorporates limits on territory size from habitat heterogeneity, or patchiness. Some approaches to modeling territorial behavior, such as Ebersole's (1980), Hixon's (1987), and Kodric-Brown's and Brown's (1978), do not express limiting conditions for territorial behavior but tacitly assume, a priori, that territoriality is economical. Understanding the limiting conditions for territorial behavior is important to understanding spacing behavior and home range variation in many species. Using economic models is a good approach to understanding limiting conditions for territoriality as long as the limiting resources do not change as conditions change (Armstrong 1992). Otherwise, the limiting resources must all be known clearly for the different conditions under which each is limiting. For example, if a small increase in the abundance of food leads to another resource becoming the limiting resource, that new limiting resource must be understood as well as the importance of food is understood. Researchers must also understand how an economic modeling approach fits into a broader picture, such as how animals use information from the environment to make decisions and how they perceive information (Stephens 1989).

When productivity of the limiting resource for an individual is very low and close to the lower limit for territoriality, the individual must maintain a territory of the maximum size possible. Such an individual should be completely territorial and not share any part of its territory. As productivity of the limiting resource for an individual approaches the upper limit for territoriality, however, its territorial behavior should change in one of two predictable ways. If necessary resources are evenly distributed in defended habitat, then the individual should maintain a smaller territory than in less productive habitat (Hixon 1982; Powell et al. 1997; Schoener 1981). If the individual's resources are distributed patchily and balanced resources cannot be found in a small territory, then it might exhibit incomplete territoriality. The individual might maintain exclusive access only to the parts of its home range with the most important resources. Coyotes (*Canis latrans,* Person and Hirth 1991), European red squirrels (*Sciurus vulgaris,* Wauters et al. 1994), and perhaps red-cockaded woodpeckers (Barr 1997) exhibit just such a pattern of partial territoriality and defend only home range cores in some habitats. Alternatively, an individual might allow territory overlap with a member of the opposite sex (Powell 1993a, 1994).

Food appears to be the limiting resource that stimulates territorial behavior by many animals and territorial defense decreases in those individuals as productivity or availability of food increases. Much research has been done on nectarivorous birds (Carpenter and MacMillen 1976; Hixon 1980; Hixon et al. 1983; Powers and McKee 1994), voles (Ims 1987; Ostfeld 1986; Saitoh 1991, reviewed by Ostfeld 1990), and mammalian carnivores (Palomares 1994; Powell et al. 1997; Rogers 1977, 1987). Black bears and nectarivorous birds (Carpenter and MacMillen 1976; Hixon 1980; Powell et al. 1997) switch quickly between territorial and nonterritorial behavior when productivity of food moves across the lower or upper limits for territoriality, respectively. For large mammals, I suspect that variation in territorial behavior around the upper limit of food production varies only over long time scales of many years (Powell et al. 1997).

Territorial behavior by members of several species (e.g., black bears, Powell et al. 1997; nectarivorous birds, Carpenter and MacMillen 1976; Hixon 1980; Hixon et al. 1983) can be predicted from variation in the productivity of food, which is good evidence that food is the limiting resource that stimulates territorial behavior for those animals. For European badgers (*Meles meles*), territory configuration can be predicted from positions of dens without reference to food (Doncaster and Woodroffe 1993), indicating that the limiting resource is den sites. However, no studies have rejected all other possible limiting resources. Wolff (1993, personal communication) argued strongly that only *offspring* are important enough, and can be defended well enough, to be the resource stimulating territorial behavior. For the black bears I have studied, adult females with and without young and adult and juvenile bears all responded in the same manner to changes in food productivity and also responded in the same manner to home range overlap with other female bears. Were Wolff correct, adult female bears with young would exhibit significantly different responses to food and to other females than do nonreproductive females. In addition, adult female black bears would be territorial in North Carolina, as they are in Minnesota. For the nectarivorous birds studied by Hixon (1980; Hixon et al. 1983), birds defended territories in the fall after reproduction but before and during migration. Were Wolff correct, hummingbirds would not defend territories after reproduction has ceased for the year. If bears, hummingbirds, and other animals use food as an index for the potential to produce offspring, then food can legitimately be considered to be at least a proximately limiting resource. Fitness is the ultimate currency in biology, and fitness may be affected by one or more limiting resources that need not be offspring or other direct components of reproduction. Evolution via natural selection requires heritable variation that affects reproductive output among individuals in a population. The effects can be via offspring, or they can be via food, nest sites, tunnel systems, or other potentially limiting resources.

## ■ Estimating Animals' Home Ranges

Added to conceptual problems of understanding an animal's home range are problems in estimating and quantifying that home range. We may never be able to find completely objective statistical methods that use location data to yield biologically significant information about animals' home ranges (Powell 1987). Nonetheless, our goal must be to develop methods that are as objective and repeatable as possible while being biologically appropriate. When analyzing data, we must use a home range estimator that is appropriate for the hypotheses being tested and appropriate for the data.

Reasons for estimating animals' home ranges are as diverse as research and management questions. Knowing animals' home ranges provides significant insight into mating patterns and reproduction, social organization and interactions, foraging and food choices, limiting resources, important components of habitat, and more. A home range estimator should delimit where an animal can be found with some level of predictability, and it should quantify the animal's probability of being in different places or the importance of different places to the animal.

Quantifying an animal's home range is an act of using data about the animal's use of space to deduce or to gain insight into the animal's cognitive map of its home (Peters 1978). These data are usually in the form of observations, trapping or telemetry locations, or tracks. Because at present we have no way of learning directly how an animal perceives its cognitive map of its home, we do not have a perfect method for quantifying home ranges. Even if we could understand an animal's cognitive map, we would undoubtedly find it difficult to quantify. Many methods for quantifying home ranges provide little more than crude outlines of where an animal has been located. For some research questions, no more information is needed. For questions that relate to understanding *why* an animal has chosen to live where it has, estimators are needed that provide more complex pictures. An animal's cognitive map will have incorporated into it the importance to the animal of different areas. The most commonly used index of that importance is the amount of time the animal spends in the different areas in its home range. For some animals, however, small areas within their home ranges may be critically important but not used for long periods of time, such as water holes. No standard approach exists to weight use of space by a researcher's understanding of importance. Therefore, to estimate importance to an animal of different areas of its home range, using any home range estimator currently available, one must assume that importance is positively associated with length or frequency of use, which are measures of time.

## UTILITY DISTRIBUTIONS

From location data such as those shown in figure 3.2, most home range estimators produce a utility distribution describing the intensity of use of different areas by an animal. The utility distribution is a concept borrowed from economics. A function, the utility function, assigns a value (the utility, which can be some measure of importance) to each possible outcome (the outcome of a decision, such as the inclusion of a place within an animal's home range; Ellner and Real 1989). If the utility distribution maps intensity of use, then it can be transformed to a probability density function that describes the probability of an animal being in any part of its home range (Calhoun and Casby 1958; Hayne 1949; Jennrich and Turner 1969; White and Garrott 1990; van Winkle 1975), as shown in figure 3.2. Utility distributions need not be probability density functions, although they usually are. A utility distribution could map the fitness an animal gains from each place in its home range, or it could map something else of importance to a researcher.

The approach using a utility distribution as a probability density function provides one objective way to define an animal's normal activities. A probability level criterion can be used to eliminate Burt's (1943) occasional sallies. Including in an animal's home range the area in which it is estimated to have a 100 percent probability of having spent time would include occasional sallies. Including only, say, the smallest area in which the animal spent 95 percent of its time could exclude occasional sallies or areas the animal will never visit again. Using a utility distribution, one can arbitrarily but operationally define the home range as the smallest area that accounts for a specified proportion of the total use. Most biologists use 0.95 (i.e., 95 percent) as their arbitrary but

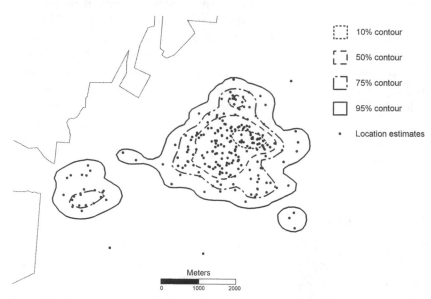

**Figure 3.2**  Location estimates (circles) and contours for the probability density function for adult female black bear 87 studied in 1985. The lightly dotted black line marks the study area border.

repeatable probability level; the smallest area with a probability of use equal to 0.95 is defined as an animal's home range. No strong biological logic supports the choice of 0.95 except that one assumes that exploratory behavior would be excluded by using this probability level; to my knowledge, this assumption has never been tested. An alternative approach is to exclude from consideration the 5 percent of the locations for an animal that lie furthest from all others. Eliminating these locations might also eliminate occasional sallies. A strong statistical argument exists for excluding some small percentage of the location data, the utility distribution, or both; extremes are not reliable and tend not to be repeatable. However, this argument does not specify that precisely 5 percent should be excluded. Using 95 percent home ranges may be widely accepted because it appears consistent with the use of 0.05 as the (also) arbitrary choice for the limiting $p$-value for judging statistical significance.

Once home range has been defined as a utility distribution, a reliable method must be sought to estimate the distribution. Estimating utility distributions has been problematic because the distributions are two- or three-dimensional, observed utility distributions rarely conform to parametric models, and data used to estimate a distribution are sequential locations of an individual animal and may not be independent observations of the true distribution (Gautestad and Mysterud 1993, 1995; Gautestad et al. 1998; Seaman and Powell 1996; Swihart and Slade 1985a, 1985b). However, lack of independence of data may not be a great problem for some analyses (Andersen and Rongstad 1989; Gese et al. 1990; Lair 1987; Powell 1987; Reynolds and Laundrè 1990). After all, data that are not statistically autocorrelated are nonetheless biologically autocorrelated because animals use knowledge of their home ranges to determine future movements. Boulanger and White (1990), Harris et al. (1990), Powell et al. (1997), Seaman and Powell (1996), and White and Garrott (1990) reviewed many home range estimators and Larkin and Halkin (1994) summarized computer software packages for home range estimators.

## GRIDS

To avoid assuming that data fit some underlying distribution (for example, that an animal's use of space is bivariate normal in nature), many researchers superimpose a grid on their study areas and represent a home range as the cells in the grid having an animal's locations (Horner and Powell 1990; Zoellick and Smith 1992). Each cell can have a spike as high as the number or proportion of times the animal was known or estimated to have been within that cell (figure 3.3) and the resultant surface is an estimate of the animal's utility distribution. For small sample sizes of animal locations, or for finely scaled grids, a home range can be estimated to have several disjunct sections (see especially figure 3.3b). The resident animal traversed the areas between the disjunct sections too rapidly, or the interval between locations was too long, for the animal to be found in intervening cells. These areas were not used for occasional sallies and therefore should probably be included within the animal's home range. One can include in the home range all cells between sequential locations, but no objective method exists to incorporate these cells into the estimated utility distribution. If possible, one should collect data until the animal has been found at least once in each cell connecting formerly disjunct locations. Using this approach to estimate home ranges, a researcher risks not including significant areas in an animal's home range.

Doncaster and Macdonald (1991) estimated the home ranges of foxes (*Vulpes vulpes*) as a retrospective count of the grid cells known to be visited at any one time. This approach is equivalent to treating the cells as marked individuals for a mark–recapture study and estimating home range size (population size of the cells) from a minimum number known alive approach (Krebs

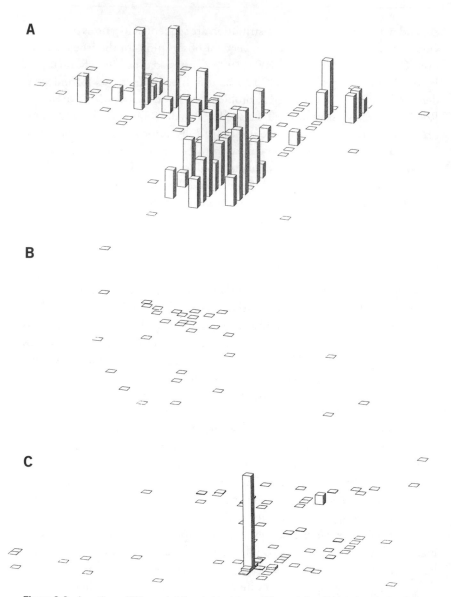

**Figure 3.3** Locations of (A) an adult female black bear, (B) an adult wolf (data from L. David Mech, personal communication), and (C) an adult male stone marten (data from Piero Genovesi, personal communication), presented as bars within grid cells. The height of each bar is proportional to the number of times the animal's location was estimated to be in that cell.

1966). Calculating back to any time, a fox's home range included the cells that had been visited before that time and that would be visited again. This approach allowed Doncaster and Macdonald to follow foxes' home ranges as they drifted across the landscape. More sophisticated survival estimators could be applied to estimate the rates at which cells were lost from home ranges and new cells added (Doncaster and Macdonald 1996). With this approach, occasional sallies are easily identified as cells visited only once.

Vandermeer (1981) cogently discussed how choosing the size of cells is a major problem for most analyses using grids. For data on animal locations, cell size should incorporate, in some objective way, information about error associated with location estimates for telemetry data, information about the radius of attraction for trapping data, information about the radius of an animal's perception and knowledge for all location data, and knowledge of the appropriate scale for the hypotheses being tested. For some comparisons, cell size must be equal for all animals; for others, cell size relative to home range size must be equal. However, changing cell size can change results of analyses (Lloyd 1967; Vandermeer 1981), often because cell size is related to the scale of the behaviors being studied.

## MINIMUM CONVEX POLYGON

The oldest and mostly commonly used method of estimating an animal's home range is to draw the smallest convex polygon possible that encompasses all known or estimated locations for the animal (Hayne 1949). This minimum convex polygon is conceptually simple, easy to draw, and not constrained by assuming that animal movements or home ranges must fit some underlying statistical distribution. However, problems with the method are myriad (Horner and Powell 1990; Powell 1987; Powell et al. 1997; Seaman 1993; Stahlecker and Smith 1993; White and Garrott 1990; van Winkle 1975; Worton 1987). Minimum convex polygons provide only crude outlines of animals' home ranges, are highly sensitive to extreme data points, ignore all information provided by interior data points, can incorporate large areas that are never used, and approach asymptotic values of home range area and outline only with large sample sizes (100 or more animal location estimates; Bekoff and Mech 1984; Powell 1987; White and Garrott 1990). Because all information about use of a home range within its borders is ignored using a minimum convex polygon, most analyses using this method implicitly assume that animals use their home ranges evenly (use all parts with equal intensity), which is clearly not the case. One can calculate a minimum convex polygon using the

95 percent of the data points that form the smallest polygon, but this does not avoid the flaws inherent in the method other than the problem with extreme data points. To construct a minimum convex polygon, a researcher discards 90 percent of the data he or she worked so hard to collect and keeps only the extreme data points. This method, more than any other, emphasizes only the unstable, boundary properties of a home range and ignores the internal structures of home ranges and central tendencies, which are more stable and are important for most critical questions about animals.

## CIRCLE AND ELLIPSE APPROACHES

Hayne (1949) suggested that to estimate an animal's home range from point location data one should use a circle; Jennrich and Turner (1969) and Dunn and Gipson (1977) generalized the circle to an ellipse. Circle and ellipse approaches assume that animals use space in a fashion conforming to an underlying bivariate normal distribution. Using a circle to represent an animal's home range assumes that each animal has a single center of activity that is the very center, or the two-dimensional arithmetic mean, of all locations. Using an ellipse assumes that each animal has two such centers of activity that are the foci of the ellipse. An ellipse can be drawn around the two centers of activity for an animal such that it contains 95 percent of the location data. This 95 percent ellipse can also be used as an estimate of the animal's home range. Dunn and Gipson's (1977) approach incorporates time data for animal location estimates but time data must conform to a highly restrictive pattern, which is usually impossible for field research. Because animals do not use space in a bivariate normal fashion, any estimator of animal home ranges that assumes such use will estimate utility distributions poorly. de Haan and Resnick (1994) recently developed a home range estimator based on polar coordinates that incorporates the time sequential aspect of location data. However, their estimator appears not to be broadly applicable to real animal location data because data must be of a restricted type and outliers (sampling errors) must be identifiable. All ellipse estimators include within an estimated home range many areas not actually used by an animal.

## FOURIER SERIES

In statistics, Fourier series are often used to smooth data, so Anderson (1982) developed a home range estimator based on the bivariate Fourier series. Each animal location estimate is treated as a spike in the third dimension above an

$x$–$y$ plane. The Fourier transform estimator smooths the spikes into a surface that estimates an animal's utility distribution. I developed a similar method using spline smoothing techniques (Powell 1987). Both of these estimators accurately show multiple centers of activity that may be considerably removed from the arithmetic mean of the $x$ and $y$ data, but both behave poorly near the edges of home ranges, probably because the location data do not meet assumptions needed to make the transformations. To address the problem of poor estimates of home range peripheries, Anderson (1982) recommended using animals' 50 percent home ranges (the smallest area encompassing a 50 percent probability of use) rather than 95 percent home ranges. Fifty percent is no less arbitrary than 95 percent, but it departs completely from the basic concept of a home range (Burt 1943) or stretches that concept to its limit by assuming that an animal is on an "occasional sally" 50 percent of the time.

## HARMONIC MEAN DISTRIBUTION

Human population densities fall in an inverse harmonic mean fashion from centers of urban areas through rural areas. Consequently, Dixon and Chapman (1980) proposed using a harmonic mean distribution to describe animal home ranges. Contours for a utility distribution are developed from the harmonic mean distance from each animal location to each point on a superimposed grid. The harmonic mean estimator may accurately show multiple centers of activity, but each estimated utility distribution is unique to the position and spacing of the underlying grid. Spencer and Barrett (1984) modified the method to reduce the problem of grid placement but a large problem with grid size remains. When a very fine grid is used, the resulting utility distribution becomes a series of sharp peaks at each animal location. When a coarse grid is used, the utility distribution lacks local detail and is overly smoothed. For many data sets, the harmonic mean estimator actually appears both to exaggerate peaks at animal locations and to oversmooth elsewhere. In addition, the estimator calculates values for all grid points, provides no outline for a home range, and does not provide a utility distribution. Most researchers choose for the home range outline the contour equal to the largest harmonic mean distance from an animal location to all other animal locations (Ackerman et al. 1988) and from this a utility distribution can be calculated. Although this is an objective criterion, it is affected by sample size. Finally, for animal home ranges that have geographic constraints that confine shapes (e.g., lakes, mountains; Powell and Mitchell 1998; Reid and Weatherhead 1988; Stahlecker and Smith 1993), much area not actually in an animal's home range will be included in the harmonic mean estimate. Boulanger and White (1990) used Monte Carlo simulations and tested the performance of the harmonic mean estimator against the other estimators just discussed. Despite its problems, the harmonic mean estimator was the best of the lot. Luckily, better estimators have since been developed.

One set of home range estimators, kernel estimators, appears best suited for estimating animals' utility distributions, and hence home ranges. Another set, fractal estimators, may have promise.

## FRACTAL ESTIMATORS

Bascompte and Vilà (1997), Gautestad and Mysterud (1993, 1995), and Loehle (1990) modeled animal movements as multiscale random walks and analyzed the patterns of locations as fractals. Bascompte and Vilà (1997) explained that $D$, the fractal dimension, can be estimated as

$$D = \frac{\log(n)}{\log(n) + \log(d/L)}$$

where $n$ is the number of steps along a trace of an animal's movements (1 less than the number of locations), $L$ is the sum of the lengths of all steps (total length of the movement), and $d$ is the planar diameter, which can be estimated as the greatest distance between two locations. For a movement that is a straight line, $d = L$, so $D = 1$; a line has one dimension. For a random walk, $D = 2$; a random walk spreads over a plane and has two dimensions.

For the animals studied by Bascompte and Vilà (1997) and Gautestad and Mysterud (1993, 1995), the fractal dimensions, $D$, for movements averaged less than 2. Finding $D < 2$ means that as they scrutinized their animal location data on smaller and smaller scales, they found clumps of locations within clumps within clumps ad infinitum. The movements of the animals did not spread randomly across the landscape. Gautestad and Mysterud (1993, 1995) argued, therefore, that animals use their home ranges in a multiscale manner, which makes ultimate sense. Optimality modeling (giving up time) and empirical data show that animals who forage in patchy environments are predicted to and, indeed, do change their movements dependent on both fine-scale and large-scale characteristics of food availability (Curio 1976; Krebs and Kacelnik 1991). Thus an animal's decision to remain in or to leave a food patch depends not just on the availability of food within the patch but also on the availability of food across its home range and on the locations of the other patches of food.

In addition, Gautestad and Mysterud (1993, 1995) showed that if animals move in a manner described by a multiscale random walk that incorporates the multiscale, fractal nature of animal movements, then the estimated home range area should increase infinitely in proportion with the square root of the number of location estimates used to estimate the area of the home range using a minimum convex polygon. Indeed, the home ranges of several species, quantified using minimum convex polygons, do appear to increase in area as predicted (Gautestad and Mysterud 1993, 1995; Gautestad et al. 1998). The predicted relationship between home range area ($A_{MCP}$, for minimum convex polygons) and the number of locations ($n$) is

$$A_{MCP} = C \cdot Q(n) \cdot n^{1/2} \qquad (3.1)$$

where $C$ is the constant of proportionality, or the scaling factor, and $Q(n)$ is a function that adjusts the relationship for underestimates of $A_{MCP}$ because of small sample size. Curve fitting indicates that

$$Q(n) = \exp(6/n^{0.7})$$

for $n \geq 5$. When not calculating home range area from minimum convex polygons, $Q(n)$ should not be used.

Gautestad and Mysterud (1993) interpret $C$ to be a measure of how an animal perceives the grain of its environment. When a grid is superimposed over a plot of an animal's locations, $C$ can be calculated for each cell and $1/C$ is a descriptor of the intensity of use for each cell (Gautestad 1998).

$1/C$ can be calculated in two ways. Superimpose a grid on a map of a study area such that no cells have fewer than five locations for a target animal (cells with fewer than five locations might alternatively be ignored). Calculate the area of the minimum convex polygon formed by all locations within each cell and use that for $A_{MCP}$ in equation 3.1. Calculate $1/C$ as

$$1/C = [Q(n) \cdot n^{1/2}]/A_{MCP}$$

Alternatively, $1/C$ can be calculated in a manner that uses different scales. Superimpose a grid on a map of a study area with cell size such that one cell contains all the locations of given animal. The area of the single can be considered as $A$ and $1/C = n^{1/2}/A$. Now divide the single cell into four equal cells and calculate $1/C$ for each cell, letting $A$ be the area of each new cell and $n$ the number of locations in each new cell. The cells can be divided again each into four equal cells and the new $1/C$ calculated for each. In either of these approaches, a utility distribution can be calculated on different scales appropriate for different questions.

Gautestad and Mysterud (1993:526) also argued that the fractal approach to animal movements shows that "it is just as meaningless to calculate [home range] areas or perimeters as it is to calculate specific lengths of a rugged coastline." They concluded that home range areas cannot be measured because the number of data points needed for an accurate estimate exceeds the number that can be collected on most studies. Unfortunately, Gautestad and Mysterud overstate their point. Clearly, home range boundaries and areas are simple and usually poor measures of animals' home ranges. The important aspects an animal's home range relate to the intensity of use and the importance of areas on the *interior* of the home range (Horner and Powell 1990). So Gautestad and Mysterud are correct in playing down the importance of boundaries and areas.

Nonetheless, boundaries and areas can be estimated. Animals' home ranges have indistinct boundaries, just as the coastline of an island becomes indistinct when viewed using several different scales. But an island whose perimeter cannot be measured accurately nonetheless has a finite limit to its area, and that limit can be estimated. Likewise, animals who confine their movements to local areas (exhibit site fidelity) do have home ranges whose areas can be estimated, even if those areas must be estimated as a range between upper and lower limits, and even if the home range boundaries may never be known precisely. In addition, a useful estimate of the internal structure of a home range may be estimated with fewer data than needed to obtain reasonable estimates of the home range boundary or area.

In fact, during a finite period of time, an animal *must* confine its movements to a finite area and limits to that area can be estimated. The black bears I have studied do confine their movements to finite areas. Fixed kernel estimates of the areas of the annual home ranges of all bears located more than 300 times reached asymptotes after at most 300 chronological locations ($131 \pm 90$, mean $\pm$ SD, $n = 7$; Powell, unpublished data; asymptote at 300 for a bear located more than 450 times, 95 percent home ranges). However, equation 3.1 states that the estimated home range area must increase infinitely as the number of location data points used to estimate the home range increases. Clearly, this is a contradiction. The solution to the contradiction lies, I believe, with whether one includes unused areas within an animal's home range and whether one uses stable measures of the interiors of home ranges or uses unstable measures of the periphery.

Gautestad and Mysterud (1993, 1995) appear to have run their simulations using simulated utility distributions so large that their simulated animals could not use their whole "home ranges" within biological meaningful time periods. When this is the case, estimates of home ranges should increase in size as more and more simulated data points are used for the estimates. Indeed, after thousands of data points were used, the estimated home range areas do reach asymptotes at the areas of the utility distributions (Gautestad and Mysterud, personal communication), but note that this implies that equation 3.1 is not accurate for large *n*.

Some real animals may not use within a single year (or within some other biologically meaningful period) all the areas with which they are familiar. This raises the question of whether areas not used by an animal during a biologically meaningful period of time should be included in the estimate of its home range. Perhaps Gautestad and Mysterud's simulated utility distributions actually represent animals' cognitive maps. Is an animal's cognitive map its home range? Or is its home range only the areas with which it is familiar *and* that it uses? No definitive answers exist for these questions. Equation 3.1 may be true for some animals. It is most likely to be true for animals that are familiar with areas far larger than they can use in a biologically meaningful period of time. And if equation 3.1 is true, then the time periods over which we estimate home ranges may be as important as the numbers of locations. The time periods must be biological meaningful periods. To obtain accurate estimates of animals' home ranges, we may need to collect as many data as possible, organized into biologically meaningful time periods.

Another solution exists to the contradiction (not necessarily an independent solution). Gautestad and Mysterud estimated home range areas using 100 percent minimum convex polygons (but using the fudge factor $Q(n)$), which use only extreme, unstable data and must increase whenever an animal reaches a new extreme location. They purposefully incorporated occasional sallies into their model but did not exclude them from their home range calculations. Small changes in sampling points at the extremes of animals' home ranges can lead to huge differences in calculated home range areas although the animals may not have changed use of the interiors of their home ranges. I calculated home ranges areas for black bears using a kernel estimator, which emphasizes central tendencies, which are stable; home range estimates from kernel estimators do not change each time an animal explores a new extreme location.

Finally, Gautestad and Mysterud's model may be unrealistic. Any model of animal movement must be a simplification, so Gautestad and Mysterud's model does simplify animal movements. It does incorporate multiscale aspects of movement and appears to be a better model than, say, random walk models. Nonetheless, the multiscale random walk model still lacks important characteristics of true animal movements, and may thereby cause equation 3.1 to give a false prediction.

Even if equation 3.1 is false, the fractal utility distribution based on $1/C$ may still provide insight into use of space by animals. Unfortunately, by calculating $C$ for each cell in a grid, one loses multiscale information that is available from an entire data set. In addition, $1/C$ provides no insight into estimated use of interstitial cells because it is only a transformation of the frequencies per cell ($n^{1/2}$ instead of $n$). Finally, Vandermeer's (1981) cautions concerning grid dimensions must be addressed. One gains equal insight by calculating kernel home ranges and examining the probabilities for animals to be in cells of different sizes (scales), and kernel estimators are free of grid size constraints.

Fractal approaches to animal movements may provide new insights into animals' home ranges, but their utility is still uncertain.

## KERNEL ESTIMATORS

I believe that the best estimators available for estimating home ranges and home range utility distributions are kernel density estimators (Powell et al. 1997; Seaman 1993; Seaman et al. 1999; Seaman and Powell 1996; Worton 1989). Nonparametric statistical methods for estimating densities have been available since the early 1950s (Bowman 1985; Breiman et al. 1977; Devroye and Gyorfi 1985; Fryer 1977; Nadaraya 1989; Silverman 1986; Tapia and Thompson 1978) and one of the best known is the kernel density estimator (Silverman 1986). The kernel density estimator produces an unbiased density estimate directly from data and is not influenced by grid size or placement (Silverman 1986). Worton (1989) suggested that a kernel density estimator could be used to estimate home ranges of animals but little work (Worton 1995) had been published on the method as a home range estimator before Seaman's (1993; Powell et al. 1997; Seaman et al. 1999; Seaman and Powell 1996; Seaman et al. 1998) work, which is elaborated here.

Kernel estimators produce a utility distribution in a manner that can be visualized as follows. On an $x$–$y$ plane representing a study area, cover each location estimate for an animal with a three-dimensional "hill", the kernel, whose volume is 1 and whose shape and width are chosen by the researcher. The width of the kernel, called the band width (also called window width or $h$), and the kernel's shape might hypothetically be chosen using location error, the radius of an animal's perception, and other pertinent information. Luckily, kernel shape has little effect on the output of the kernel estimators, as long as the kernel is hill-shaped and rounded on top (Silverman 1986), not sharply peaked (deduced from criticisms by Gautestad and Mysterud, personal communication). Although no objective method exists at present to tie band width to biology or to location error, except that band width should be greater than location error (Silverman 1986), objective methods do exist for choosing a band width that is consistent with statistical properties of the data on animal locations. Band width can be held constant for a data set (fixed kernel). Or band width can be varied (adaptive kernel) such that data points are covered with kernels of different widths ranging from low, broad kernels for widely spaced points to sharply peaked, narrow kernels for tightly packed points. Although adaptive kernel density estimators have been expected, intuitively, to perform better than fixed kernel estimators (Silverman 1986), this has not been the case (Seaman 1993; Seaman et al. 1999; Seaman and Powell 1996). The utility distribution is a surface resulting from the mean at each point of the values at that point for all kernels. In practice, a grid is superimposed on the data and the density is estimated at each grid intersection as the mean at that point of all kernels. The probability density function is calculated by multiplying the mean kernel value for each cell by the area of each cell.

Choosing band width is one of the most important and yet the most difficult aspects of developing a kernel estimator for animal home ranges (Silverman 1986). Narrow kernels reveal small-scale details in the data, and, consequently, tend also to highlight measurement error (telemetry error or trap placement, for example). Wide kernels smooth out sampling error but also hide local detail. The optimal band width is known for data that are approximately normal but, unfortunately, animal location data seldom approximate bivariate, normal distributions (Horner and Powell 1990; Seaman and Powell 1996). For distributions that are not normal, a band width more appropriate than that for a normal distribution can be chosen using least squares cross validation. This process chooses various band widths and selects the one that provides the minimum estimated error. Seaman (1993; Seaman and Powell 1996) found that cross-validation chooses band widths that estimate known utility distributions better than do band widths appropriate for bivariate normal distributions.

Using computer simulations and telemetry data for bears, Seaman (Seaman 1993; Seaman et al. 1999; Seaman and Powell 1996) explored the accuracy of both fixed and adaptive kernel home range estimators and compared their accuracies to the harmonic mean estimator. He used simulated home ranges that looked much like real home ranges but he knew the utility distributions for the simulated home ranges. Seaman chose points randomly within the simulated home ranges, simulating the collection of telemetry or trapping or sighting location data, and then he estimated the simulated home ranges from the "location" data points. He then compared the kernel estimators to the harmonic mean estimator because the harmonic mean estimator was widely used into the early 1990s, it appeared preferable to most well-known nonkernel estimators (Boulanger and White 1990), and Seaman's comparisons can be extrapolated to other home range estimators through Boulanger and White's (1990) results. Seaman found that the different home range estimators varied greatly in accuracy of estimating both home range areas and utility distributions (figure 3.4).

The fixed kernel estimator, using cross-validation to choose band width,

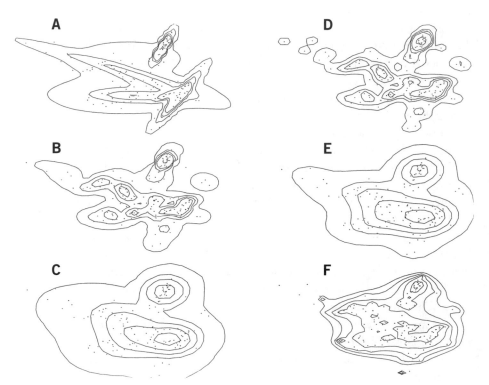

**Figure 3.4** A complex, simulated home range. (A) True density contours. (B) Fixed kernel density estimate with cross-validated band width choice. (C) Adaptive kernel density estimate with cross-validated band width choice. (D) Fixed kernel density estimate with ad hoc band width choice. (E) Adaptive kernel density estimate with ad hoc band width choice. (F) Harmonic mean estimate. Modified from Powell et al. (1997).

yielded the most accurate estimates of home range areas and had the smallest variance. These estimates averaged 0.7 percent smaller than the true areas of the simulated home ranges, whereas the adaptive kernel estimates averaged about 25 percent larger than true. The harmonic mean estimator overestimated true home range area by about 20 percent. The cross-validated, fixed kernel estimator also estimated the shapes of the utility distributions the best (figure 3.4). Figure 3.2 depicts the utility distribution isoclines for the home range of an adult female black bear. In addition, for simple, simulated home ranges, the fixed and adaptive kernel estimators generate consistent 95 percent home range areas with as few as 20 location estimates (Noel 1993; Seaman et al. 1999). However, the harmonic mean estimator requires 125 location estimates or more.

The adaptive kernel estimators performed slightly worse than the fixed kernel estimators in all of the tests, apparently through overestimation of peripheral use (Seaman 1993; Seaman et al. 1999; Seaman and Powell 1996). Adaptive kernel estimators also appear sensitive to autocorrelation within data sets. The amount of kernel variation can be adjusted for adaptive kernel estimators, but Seaman has found no consistent or predictable pattern of adjustment that minimizes error for these estimators (Seaman et al. 1999). Consequently, the best estimators at present are fixed kernel estimators with band width chosen via least-squares cross-validation (Seaman 1993; Seaman et al. 1999; Seaman and Powell 1996).

Kernel estimators share three shortcomings with most other home range estimators. First, they ignore time sequence information available with most data on animal locations (White and Garrott 1990). All estimators assume that all location data points are independent and that time sequence information is irrelevant. Future kernel estimators will incorporate brownian bridges between consecutive location estimates, with the heights, widths, and shapes of the bridges dependent on the time and distance between locations, as developed by Bullard (1999). Second, kernel estimators estimate the probability that an animal will be in any part of its home range; therefore, they sometimes produce 95 percent home range outlines that have convoluted shapes or disjunct islands of use. For example, figure 3.5 shows the 95 percent fixed kernel home range for an adult female black bear, bear 61, whom I studied in 1983–1985. Bear 61's home range in 1983 nearly surrounds a large area not designated as her home range. Surely, this bear was familiar with the surrounded area and included it on her cognitive map; however, she chose not to use that area regularly in 1983. In other years, she did use that area (figure 3.1). The fixed kernel estimate of bear 61's home range accurately quantifies the

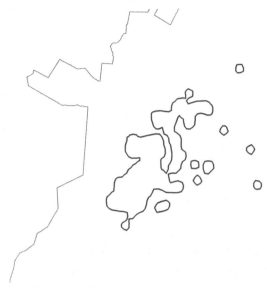

### Adult Female Black Bear 61
Home Range in 1983

**Figure 3.5** The 95% fixed kernel home range for adult female black bear 61 in 1983. Bear 61's home range nearly surrounds a large, central area not designated as her home range. The thin dotted line marks the study area border.

area that she *used* in 1983 but may not accurately define the area that she actually *considered* to be her home range in 1983.

Third, related to the second problem, kernel estimators estimate the probability that an animal will be in any part of its home range but do not estimate how *important* that part of the home range is to the animal. For researchers asking questions related to time or studying animals for whom time and importance coincide, no problem exists because home range estimators provide probabilities for use of space or, alternatively, probabilities for extent of time in given areas. This aspect of home range estimators *is* a problem for researchers interested in the underlying importance of habitats or landscape characteristics when time and importance may not coincide. Some parts of a home range that are used little may be very important because they contain a limiting resource needed only at low levels of use. If time and importance do not coincide, kernel estimators (and all other estimators) do not estimate importance accurately.

## HOME RANGE CORE

Particular parts of an animal's home range must be more important than other parts. In general, foods and other resources are patchily distributed (Curio 1976; Frafjord and Prestrud 1992; Goss-Custard 1977; Mitchell 1997; Powell et al. 1997), so the parts of a home range with greater density of critical resources ought, logically, to be more important than areas with few resources. For years, biologists have conceived the core as the part of an animal's home range that is most important to it (Burt 1943; Ewer 1968; Kaufmann 1962; Samuel et al. 1985; Samuel and Green 1988). To understand home ranges well, identifying cores is important, if cores do indeed exist.

My understanding of a home range core has two major parts (Powell et al. 1997; Seaman and Powell 1990). First, a core must be used more heavily than the apparent clumps of heavy use that occur from uniform random use of space within a home range. Random use of space leads to some areas being used more than others, even though no place is more important to the animal than any other. Therefore, random use of space leads to apparent clumps of use in some places and little use of other places. Consequently, the core of a home range must be used more than expected by random use, which means that for a home range to have a core, use of space within that home range must be statistically clumped and not random or even. Testing for clumped versus random (or for clumped versus even) use of space is usually straightforward (Horner and Powell 1990; Mitchell 1997). In a uniform random distribution, the mean equals the variance. If an animal uses space at random, then the mean number of locations in each cell in its home range equals the variance. If the variance is significantly greater than the mean, then use is clumped; if the variance is less than the mean, then use is even. Note that many nonrandom distributions have means equal to their variances. Therefore, equal mean and variance does not prove uniform random use of space, but unequal mean and variance does disprove random use of space.

Second, a core must not be strictly determined by home range area. Animals with home ranges of equal size but with different patterns of home range use (e.g., central place foragers, strongly territorial animals, extensive wanderers) should have differently sized cores. Any technique developed to identify the core of a home range must reflect this biological understanding of what a core is.

Most definitions of cores have been ad hoc or subjective. Many define the core as the smallest area with an arbitrary probability of use (e.g., the smallest area enclosing 25 percent of total use). A crucial problem with such definitions, beyond their subjectivity, is that the cores so defined are not strictly tied to intensity of use of space. In fact, animals that use their home ranges randomly or in an even fashion have cores by these definitions. Samuel et al. (1985) developed an objective method for identifying the maximum possible core: all parts of the home range used more heavily than would result from evenly distributed use. Although this definition is objective, it is still arbitrary, it allows cores to incorporate space used little more than adjacent space, and it defines cores for animals that use space randomly. Seaman and I (1990; Powell et al. 1997) introduced a technique for identifying cores that is objective, not arbitrary, and that allows the animals themselves to define their cores. Our technique is based on the logic used to identify behavior bouts (Fagen and Young 1978; Slater and Lester 1982). Bingham and Noon (1997) used the same method and Harris et al. (1990) may have, but their explanation is not clear.

If an animal's use of space is random within its home range, a plot of home range area at a certain percentage use versus probability of use should yield a straight line going from the 100 percent home range at probability of use equal to 0 (100 percent of the home range has probability of use 0 or above) to the 0 percent home range at probability of use greater than the maximum probability of use (0 percent of the home range has probability of use greater than the greatest probability of use; figure 3.6a). If probability of use is transformed to percentage of largest probability of use, then both $x$ and $y$ axes on the graph range from 0 to 100 (figure 3.6a) and the descending line representing random use has a slope of $-1$. If use of space by an animal is clumped, however, its curve will sag below the line of random use (figure 3.6b) and if use of space is even (all areas used with equal intensity), the graph will remain as a high plateau from $x$ equals 0 probability of use up to $x$ equal some large probability of use and then plummet (figure 3.6c).

When use of space by animals is clumped (figure 3.6b), Seaman and I defined as an animal's core the areas in its home range that are used most intensively. The parts of the home range mapped onto the steeply descending slope of the area–probability curve along the $y$-axis are used least and constitute the periphery of the home range. The parts of the home range mapped onto the nearly horizontal slope along the $x$-axis are used most intensively and constitute the core. The curve can be divided into two pieces at the point whose tangent has a slope of $-1$, that is, whose tangent is parallel to the line for random use. This is also the point that is furthest from the line with a slope of $-1$ (figure 3.6b). Plots of actual data may not yield smooth curves; these plots can be fit with reasonable curves.

# HOME RANGE CORES

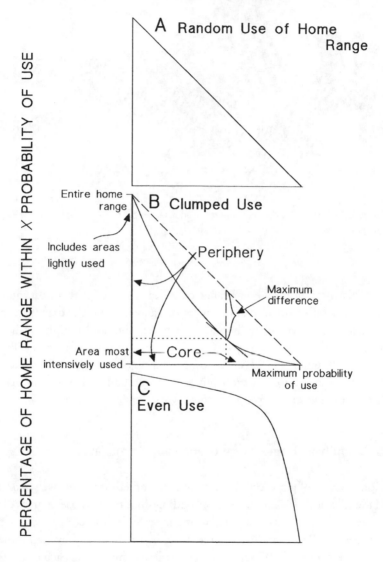

**Figure 3.6** For an animal's home range, possible relationships between probability of use and percentage of home range with the probability of use or greater. The x-axis is probability of use for areas within an animal's home range calculated as the percentage of maximum probability of use. The y-axis is the percentage of the home range with the given probability of use of higher use. (A) Relationship for random use by an animal of the area within its home range, (B) clumped or patchy use of the home range, (C) even or overdispersed use of the home range.

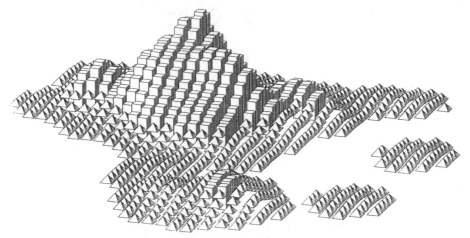

**Figure 3.7**  Core area and home range for an adult female bear. The core is shown with flat-topped symbols, the periphery with triangular symbols.

This criterion for a home range core clearly identifies the most intensively used areas within an animal's home range, and it allows the data (i.e., the animal) to decide where the boundary between core and periphery should be located. The criterion is objective and, for me, intuitive (figure 3.7; Powell et al. 1997; Seaman and Powell 1990). By this objective, each animal's core, if it has one, is at a different probability of use. In addition, some animals have large cores and some have small cores.

## ■ Quantifying Home Range Overlap and Territoriality

Home ranges of conspecifics often overlap, sometimes extensively. For a population, hypotheses can be tested regarding simultaneous use of areas of home range overlap. Subsets of a population may exhibit different patterns of simultaneous use. Relatives, for example, may use their areas of overlap more than expected from random use, whereas nonrelatives may avoid each other and use areas of overlap less than expected. Intrasexually territorial animals may spend less time in their areas of overlap with members of the opposite sex than expected.

For some species, territorial behavior has been documented objectively. Extensive experimentation with limiting resources and territorial displays and calls has defined territoriality clearly in many birds (e.g., nectarivorous birds, reviewed by Hixon 1980; Hixon et al. 1983; blackbirds, reviewed by Orians 1980). For many species, however, apparent lack of overlap of home ranges is the only evidence of territorial behavior. For many carnivorous mammals, territory defense and responses to scent marks are difficult to document. For such species, home range overlap can be quantified in an objective manner that weighs probability of use of different parts of a home range or territory. Home range overlap can then be compared statistically among populations or species that appear to differ in territorial behavior but for whom territorial behavior has not been manipulated experimentally.

Doncaster (1990) defined two types of overlap, called static and dynamic interactions. Static interaction is the spatial overlap of two home ranges and dynamic interaction involves interdependent movements of the two animals whose home ranges overlap. These types of overlap can be quantified in several different ways.

## STATIC INTERACTIONS

Area of overlap is a poor estimate of the effect or importance of the overlap on two individuals whose home ranges overlap. Areas of overlap vary in probability of use and the two individuals may have a large overlap of areas used little by each or a small overlap of areas used intensively by each. Although Genovesi and Boitani (1997) found that area overlap of minimum convex polygons correlated strongly with weighted overlap, this need not be the case for all populations or individuals.

I have used two indices of pairwise overlap of home ranges using 95 percent probability density functions for each animal's home range. Such probability density functions could be generated by a fixed kernel home range estimator or by any other estimator that produces accurate utility distributions. The first, $I_p$, is

$$I_p = \sum_{k \in 0} P_{ki} \cdot \sum_{k \in 0} P_{kj}$$

where $p_{ki}$ and $p_{kj}$ are the independent probabilities that at any arbitrary time animals $i$ and $j$ are in cell $k$ of a study area that has a grid superimposed on it and that cell $k$ is within the area of overlap, $O$, of the animals' home ranges. $I_p$ ranges from 0 to 1. This index is the simple probability that the two animals $i$

and $j$ will be in their area of home range overlap at the same time were they to move independently of each other. Of course, most animals do not move without respect to the movements of other animals. Consequently, static interactions should be studied in conjunction with dynamic interactions.

A similar index, $I_L$, is Lloyd's (1967) $\overset{*}{m}$ index of mean crowding, which Hurlbert (1978) identified as probably the least biased overlap index:

$$\overset{*}{m} = N \cdot \sum_k P_{ki} \cdot P_{kj}$$

where $N$ is the number of cells in which animal $i$ or animal $j$ (or both) has nonzero probability of use. The mean crowding index standardizes overlap by the probability of both animals being in the same cell at the same time were each to use all cells in the combined home ranges evenly and without respect to use by the other.

Seaman (1993; Seaman et al. 1999; Seaman and Powell 1996) indexed the abilities of different home range estimators to reconstruct a known utility distribution by calculating the summed squared differences between predicted and known values for each cell. This index can be used to index overlap of two home ranges as $\sum(p_{ki} - p_{kj})^2$, where $p_{ki}$ and $p_{kj}$ are defined as above, and emphasizes differences where probability of use is high. The behavior of this index has not been explored.

Doncaster (1990) indexed overlap using Spearman's coefficient of rank correlation. Spearman's $r$ is calculated for the utility distributions of two animals with overlapping home ranges, or for the frequencies of use of cells in a grid. Doncaster showed that the index behaves well and that nonlinear responses of the index outside of the area of overlap (where one individual has probability of use equal to 0) do not affect the overall usefulness of the index. This appears to be a robust index that can be used broadly to index overlap of home ranges.

I index pairwise overlap only for animals whose home ranges overlap, or are adjacent, in the same year, that is, animals for which $I_p > 0$. This criterion sometimes excludes from analyses two home ranges that are adjacent but do not overlap, possibly because of sampling error. An objective method of choosing nearest neighbors is needed. Because an animal's home range can overlap with the home ranges of several other animals, all pairwise index values within a study site are not strictly independent. Similarly, for studies that follow some individuals for more than 1 year, index values for different years may not be independent. Statistical tests must be controlled for both individuals in each pairwise overlap when testing for differences among sites, among years, or among populations of different species.

These indices of overlap and similar indices (Hurlbert 1978) can be used to compare overlap among sites, to compare changes in overlap with changes in food or other limiting resources, or to deduce territorial behavior when active defense, scent marks, or calls have not been documented.

## DYNAMIC INTERACTIONS

Several approaches can be used to quantify and test whether two individuals affect the behavior of each other. The indices test predominantly for attraction or avoidance.

Doncaster (1990), Horner and Powell (1990, citing Minta's work in preparation) and Minta (1992) used chi-square and $G$ tests to explore whether two animals located at the same time (approximately) tended to be found together. In a $2 \times 2$ contingency table, paired and unpaired distances can be labeled as near (animals together or associated) or far (not together or not associated). Doncaster labeled the foxes he studied as near when they were close enough to detect each other; Horner and Powell labeled black bears as near when they were closer than the median distance of telemetry error. Other objective criteria might be based on the animals being within an area of overlap at the same time (Minta 1992). The $N$ paired locations are a sample estimating how often the two animals are close together. By taking each location for each animal and calculating the distance to the $N-1$ locations of the other animal not taken at the same time, one obtains a sample of $N^2 - N$ distances that can also be divided into near and far and estimate how often the animals would be found near each other if the movements of each were unaffected by the other. A significant chi-square or $G$ value indicates that the animals attract or avoid each other. Minta (1992) noted that chi-square statistics behave better than $G$ with small sample sizes.

My coworkers and I used fixed kernel estimates of utility distributions to estimate the probabilities that two animals would be in their area of overlap simultaneously (Powell et al. 1987). Because $p_{ki}$ is the probability that at any given time individual $i$ will be found in cell $k$, $L_p$ is the probability that two animals are in their area of home range overlap at the same time if each uses the area without regard to use by the other. $L_p$ can be tested against the actual proportion of time (proportion of location estimates) that either animal spends in the area of overlap with the other. Because the two animals are unlikely to be located the same number of times, but must be in the overlap together the same number of times, the proportion of locations in the area of overlap will differ slightly between the two. This approach cannot be used to test whether two specific individuals attract or avoid each other but can be used to test whether classes of animals exhibit attraction or avoidance.

Minta (1992) developed further tests for attraction or avoidance to an area of overlap that allow researchers to test for more specific use of areas of overlap and that accommodate more diverse data than needed for the approaches used by Doncaster (1990) and my coworkers and I (Horner and Powell 1990; Powell et al. 1997). Minta showed how to test whether one animal of a pair is attracted but the other not, how to test for attraction when animals are not always located simultaneously, and how to test for attraction when home ranges are not known but the area of overlap is.

## TESTING FOR TERRITORIALITY

For many animals, territory defense is difficult or impossible to document but patterns of home range overlap can be documented. Such patterns of overlap can often be used to deduce territorial behavior. Table 3.1 gives overlap index results for territorial wolves in northeastern Minnesota (Mech, personal communication, unpublished data), for adult female black bears in North Carolina and in Minnesota that appear to differ in territorial behavior (Powell et al. 1997, Rogers 1977), and for intrasexually territorial stone martens (*Martes foina*) in Italy (Genovesi and Boitani 1997; Genovesi et al. 1997; Genovesi personal communication, unpublished data).

Home ranges of wolves in different packs, and pack home ranges (from combined home ranges of pack members), overlapped little compared to home ranges of wolves in the same pack ($p < 0.0001$, for each index general linear model [GLM] test). Wolves in the same pack have home ranges that overlap extensively and Spearman's $r$ is positive, showing significant correlation of the home ranges of wolves in the same pack and attraction among wolves of the same pack. Spearman's $r$ is negative for wolves of different packs and for pack home ranges, indicating avoidance by wolves of different packs. Such quantified patterns of home range overlap and lack of overlap agree with the extensive field observations that wolves defend territories (Mech 1970; Peters and Mech 1975; Peterson 1977, 1995).

Overlap of home ranges for bears in the North Carolina site was significantly greater than was overlap for bears at the Minnesota site ($p < 0.01$, GLM, for each index, table 3.1; Powell et al. 1997). Productivity of food was significantly higher for the North Carolina population (Powell et al. 1997) and anecdotal evidence indicates that the lack of home range overlap for the bears in

**Table 3.1  Simple Probability Index I$_p$, Lloyd's Index I$_L$, and Spearman's Rank Correlation r for Home Range Overlap of Adult Female Black Bears, Wolves and Wolf Packs, and Stone Martens**

| Animal, Study Site | N | Probability index, $I_p$ Mean ± SD | Lloyd's Index, $I_L$ Mean ± SD | Spearman's r Mean ± SD |
|---|---|---|---|---|
| Wolves, Minnesota, USA | | | | |
| Wolf packs[a] | 84 | 0.293 ± 0.484 | 0.34 ± 0.56 | −0.013 ± 0.097 |
| Wolves, different packs[a] | 120 | 0.077 ± 0.150 | 0.37 ± 0.70 | −0.784 ± 0.0174 |
| Wolves, same pack[a] | 17 | 0.827 ± 0.087 | 2.84 ± 0.97 | 0.219 ± 0.447 |
| Black bears, USA | | | | |
| North Carolina[b] | 46 | 0.111 ± 0.169 | 0.40 ± 0.50 | −0.722 ± 0.180 |
| Northeastern Minnesota[c] | 48 | 0.039 ± 0.076 | 0.20 ± 0.26 | −0.828 ± 0.090 |
| Stone martens, Italy | | | | |
| Same sex[d] | 11 | 0.018 ± 0.036 | 0.93 ± 1.28 | −0.702 ± 0.219 |
| | | | | −0.665 ± 0.277 |
| | | | | −0.786 ± 0.144 |
| Different sex[d] | 14 | 0.081 ± 0.106 | 3.79 ± 6.04 | −0.558 ± 0.242 |
| | | | | −0.480 ± 0.278 |
| | | | | −0.806 ± 0.077 |

[a]L. David Mech, personal communication; [b]Powell et al. 1997; [c]Rogers 1977; [d]Genovesi et al. 1997; personal communication.

Overlap was calculated only for animals having overlapping or adjacent home ranges. Home ranges of wolves in different packs and home ranges of wolf packs overlapped significantly less than did home ranges of wolves in the same pack ($p < 0.0001$, general linear model [GLM], for each index). Home ranges of adult female black bears in North Carolina overlapped significantly more than did those for bears in Minnesota ($p \leq 0.01$, GLM, for each index). Home ranges of stone martens of the same sex overlapped significantly less than home ranges of martens of the opposite sex using $I_p$ ($p \approx 0.05$, GLM) and $I_L$ ($p \approx 0.01$, GLM), but not using Spearman's r. N is the total numbers of bears, wolves, packs, or martens but not the sample size used in blocked statistical tests.

Under Spearman's r, the top figure in each row has been calculated on a 1- × 1-km grid, the second on a 500- × 500-m grid, and the third on a 125- × 125-m grid.

Minnesota was caused by territory defense (Rogers 1977). No territorial behavior was documented in North Carolina, but bears in both sites were seldom seen by the researchers. Table 3.1 provides solid evidence for differences in home range overlap and for differences in tolerance between individual adult females at sites differing in productivity of food. The difference in home range overlap between the two sites is consistent with observations of territorial behavior and with territory theory (Powell et al. 1997). Nonetheless, black bears in North Carolina have negative values of Spearman's $r$, indicating that avoidance does occur among these bears at some scale.

Home range overlap differed significantly between stone martens of the same sex and martens of opposite sex using two of the three indices for overlap (table 3.1). Intrasexual territoriality has been deduced from observations of home range overlap (Powell 1993b, 1994), but objectively quantified overlap that incorporates probability of use has been lacking (but see Genovesi and Boitani 1997). Erlinge (1977, 1979) suggested that males of *Mustela* species avoided females' territories, indicating that large overlap in space may not indicate large overlap of important space. The data for stone martens indicate that overlap between the sexes is significant. That Spearman's $r$ is negative both for members of the same sex and for members of opposite sex indicates that members of opposite sex also exhibit some avoidance but not as much as do members of the same sex. Values of Spearman's $r$ vary as the size of the cells in the grid varies and the mean difference in $r$ between the sexes is maximal at an intermediate cell size. When a cell size larger than the average home range area is used, home range overlap is similar within and between sexes. For cell size smaller than home range area, overlap among members of the same sex becomes smaller than overlap between the sexes. When cell size is much smaller than mean home range area, overlap within and between the sexes again becomes similar, indicating that members of the opposite sex with overlapping home ranges avoid each other on a fine scale. Taber et al. (1994) found similar avoidance by groups of Chacoan (*Catagonus wagneri*) and collared peccaries (*Tayassu tajacu*).

$I_p$ is little affected by cell size as long as cell size is small compared to home range size because probabilities of being in cells are summed over the area of overlap. Lloyd's index $\overset{*}{m}$ and Spearman's $r$ are both sensitive to cell size and therefore have the potential to give different results with different cell sizes. Cell size must match the grain or scale of the hypotheses being tested. Therefore, $I_p$ shows that stone martens of the same and opposite sex show different patterns of home range overlap, whereas Spearman's $r$ is able to show that spatial avoidance may occur on a scale much finer than that of individual home ranges.

## ■ Lessons

Within Burt's (1943) definition of *home range*, tremendous latitude exists for different approaches to understanding what a home range is. No consensus exists for a single, precise definition of *home range* and this lack of consensus leads to some of the confusion regarding home ranges. Many researchers may be unaware of the confusion. The major confusion appears to be between those who use a conceptual definition and those who use an operational definition. Burt's definition is conceptual and has two components: familiarity and use. Operational definitions and home range estimators quantify use and may add some estimate of familiarity. The minimum convex polygon estimator assumes that animals are familiar with all areas within the extremes of their movements but not familiar with any area outside those extremes, no matter how close. Kernel estimators assume that animals are familiar with, and have interest in, areas surrounding their movements in a fashion that decreases as does the shape of the kernels. The only quantification of a home range that excludes some guess about familiarity would include only areas over which an animal actually traveled (but what about flight by birds?).

Should areas to which an animal seldom travels but with which the animal is familiar be included in its home range? This question addresses the difference between an animal's home range and its cognitive map. An animal's cognitive map includes all areas for which the animal has information, whereas I consider its home range to include only the areas with which it maintains familiarity. Familiarity may be maintained through the senses, without actual visits. In addition, familiarity is undoubtedly both graded and of a multiscale nature, making familiarity extremely difficult to quantify. How unused areas are quantified must be considered by researchers. To some extent, how unused areas are quantified depends on research questions. However, researchers must understand how home range estimators estimate familiarity and must use estimators that are consistent with their own concepts of home range and with their research questions. No agreement exists as to how best to quantify familiarity.

What currencies are best for quantifying home ranges and territories? Home ranges should not always be quantified with respect to the time that an animal spends, or is predicted to spend, in different places. Most researchers use time to index importance, but what really counts is fitness. For some questions, home ranges may need to be quantified by probability density functions of energy expenditure or energy acquisition. Ultimately, for much research, home ranges should be quantified as probability density functions for contributions to fitness. How does a person map a home range's contributions to fitness?

My coworkers and I (Powell et al. 1997) documented a strong correlation between home range size and the size of the periphery for black bears, suggesting that peripheral parts of home ranges should receive more research attention than they do. Peripheries expand and contract as home range cores contain or fail to contain necessary resources. How are peripheries expanded? How does an animal choose where and how to expand its peripheral home range? Does the expansion depend on past knowledge and, if so, how did the animal obtain that knowledge? How important are occasional sallies? And if occasional sallies are important, should they be included in home ranges? Conversely, what critical resources are contained in a home range core?

How can estimates of animals' home ranges be improved? No universally best method can be developed for estimating home ranges of all animals because the best method for one research project may be different than that for another. Optimal methods depend on the hypotheses being tested and different hypotheses demand different analyses. Researchers must understand the strengths and limitations of each home range estimator and choose the one that provides the most accurate information related to the hypotheses being tested or to research objectives. Nonetheless, some generalities exist for improvement.

Home ranges must be quantified over biologically meaningful periods of time. Data collected over too short a time period may not sample an animal within all areas that it considers to be its home range. Data collected over too long a time period may suffer from changes in an animal's home range during the time period.

Kernels and brownian bridges incorporating time sequence information (i.e., in what direction is an animal most likely to travel next?) promise to be useful for many studies (see Bullard 1999). Future kernel estimators should also use information on telemetry error or area of trap attraction and information on an animal's radius of perception when calculating band width. Using probabilities generated by kernel estimators to explore use of space at different scales has much promise.

The fractal estimator, $1/C$, may have promise. A future fractal estimator could become an estimator of choice for some research if development allows $C$ for each cell to include time and multiscale information from the entire data set (not just from one cell), finds a way to determine optimal cell size for research questions of known scale, and solves the critical problem of $C$ depending on an unstable estimate of home range area.

All methods of estimating and quantifying animals' home ranges have problems and no method is best for all research. Because of the myriad problems associated with quantifying home ranges, especially boundaries and areas of home ranges, White and Garrott (1990) suggested that the home range concept is obsolete and that our understanding of animal ecology and behavior will advance best through research directed toward unambiguous analyses of data. Gautestad and Mysterud (1995) implied the same when they argued that boundaries of home ranges cannot be quantified precisely. Clearly, as White and Garrott stated (1990:179), "home range estimates are a poor substitute for good experimental protocol," but home range estimates and good research protocol are not mutually exclusive. Animals do have home ranges. They do not move randomly through the world but stay in confined, local areas for days, seasons, or even years. To understand why animals live where they do, why they go certain places to do certain things, and how they share or divide the locale, researchers must grapple with the concept of home range. For some questions, researchers should do as White and Garrott and as Gautestad and Mysterud suggested: Document exact animal locations, distances moved and rates of movement between locations, and so forth. Answering other questions will require estimates of home ranges, especially estimates of how animals use space within their home ranges. Exploring the concept of home range will improve our understanding of how animals conceive and perceive where they live and will further our understanding of animals' cognitive maps of the land in which they live. We have a sense of place for where we live; other animals do, too. That is what we seek to understand.

### Acknowledgments

Mike Mitchell, Consie Powell, Erran Seaman, and Luigi Boitani provided critical comments on early drafts of this manuscript. Arild Gautestad and Ivar Mysterud provided extremely helpful critical remarks. Piero Genovesi and Dave Mech graciously provided unpublished data used in table 3.1 and figure 3.3.

### Literature Cited

Ackerman, B. B., F. A. Leban, E. O. Garton, and M. D. Samuel. 1988 (2d ed). *User's manual for program home range.* Forestry, Wildlife and Range Experiment Station Technical Report. Moscow: University of Idaho.

Andersen, D. E. and O. J. Rongstad. 1989. Home-range estimates of red-tailed hawks based on random and systematic relocations. *Journal of Wildlife Management* 53: 802–807.

Anderson, D. J. 1982. The home range: A new nonparametric estimation technique. *Ecology* 63: 103–112.

Armstrong, D. P. 1992. Correlation between nectar supply and aggression in territorial honeyeaters: Causation or coincidence? *Behavioral Ecology and Sociobiology* 30: 95–102.

Barr, R. P. 1997. *Red-cockaded woodpecker habitat selection and landscape productivity in the North Carolina Sandhills.* M.S. thesis. Raleigh: North Carolina State University.

Bascompte, J. and C. Vilà. 1997. Fractals and search paths in mammals. *Landscape Ecology* 12: 213–221.

Bekoff, M. and L. D. Mech. 1984. Computer simulation: Simulation analyses of space use: Home range estimates, variability, and sample size. *Behavior Research Methods, Instruments, and Computers* 16: 32–37.

Bekoff, M. and M. C. Wells. 1981. Behavioural budgeting by wild coyotes: The influence of food resources and social organization. *Animal Behaviour* 29: 794–801.

Bingham, B. B. and B. R. Noon. 1997. Mitigation of habitat "taek": Application to habitat conservation planning. *Conservation Biology* 11: 127–139.

Boulanger, J. G. and G. C. White. 1990. A comparison of home-range estimators using Monte Carlo simulation. *Journal of Wildlife Management* 54: 310–315.

Bowman, A. W. 1985. A comparative study of some kernel-based nonparametric density estimators. *Journal of Statistical Computer Simulations* 21: 313–327.

Breiman, L., W. Meisel, and E. Purcell. 1977. Variable kernel estimates of multivariate densities. *Technometrics* 19: 135–144.

Brown, J. L. 1969. Territorial behavior and population regulation in birds: A review and re-evaluation. *Wilson Bulletin* 81: 293–329.

Bullard, F. 1999. *Estimating the home range of an animal: A brownian bridge approach.* M.S. thesis. Chapel Hill: University of North Carolina.

Burt, W. H. 1943. Territoriality and home range concepts as applied to mammals. *Journal of Mammalogy* 24: 346–352.

Calhoun, J. B. and J. U. Casby. 1958. Public Health Monograph No. 55. Washington, D.C.: U.S. Government Printing Office.

Carpenter, F. L. and R. E. MacMillen. 1976. Threshold model of feeding territoriality and test with a Hawaiian honeycreeper. *Science* 194: 634–642.

Curio, E. 1976. *The ethology of predation.* Berlin: Springer-Verlag.

Devroye, L. and L. Gyorfi. 1985. *Nonparametric density estimation, the L1 view.* New York: Wiley.

Dixon, K. R. and J. A. Chapman. 1980. Harmonic mean measure of animal activity areas. *Ecology* 61: 1040–1044.

Doncaster, C. P. 1990. Non-parametric estimates of interaction from radio-tracking data. *Journal of Theoretical Biology* 143: 431–443.

Doncaster, C. P. and D. W. Macdonald. 1991. Drifting territoriality in the red fox *Vulpes vulpes. Journal of Animal Ecology* 60: 423–439.

Doncaster, C. P. and D. W. Macdonald. 1992. Optimum group size for defending heterogenous distributions of resources: A model applied to red foxes, *Vulpes vulpes,* in Oxford City. *Journal of Theoretical Biology* 159: 189–198.

Doncaster, C. P. and D. W. Macdonald. 1996. Intra-specific variation in movement behaviour of foxes (*Vulpes vulpes*): A reply to White, Saunders and Harris. *Journal of Animal Ecology* 65: 126–127.

Doncaster, C. P. and R. Woodroffe. 1993. Den site can determine shape and size of badger territories: Implications for group-living. *Oikos* 66: 88–93.

Dunn, J. E. and P. S. Gipson. 1977. Analysis of radio telemetry data in studies of home range. *Biometrics* 33: 85–101.

Ebersole, S. P. 1980. Food density and territory size: An alternative model and test on the reef fish *Eupomacentrus leucostius*. *American Naturalist* 115: 492–509.

Ehrlich, P. R. 1975. The population biology of coral reef fishes. *Annual Review of Ecology and Systematics* 6: 211–248.

Ellner, S and L. A. Real. 1989. Optimal foraging models for stochastic environments: Are we missing the point? Comments. *Theoretical Biology* 1: 129–158.

Erlinge, S. 1977. Spacing strategy in stoat *Mustela erminea*. *Oikos* 28: 32–42.

Erlinge, S. 1979. Adaptive significance of sexual dimorphism in weasels. *Oikos* 33: 223–245.

Ewer, R. F. 1968. *Ethology of mammals*. London: Logos.

Fagen, R. M. and D. Y. Young. 1978. Temporal patterns of behavior: Durations, intervals, latencies, and sequences. In P. Colgen, ed., *Quantitative ethology*, 79–114. New York: Wiley.

Frafjord, K. and P. Prestrud. 1992. Home range and movements of arctic foxes *Alopex lagopus* in Svalbard. *Polar Biology* 12: 519–526.

Fretwell, S. D. and H. L. Lucas, Jr. 1970. On territorial behavior and other factors influencing habitat distribution in birds. I. Theoretical development. *Acta Biotheoretica* 19: 16–36.

Fryer, M. J. 1977. A review of some non-parametric methods of density estimation. *Journal of the Institute of Mathematics Applications* 20: 335–354.

Garshelis, D. L. and M. R. Pelton. 1980. Activity of black bears in the Great Smoky Mountains National Park. *Journal of Mammalogy* 61: 8–19.

Garshelis, D. L. and M. R. Pelton. 1981. Movements of black bears in the Great Smoky Mountains National Park. *Journal of Wildlife Management* 45: 912–925.

Gautestad, A. O. 1998. *Site fidelity and scaling complexity in animal movement and habitat use: The multiscaled random walk model*. Ph.D. thesis, University of Oslo.

Gautestad, A. O. and I. Mysterud. 1993. Physical and biological mechanisms in animal movement processes. *Journal of Applied Ecology* 30: 523–535.

Gautestad, A. O. and I. Mysterud. 1995. The home range ghost. *Oikos* 74: 195–204.

Gautestad, A. O., I. Mysterud, and M. R. Pelton. 1998. Complex movement and scale-free habitat use: Testing the multiscale home range model on black bear telemetry data. *Ursus* 10: 219–234.

Genovesi, P. and L. Boitani. 1997. Social ecology of the stone marten in central Italy. In G. Proulx, H. N. Bryant, and P. M. Woodard, eds., *Martes: Taxonomy, ecology, techniques, and management*, 110–120. Edmonton: Provincial Museum of Alberta.

Genovesi, P., I. Sinibaldi, and L. Boitani. 1997. Spacing patterns and territoriality of the stone marten. *Canadian Journal of Zoology* 75: 1966–1971.

Gese, E. M., D. E. Andersen, and O. J. Rongstad. 1990. Determining home-range size of resident coyotes from point and sequential locations. *Journal of Wildlife Management* 54: 501–506.

Goss-Custard, J. D. 1977. Optimal foraging and the size selection of worms by redshank *Tringa totanus*. *Animal Behavior* 25: 10–29.

Haan, L. de and S. Resnick. 1994. Estimating the home range. *Journal of Applied Probability* 31: 700–720.

Harris, S., W. J. Cresswell, P. G. Forde, W. J. Trewhella, T. Wollard, and S. Wray. 1990. Home-range analysis using radio-telemetry data: A review of problems and techniques particularly as applied to the study of mammals. *Mammal Review* 20: 97–123.

Hayne, D. W. 1949. Calculation of size of home range. *Journal of Mammalogy* 30: 1–18.

Hixon, M. A. 1980. Food production and competitor density as the determinants of feeding territory size. *American Naturalist* 115: 510–530.

Hixon, M. A. 1982. Energy maximizers and time minimizers: Theory and reality. *American Naturalist* 119: 595–599.

Hixon, M. A. 1987. Territory area as a determinant of mating systems. *American Zoologist* 27: 229–247.

Hixon, M. A., F. L. Carpenter, and D. C. Paton. 1983. Territory area flower density, and time budgeting in hummingbirds: An experimental and theoretical analysis. *American Naturalist* 122: 366–391.

Horner, M. A. and R. A. Powell. 1990. Internal structure of home ranges of black bears and analyses of home-range overlap. *Journal of Mammalogy* 71: 402–410.

Hurlbert, S. H. 1978. The measurement of niche overlap and some relatives. *Ecology* 59: 67–77.

Ims, R. A. 1987. Male spacing systems in microtine rodents. *American Naturalist* 130: 74–484.

Jenkins, S. H. and P. E. Busher. 1979. *Castor canadensis*. *Mammalian Species* 120: 1–8.

Jennrich, R. I. and F. B. Turner. 1969. Measurement of non-circular home range. *Journal of Theoretical Biology* 22: 227–237.

Kaufmann, J. H. 1962. Ecology and social behavior of the coati, *Nasua narica*, on Barro Colorado Island, Panama. *University of California Publications in Zoology* 60: 95–222.

Kodric-Brown, A. and J. H. Brown. 1978. Influence of economics, interspecific competition, and sexual dimorphism on territoriality of migrant rufous hummingbirds. *Ecology* 59: 285–296.

Krebs, C. J. 1966. Demographic changes in fluctuating populations of *Microtus californicus*. *Ecological Monographs* 36: 239–273.

Krebs, J. R. and A. Kacelnik. 1991. Decision-making. In J. R. Krebs and N. B. Davies, eds., *Behavioural ecology: An evolutionary approach*, 105–136. Oxford, U.K.: Blackwell Scientific.

Kruuk, H. 1972. *The spotted hyena*. Chicago: University of Chicago Press.

Kruuk, H. 1989. *The social badger*. Oxford, U.K.: Oxford University Press.

Kruuk, H. and T. Parish. 1982. Factors affecting population density, group size and territory size of the European badger, *Meles meles*. *Journal of Zoology* 196: 31–39.

Lair, H. 1987. Estimating the location of the focal center in red squirrel home ranges. *Ecology* 68: 1092–1101.

Larkin, R. P. and D. Halkin. 1994. Wildlife software: A review of software packages for estimating animal home ranges. *Wildlife Society Bulletin* 22: 274–287.

Leopold, A. 1949. *A Sand County almanac and sketches here and there.* New York: Oxford University Press.

Lewis, M. A. and J. D. Murray. 1993. Modelling territoriality and wolf–deer interactions. *Nature* 366: 738.

Loehle, C. 1990. Home range: A fractal approach. *Landscape Ecology* 5: 39–52.

Lloyd, M. 1967. Mean crowing. *Journal of Animal Ecology* 36: 1–30.

Macdonald, D. W. 1981. Resource dispersion and the social organisation of the red fox, *Vulpes vulpes.* In J. A. Chapman and D. Pursley, eds., *Proceedings of the Worldwide Furbearer Conference,* 918–949. Baltimore, Md.: Worldwide Furbearer Conference.

Macdonald, D. W. 1983. The ecology of carnivore social behaviour. *Nature* 301: 379–384.

Macdonald, D. W. and G. M. Carr. 1989. Food security and the rewards of tolerance. In V. Standen and R. A. Foley, eds., *Comparative socioecology: The behavioural ecology of humans and other mammals,* 75–99. Oxford, U.K.: Blackwell Scientific.

Maynard Smith, J. 1976. Evolution and the theory of games. *American Scientist* 64: 41–45

Mech, L. D. 1970. *The wolf: Ecology and behavior of an endangered species.* Garden City, N.J.: Natural History Press.

Minta, S. C. 1992. Tests of spatial and temporal interaction among animals. *Ecological Applications* 2: 178–188.

Mitchell, M. S. 1997. *Optimal home ranges and application to black bears.* Ph.D. thesis. Raleigh: North Carolina State University.

Nadaraya, E. A. 1989. *Nonparametric estimation of probability densities and regression curves.* Dordrecht, The Netherlands: Kluwer.

Noel, J. 1993. *Food productivity and home range size in black bears.* M.S. thesis. Raleigh: North Carolina State University.

Orians, G. H. 1980. *Some adaptations of march-nesting blackbirds.* Princeton, N.J.: Princeton University Press.

Ostfeld, R. S. 1986. Territoriality and mating system of California voles. *Journal Animal Ecology* 55: 691–706.

Ostfeld, R. S. 1990. The ecology of territoriality in small mammals. *Trends in Ecology and Evolution* 5: 411–415.

Palomares, F. 1994. Site fidelity and effects of body mass on home-range size of Egyptian mongooses. *Canadian Journal of Zoology* 72: 465–469.

Person, D. K. and D. H. Hirth. 1991. Home range and habitat use of coyotes in a farm region of Vermont. *Journal of Wildlife Management* 55: 433–441.

Peters, R. 1978. Communication, cognitive mapping, and strategy in wolves and hominids. In R. L. Hall and H. S. Sharp, eds., *Wolf and man: Evolution in parallel,* 95–108. New York: Academic Press.

Peters, R. and L. D. Mech. 1975. Scent-marking in wolves. *American Scientist* 63: 628–637.

Peterson, R. O. 1977. *Wolf ecology and prey relationships on Isle Royale.* Washington D.C.: U.S. National Park Service, Fauna Series No. 11.

Peterson, R. O. 1995. *The wolves of Isle Royale: A broken balance.* Minoqua, Wisc.: Willow Creek Press.

Powell, R. A. 1979. Mustelid spacing patterns: Variations on a theme by *Mustela. Zeitschrift für Tierpsychologie* 50: 153–165.

Powell, R. A. 1987. Black bear home range overlap in North Carolina and the concept of home range applied to black bears. *International Conference on Bear Research and Management* 7: 235–242.

Powell, R. A. 1989. Effects of resource productivity, patchiness and predictability on mating and dispersal strategies. In V. Standen and R. A. Foley, eds., *Comparative socioecology of mammals and humans,* 101–123. Oxford, U.K.: Blackwell Scientific.

Powell, R. A. 1993a. Why do some forest carnivores exhibit intrasexual territoriality and what are the consequences for management? *Proceedings of the International Union of Game Biologists* 21: 268–273.

Powell, R. A. 1993b (2d ed). *The fisher: Ecology, behavior and life history.* Minneapolis: University of Minnesota Press.

Powell, R. A. 1994. Structure and spacing of *Martes* populations. In S. W. Buskirk et al., eds., *Martens, sables and fishers: Biology and conservation,* 101–121. Syracuse, N.Y.: Cornell University Press.

Powell, R. A. and J. J. Fried. 1992. Helping by juvenile pine voles (*Microtus pinetorum*), growth and survival of younger siblings, and the evolution of pine vole sociality. *Behavioral Ecology* 4: 325–333.

Powell, R. A. and M. S. Mitchell. 1998. Topographical constraints on home range quality. *Ecography* 21: 337–341.

Powell, R. A., J. W. Zimmerman, and D. E. Seaman. 1997. *Ecology and behaviour of North American black bears: Home ranges, habitat and social organization.* London: Chapman & Hall.

Powers, D. R. and T. McKee. 1994. The effect of food availability on time and energy expenditure of territorial and non-territorial hummingbirds. *Condor* 96: 1064–1075.

Price, K., S. Boutin, and R. Ydenberg. 1990. Intensity of territorial defense in red squirrels: An experimental test of the asymmetric war of attrition. *Behavioral Ecology and Sociobiology* 27: 217–222.

Pulliainen, E. 1984. Use of the home range by pine martens (*Martes martes* L.). *Acta Zoologica Fennica* 171: 271–274.

Pyke, G. H. 1984. Optimal foraging theory: A critical review. *Annual Review of Ecology and Systematics* 15: 523–575.

Pyke, G. H., H. R. Pulliam, and E. L. Charnov. 1977. Optimal foraging: A selective review of theory and tests. *Quarterly Review of Biology* 52: 137–154.

Reid, M. L. and P. J. Weatherhead. 1988. Topographic constraints on competition for territories. *Oikos* 51: 115–117 plus *erratum Oikos* 53: 143.

Reynolds, R. D. and J. W. Laundrè. 1990. Time intervals for estimating pronghorn and coyote home ranges and daily movements. *Journal of Wildlife Management* 54: 316–322.

Rogers, L. L. 1977. *Social relationships, movements, and population dynamics of black bears in northeastern Minnesota.* Ph.D. thesis. Minneapolis: University of Minnesota.

Rogers, L. L. 1987. Effects of food supply and kinship on social behavior, movements, and population growth of black bears in northeastern Minnesota. *Wildlife Monographs* 97: 1–72.

Rood, J. P. 1986. Ecology and social evolution in the mongooses. In D. I. Rubenstein and R. W. Wrangham, eds., *Ecological aspects of social evolution,* 131–152. Princeton, N.J.: Princeton University Press.

Saitoh, T. 1991. The effects and limits of territoriality on population regulation in grey red-backed voles, *Clethrionomys rufocanus bedfordiae*. *Research on Population Ecology* 33: 367–386.

Samuel, M. D. and R. E. Green. 1988. A revised test procedure for identifying core areas within the home range. *Journal of Animal Ecology* 57: 1067–1068.

Samuel, M. D., D. J. Pierce, and E. O. Garton. 1985. Identifying areas of concentrated use within the home range. *Journal of Animal Ecology* 54: 711–719.

Schoener, T. W. 1981. An empirically based estimate of home range. *Theoretical Population Biology* 20: 281–325.

Seaman, D. E. 1993. *Home range and male reproductive optimization in black bears.* Ph.D. thesis. Raleigh: North Carolina State University.

Seaman, D. E., J. J. Millspaugh, B. J. Kernohan, G. C. Brundige, K. J. Raedeke, and R. A. Gitzen. 1999. Effects of sample size on kernel home range estimates. *Journal of Wildlife Management* 63.

Seaman, D. E., B. Griffith, and R. A. Powell. 1998. KERNELHR: A program for estimating animal home ranges. *Wildlife Society Bulletin* 26: 95–100.

Seaman, D. E. and R. A. Powell. 1990. Identifying patterns and intensity of home range use. *International Conference on Bear Research and Management* 8: 243–249.

Seaman, D. E. and R. A. Powell. 1996. Accuracy of kernel estimators for animal home range analysis. *Ecology* 77: 2075–2085.

Silverman, B. W. 1986. *Density estimation for statistics and data analysis.* London: Chapman & Hall.

Slater, P. J. B. and N. P. Lester. 1982. Minimising errors in splitting behaviour into bouts. *Behaviour* 79: 153–161.

Smith, C. C. 1968. The adaptive nature of social organization in the genus of tree squirrel *Tamiasciurus*. *Ecological Monographs* 38: 31–63.

Spencer, A. R., G. N. Cameron, and R. K. Swihart. 1990. Operationally defining home range: Temporal independence exhibited by hispid cotton rats. *Ecology* 71: 1817–1822.

Spencer, W. D. and R. H. Barrett. 1984. An evaluation of the harmonic mean measure for defining carnivore activity areas. *Acta Zoologica Fennica* 171: 277–281.

Stahlecker, D. W. and T. G. Smith. 1993. A comparison of home range estimates for a bald eagle wintering in New Mexico. *Journal of Raptor Research* 27: 42–45.

Stamps, J. 1995. Motor learning and the value of familiar space. *American Naturalist* 146: 41–58.

Stephens, D. W. 1989. Variance and the value of information. *American Naturalist* 134: 128–140.

Swihart, R. K. and N. A. Slade. 1985a. Testing for independence of observations in animal movements. *Ecology* 66: 1176–1184.

Swihart, R. K. and N. A. Slade. 1985b. Influence of sampling interval on estimates of home range size. *Journal of Wildlife Management* 49: 1019–1025.

Taber, A. B., C. P. Doncaster, N. N. Neris, and F. Colman. 1994. Ranging behaviour and activity patterns of two sympatric peccaries, *Catagonus wagneri* and *Tayassu tajacu,* in the Paraguayan chaco. *Mammalia* 58: 61–71.

Tapia, R. A. and J. R. Thompson. 1978. *Nonparametric probability density estimation.* Baltimore, Md.: Johns Hopkins University Press.

Vandermeer, J. 1981. *Elementary mathematical ecology.* New York: Wiley.

Walters, J. R., C. K. Copeyon, and J. H. Carter, III. 1992. Test of the ecological basis of cooperative breeding in red-cockaded woodpecker. *Auk* 109: 90–97.

Walters, J. R., P. D. Doerr, and J. H. Carter, III. 1988. The cooperative breeding system of the red-cockaded woodpecker. *Ethology* 78: 275–305.

Watson, A. and R. Moss. 1970. Dominance, spacing behaviour and aggression in relation to population limitation in vertebrates. In A Watson, ed., *Animal populations in relation to their food resources,* 167–222. Oxford, U.K.: Blackwell.

Wauters, L., P. Casale, and A. A. Dhondt. 1994. Space use and dispersal of red squirrels in fragmented habitats. *Oikos* 69: 140–146.

White, G. C. and R. A. Garrott. 1990. *Analysis of wildlife radio-tracking data.* New York: Academic Press.

Winkle, W., van. 1975. Comparison of several probabilistic home-range models. *Journal of Wildlife Management* 39: 118–123.

Wolff, J. O. 1989. Social behavior. In G. L. Kirkland, Jr. and J. N. Layne, eds., *Advances in the study of Peromyscus (Rodentia),* 271–291. Lubbock: Texas Tech University Press.

Wolff, J. O. 1993. Why are female small mammals territorial? *Oikos* 68: 364–369.

Worton, B. J. 1987. A review of models of home range for animal movement. *Ecological Modeling* 38: 277–298.

Worton, B. J. 1989. Kernel methods for estimating the utility distribution in home-range studies. *Ecology* 70: 164–168.

Worton, B. J. 1995. Using Monte Carlo simulation to evaluate kernel-based home range estimators. *Journal of Wildlife Management* 59: 794–800.

Zoellick, B. W. and N. W. Smith. 1992. Size and spatial organization of home ranges of kit foxes in Arizona. *Journal of Mammalogy* 73: 83–88.

## Chapter 4

# Delusions in Habitat Evaluation: Measuring Use, Selection, and Importance

DAVID L. GARSHELIS

Management of wildlife populations, whether to support a harvest, conserve threatened species, or promote biodiversity, generally entails habitat management. Habitat management presupposes some understanding of species' needs. To assess a species' needs, researchers commonly study habitat use and, based on the results, infer selection and preference. Presumably, species should reproduce or survive better (i.e., their fitness should be higher) in habitats that they tend to prefer. Thus, once habitats can be ordered by their relative preference, they can be evaluated as to their relative importance in terms of fitness. Managers can then manipulate landscapes to contain more high-quality habitats and thus produce more of the targeted species. Habitat manipulations specifically intended to produce more animals have been conducted since at least the days of Kublai Khan (A.D. 1259–1294; Leopold 1933).

However, the processes of habitat evaluation are fraught with problems. Some problems are specific to the methods used in the data collection or analyses. Many of these problems have already been recognized, and discussions about them in the literature have prompted a host of evolving techniques. Other problems are inherent in the two most basic assumptions of this approach: that researchers can discern habitat selection or preference from observations of habitat use and that such selection, perceived or real, relates to fitness and hence to population growth rate.

My goal is to illuminate the scope of the problems involved in habitat evaluation. Assessments of habitat selection and presumed importance are done so often, and study methods have become so routine, that it is apparent that researchers and managers tend to believe that the major problems have, for the most part, been overcome. I contend that this view is overly sanguine and propose a reconsideration of the ways in which habitat evaluations are conducted.

## ■ Terminology

The word *habitat* has two distinct usages. The true dictionary definition is the type of place where an animal normally lives or, more specifically, the collection of resources and conditions necessary for its occupancy. Following this definition, habitat is organism specific (e.g., deer habitat, grouse habitat). A second definition is a set of specific environmental features that, for terrestrial animals, is often equated to a plant community, vegetative association, or cover type (e.g., deer use different habitats or habitat types in summer and winter). *Nonhabitat* could mean either the converse of *habitat* in the first sense (a setting that an animal does not normally occupy) or the second (a specific vegetative type that the animal views as unsuitable); here, the two meanings of *habitat* converge (see also pages 392–396 in this volume).

Hall et al. (1997) argue that only the first definition of *habitat* is correct and that the second represents a confusing misuse of the term. They reviewed 50 articles dealing with wildlife–habitat relationships and, based on their definition, found that 82% discussed habitat vaguely or incorrectly. I suggest that given the prevalent use of *habitat* to mean habitat type, this alternative definition is legitimate and well understood in the wildlife literature. Moreover, this common usage of the term is consistent with the normally accepted meaning of *habitat use:* the extent to which different vegetative associations are used. Hall et al. (1997:175) define *habitat use* as "the *way* an animal uses . . . a collection of physical and biological components (i.e., resources) *in* a habitat" (emphasis mine), which seems difficult to measure.

Habitat selection and preference are also more easily understood in terms of differential use of habitat types. *Selection* and *preference* are often used interchangeably in the wildlife literature; however, they have subtly different meanings. I will adopt the distinction posed by Johnson (1980), who defined *selection* as the process of choosing resources and *preference* as the likelihood of a resource being chosen if offered on an equal basis with others. Peek (1986) suggested that innate preferences exist even for resources not actually available. Furthering this concept, Rosenzweig and Abramsky (1986) characterized preferred habitats as those that confer high fitness and would therefore support a high equilibrium density (in the absence of other confounding factors, such as competitors). Thus use results from selection, selection results from preference, and preference presumably results from resource-specific differential fitness. In controlled experiments, preferences can be assessed directly by offering equal portions of different resources and observing choices that are made

(Elston et al. 1996). In the wild, however, preferences must be inferred from patterns of observed use of environments with disparate, patchy, and often varying resources.

Generally, the purpose for determining preferences is to evaluate habitat *quality* or *suitability*, which I define as the ability of the habitat to sustain life and support population growth. *Importance* of a habitat is its quality relative to other habitats—its contribution to the sustenance of the population. Assessments of habitat quality and importance (i.e., habitat *evaluation*) are thus based on the presumption that preference, and hence selection, are linked to fitness (reproduction and survival) and that preference can be gleaned from patterns of observed use.

Use of habitat is generally considered to be selective if the animal makes choices rather than wandering haphazardly through its environment. Typically, the disproportionate use of a habitat compared to its availability is taken as prima facie evidence of selection. Although technically resource *availability* encompasses accessibility and procurability (Hall et al. 1997), these attributes are difficult to measure, so it seems reasonable to equate habitat availability with abundance (typically measured in terms of area), as is normally done in habitat selection studies. A habitat that is used more than its availability is considered to be selected for. Conversely, a habitat that is used less than its availability is often referred to as being *selected against*, or even *avoided*. This is poor terminology, however, in that it suggests that the animal preferred not to be in that habitat at all, but occasionally just ended up there. Use that is proportional to availability is generally taken to be indicative of lack of selection, which is also unfortunate terminology, as illustrated by the following examples.

Consider an animal living in an area with only two habitats and using each in proportion to its availability; from this we might assume that the animal was not exhibiting habitat selection. However, unless the animal was a very low life form, it certainly made choices as to when it visited each habitat and what it did when it got there; anytime it made a choice, and either stayed or moved, it selected one habitat over the other. Arguably, if one analyzed these movements on a short enough time scale, habitat use would be disproportionate to availability, enabling detection of habitat selection. As the time scale is shortened, though, the sheer physical constraint of moving between the two habitats (i.e., the distance between them) also affects their relative use.

On the flip side, imagine a dispersing animal attempting to traverse an area with no regard for habitat. If its route was frequently diverted by the presence of other, more dominant resident animals, living in their presumably preferred habitats, the disperser's movements would appear to reflect habitat selection (i.e., selection for habitats not preferred by the residents). Indeed, one could reasonably assert that this represents true habitat selection as defined earlier, in that the disperser chose to avoid habitats with dominant conspecifics and thereby improved its chance of obtaining resources and not getting killed; however, one could also legitimately contend that the disperser was simply exhibiting avoidance of conspecifics, and used whatever cues, including markings, droppings, and possibly habitat characteristics, to do so.

These are not trivial complications, but rather examples of the intrinsic ambiguities associated with the application of these concepts. Terms such as *selection* and *preference* can be clearly defined, but not easily measured in the real world. Moreover, as I will show later, the link between selection, preference, and habitat-related fitness may be tenuous.

## ■ Methods for Evaluating Habitat Selection, Preference, and Quality

Three general study designs have been used to infer habitat quality. The first, generally called the use–availability design, compares the proportion of time that an animal spends in each available habitat type (generally judged by the number of locations, or less commonly, by the distance traveled; e.g., Salas 1996) to the relative area of each type. The second, which I call the site attribute design, compares habitat characteristics of sites used by an animal to unused or random sites. These two designs generate measures of selection for various habitats or habitat attributes, and habitat quality or importance is inferred from the magnitude of this apparent selection. The third method, which I call the demographic response design, uses a more direct approach for assessing habitat quality by comparing the demographics (density, reproduction, or survival) of animals living in different habitats. This design thus circumvents the need to interpret animal behavior (habitat choices).

### USE–AVAILABILITY DESIGN

Among studies of birds and mammals, the use–availability design is the most popular. I reviewed habitat-related papers dealing with birds and mammals published in the *Journal of Wildlife Management* during 1985–1995 and found that most (90 of 156, or 58 percent) relied on a use–availability study design to assess habitat selection, preference, or quality. Thomas and Taylor (1990) further categorized use–availability studies into three approaches: one in which habitat-use data are collected on animals that are not individually recognizable (e.g., visual sightings or sign), one in which data are collected on individuals (e.g., radiocollared animals) but habitat availability is considered the same for all individuals (so individuals are typically pooled for analysis), and one in which use and availability are measured and compared for each individual. They also reviewed papers published in the *Journal of Wildlife Management* (1985–1988) and found that nearly twice as many studies collected data on individuals but pooled them for analysis than either of the other two approaches.

Studies that pooled animals for analysis have commonly compared frequencies of use and availability for an array of habitats using a chi-square test. Two-thirds of the use–availability studies that I reviewed (61 of 90) did this. Determination of which habitat types were used more or less than expected is generally made by comparing availability of each habitat type to Bonferroni confidence intervals around the percentage use of each type. This procedure was described initially by Neu et al. (1974) and clarified by Byers et al. (1984), although a more accurate method of constructing such confidence intervals was recently proposed by Cherry (1996). If the areas of available habitats are estimated (e.g., from sampling) rather than measured (e.g., from a map), use and availability should be compared with the chi-square test for homogeneity rather than goodness-of-fit (Marcum and Loftsgaarden 1980). A chi-square goodness-of-fit test assumes that the availabilities are known constants against which use is compared, so if availabilities are actually estimated, with some sampling error, this test is more prone to indicate selection when there is none (type I error) (Thomas and Taylor 1990).

Various other methods of comparing use and availability have been advanced but less often used in wildlife habitat studies. Ivlev (1961) proposed an electivity index to measure relative selection of food items on a scale from −1 to 1; this has since been adopted for some habitat selection studies. However, Chesson (1978, 1983) noted that Ivlev's index may yield misleading results because it varies with availability even if preference is unchanged, and advocated use of a 0 to 1 index originally proposed by Manly et al. (1972), also for feeding preference studies. This Manly–Chesson index is simply the proportional use divided by the proportional availability of each habitat, standardized so the values for all habitats sum to 1. As adapted to habitat studies, it is interpretable as the relative expected use of a habitat had all types been equally available (i.e., preference). Thus in an area with four habitats, an index of 0.25 for each habitat would indicate no preference, whereas deviations from this would indicate relative preference for or against certain habitat types. Heisey (1985) and Manly et al. (1993) extended this method to test for differences in habitat preference among individuals or sex–age groups, and also showed how to test for statistically significant differences among preferences for different habitat types. Kincaid and Bryant (1983) and Kincaid et al. (1983) offered an alternative method that scores relative differences between use and availability for habitats defined as geometric vectors.

Most studies using these tests pooled data among individuals, so that animal captures, sightings, radiolocations, and so on represented the sample units. Aebischer et al. (1993b) pointed out that this constitutes pseudoreplication (Hurlbert 1984) and advised comparing use to availability for each animal individually (i.e., so individuals are the sample units). Several methods have been developed specifically to do this. Of these, the most commonly used is Johnson's (1980), which is based on the difference between the rankings of habitat use and the rankings of habitat availability. This method also provides a means of detecting statistically significant differences among habitats, not just a relative ordering of their selection. Moreover, because comparisons are made on an individual-animal basis, habitat availability can be considered either within each individual home range, or within the study area as a whole. Johnson (1980) defined first-order selection as that which distinguishes the geographic distribution of a species, second-order selection as that which determines the composition of home ranges within a landscape, and third-order selection as the relative use of habitats within a home range. Thus, both second-order and third-order selection can be addressed with Johnson's (1980) technique; with chi-square tests it is possible (Gese et al. 1988; Carey et al. 1990; Boitani et al. 1994) but more difficult (because of sample size constraints) to consider both of these levels of selection.

Alldredge and Ratti (1986, 1992) compared four methods (including the chi-square, Johnson's, and two others based on individual-animal comparisons) in simulated conditions and found that none performed (with regard to type I and type II error rates) consistently better than the others. However, some methods are better suited for given situations. For example, because data for all animals are generally pooled for chi-square tests, unequal sampling among individuals could strongly affect the results if all individuals did not make similar selections. Conversely, the methods that weight animals equally, regardless of the amount of data collected on each, may be subject to spurious results caused by small sample sizes and variability among individuals. McClean et al. (1998) used real data on young turkeys (*Meleagris gallopavo*), which have fairly narrow and well-known habitat requirements, to compare results of six analytical techniques for assessing habitat selection. In this case, the methods that treat individuals as sample units tended to be less apt to detect habitat selection.

Aebischer et al. (1993b) offered what appears to be an improved procedure for comparing use with availability on an individual animal basis (although it performed poorly in McClean et al.'s 1998 evaluation). This method (compositional analysis) has become increasingly popular because it enables assessment of both second-order and third-order selection and yields statistical comparisons (rankings) among habitats (Donázar et al. 1993; Carroll et al. 1995; Macdonald and Courtenay 1996; Todd et al. 2000). Additionally, because the data are arranged analogous to an ANOVA, in which between-group differences can be tested against within-group variation among individuals, it provides a means of testing for differences among study sites (e.g., with different habitats, different animal density, or different predators or competitors), seasons or years (e.g., with different food conditions), sex–age groups, or groups of animals with different reproductive outputs or different fates (Aebischer et al. 1993a; Aanes and Andersen 1996).

## SITE ATTRIBUTE DESIGN

Site attribute studies differ from use–availability studies in that they measure a multitude of habitat-related variables at specific sites and attempt to identify the variables and the values of those variables that best characterize sites that are used (often for a specific activity). With this design, the dependent variable is not the amount of use (as with use–availability studies) but simply whether each site was used or unused (or a random location with unknown use); the independent variables can be many and varied. Use–availability studies generally just deal with broad habitat types, or if more variables are considered, they are analyzed individually (Gionfriddo and Krausman 1986; Armleder et al. 1994).

   A site attribute design was used in 45 (29 percent) of the habitat selection studies I reviewed. Of these, 28 were on birds and 17 on mammals. This design requires measurement of habitat variables at some defined site, usually one that serves some biological importance to the animal. Nest sites of birds are easily defined and biologically important, and hence are often the subject of studies of this nature. Habitat characteristics of breeding territories (Gaines and Ryan 1988; Prescott and Collister 1993), drumming sites (Stauffer and Peterson 1985; Thompson et al. 1987), and roosting sites (Folk and Tacha 1990) also have been investigated. Among mammals, studies have focused on characteristics of feeding sites (e.g., as evidenced by browsed or grazed vegetation; Edge et al. 1988), food storage sites (e.g., squirrel middens; Smith and Mannan 1994), resting sites (e.g., deer beds; Huegel et al. 1986; Ockenfels and Brooks 1994), shelters (such as cliff overhangs, cavities, burrows, lodges, or dens; Lacki et al.

1993; Loeb 1993; Nadeau et al. 1995), wintering areas (Nixon et al. 1988), or areas recolonized by an expanding population (Hacker and Coblentz 1993). Other studies have compared habitat characteristics of randomly located sites to sites where birds or mammals were observed, radiolocated, or known to have been from remaining sign (Dunn and Braun 1986; Krausman and Leopold 1986; Beier and Barrett 1987; Edge et al. 1987; Lehmkuhl and Raphael 1993; Flores and Eddleman 1995).

The statistical procedures used in such studies vary. Most have used multivariate analyses to differentiate combinations of variables that tend to be associated with the used sites. Discriminant function analysis (DFA) is the most popular of these. Logistic regression is an alternative, and is especially useful when the data consist of both discrete and continuous variables (Capen et al. 1986) or are related to site occupancy in a nonlinear fashion (Brennan et al. 1986; Nadeau et al. 1995).

### DEMOGRAPHIC RESPONSE DESIGN

Ideally, studies should identify relationships between habitat characteristics and the animal's fitness. Studies employing use–availability and site attribute designs assume that certain habitat features are selected because they improve fitness. Demographic response designs attempt to test this more directly. However, although I refer to the measured demographic parameters in these studies as response variables, they really only represent correlates with given habitats.

I identified 39 studies among those that I reviewed (25 percent) that measured an association between a demographic parameter and habitat (note that percentages for the three designs total more than 100 percent because some studies used more than one design). Most of these investigated differences in animal density among habitats. Fourteen studies, all on birds, related reproduction (i.e., nesting success) to habitat of nest sites. Three studies, two on birds and one on mammals, attempted to find an association between habitat and survival (Hines 1987; Klinger et al. 1989; Loegering and Fraser 1995), but only one (Loegering and Fraser 1995) detected such a relationship.

## ■ Problems with Use–Availability and Site Attribute Designs

### DEFINING HABITATS

The first prerequisite for assessing habitat selection is that habitats be defined as discrete entities. For use–availability studies in particular, the defined number of habitats can directly affect the results. Yet habitat distinctions often are not clear-cut. A researcher might distinguish two general forest types, uplands and lowlands, or might classify habitats by dominant overstory, or might divide these further by stand age or understory, and so on. As more types are defined, sample sizes are reduced for observed use of each type, thereby diminishing the power of the statistical tests to distinguish differences between use and availability. Also, because the proportional use and availability of all habitats each sum to 1, the number of habitats distinguished affects all of these proportions. Aebischer et al. (1993a, 1993b) observed that this unit–sum constraint renders invalid many of the statistical tests often employed to compare use and availability because the proportions are not independent. That is, if one habitat type has a low proportional use, others will have a correspondingly high use, and if there are only a few types, then the infrequent use of one type will lead to the apparent selection for another. Aebischer et al.'s (1993a, 1993b) method of compositional analysis was developed specifically to circumvent this problem.

Not just the number of types, but the criteria used to partition types may greatly affect results. Knight and Morris (1996) were able to visually differentiate 13 habitat types on landscape photographs of their study area, but postulated that only two broad classifications were distinguished by red-backed voles (*Clethrionomys gapperi*), the subject of their study. After analysis of their data, however, it became clear that from the voles' perspective, at least three functional habitats existed.

Another problem is the scale at which habitats are viewed. For example, an animal might appear to select for a certain habitat type, defined by a dominant cover type, whereas in reality it selected for certain specific kinds of sites that just happened to occur more commonly in that cover type than in others. An animal's choice of habitat type is often called macrohabitat selection and the choice of specific sites or patches within habitats is called microhabitat selection. These may be perfectly hierarchical in that the most preferred microhabitats always occur within the same macrohabitat, in which case an animal may really select initially at the scale of macrohabitat, and then focus on specific sites within it. Schaefer and Messier (1995) observed this sort of nested hierarchy across a range of scales for foraging muskoxen (*Ovibos moschatus*) in the Canadian High Arctic. Alternatively, the distribution of preferred microhabitats could be largely unrelated to the broader habitats defined by the biologist; in this case, a site attribute study might identify characteristics related to preferred microhabitats, whereas a use–availability study would detect no selection at the level of habitat type. This situation was apparently the case for wood mice (*Apodemus sylvaticus*) inhabiting arable lands in Great Britain: The mice seemed not to select (based on a use–availability study) from among three types of croplands (macrohabitats), but within each of these croplands they chose microhabitats with a high abundance of certain plants (Tew et al. 2000; Todd et al. 2000).

In sum, significant challenges in defining habitats include: partitioning them in terms of the features that the animals are selecting for, which are not necessarily the ones we most easily discern; delineating sufficient habitat categories to ensure that the truly important types are not lumped with and thus diluted by less important types; and not diminishing the power to discern selection by parceling out too many types.

## MEASURING HABITAT USE

Sample bias is an obvious potential problem in measuring habitat use. Interpretations of habitat use from visual observations of animals or their sign can vary among observers (Schooley and McLaughlin 1992) and sightability can vary among types of habitats (e.g., because of differing vegetative density; Neu et al. 1974), both of which can introduce biases in the data. For example, Powell (1994) noted that fisher (*Martes pennanti*) tracks in snow were difficult to follow in habitats with dense vegetation, especially where fishers followed trails of snowshoe hares (*Lepus americanus*); in this case the bias against observing tracks in dense vegetation merely detracted from the overall conclusion that densely vegetated habitats were frequently used.

Counts of pellet groups (e.g., from ungulates or lagomorphs) may poorly reflect habitat use because defecation rates often vary with the food source, and hence the habitat type (Collins and Urness 1981, 1984; Andersen et al. 1992). Capture locations may be a poor indicator of habitat use because baits and other trap odors (e.g., from captures of other animals) may affect behaviors in an unpredictable way (Douglass 1989).

Telemetry also may yield biased data on habitat use because the detection of an animal's radio signal may depend on the habitat it is in (e.g., GPS collars; Moen et al. 1996), and location data obtained by triangulation have inherent associated errors. Intuitively, and as shown in computer simulations by White and Garrott (1986), errors in determining habitat use increase with increased habitat complexity and decreased precision in the telemetry system. Errors do not necessarily introduce bias, but can if patch size differs among habitats (detected use would be underrepresented in habitat types that tend to occur as small patches) or if the animal preferentially used the edge of some habitat types but not others. Powell (1994) reported different perceptions of habitat use of fishers between his study, where he followed tracks in the snow, and another nearby radiotelemetry study; he attributed the difference to error in the telemetry system and consequent incorrect habitat categorization for animals near edges. Nams (1989) showed that simply discarding locations because of large telemetry error, as is common practice, exacerbates bias; he offered a procedure for circumventing it, but few studies have used it. Kufeld et al. (1987) suggested using the habitat composition of error polygons formed by the triangulation of radio bearings, but this would not alleviate bias. Chapin et al. (1998) solved the problem a different way. In a study of habitat use of American martens (*Martes americanus*), which have a documented affinity for mid- to late-successional forests, they classified telemetry locations that were outside small patches of forest but within telemetry error of the edge as representing use of those forest patches.

Even if habitat use can be measured accurately, biases may result from sampling or analytical procedures. Habitat use may vary by individual, sex–age group, social status, time of day, season, and year, yet many (most) studies pool individuals and do not sample adequately. Schooley (1994) reviewed habitat studies published in the *Journal of Wildlife Management* and found that most lasted only 2 years, and most pooled results across years without testing for annual variation. He used results of a black bear (*Ursus americanus*) study to show that habitat use can vary annually, and that the data pooled across years can yield misleading results. Beyer and Haufler (1994) found that most published studies that they reviewed collected data only during daylight hours; in their study of elk (*Cervus elaphus*), habitat use differed between day and night. Similarly, Arthur and Schwartz (1999) reported diurnal and nocturnal differences in habitat use for brown bears (*Ursus arctos*) that fed at a salmon stream that was used by people during the day; this difference was detected with data from GPS collars, but was not apparent from conventional diurnal telemetry data. Ostfeld et al. (1985) and Belk et al. (1988) observed sex-related differences in habitats used by ground-dwelling rodents; Belk et al. remarked that combining the two sexes would produce a false perception of habitat use. Paragi et al. (1996) observed differences in habitat use of resident and transient martens. Boitani et al. (1994) and Macdonald and Courtenay (1996) observed individual differences in habitat use, apparently related to social status. Bowers (1995:18) found that habitat use of eastern chipmunks (*Tamias striatus*) varied significantly with distance from their burrows, a finding noticeable only by considering the data on an individual basis. "It is time," Bowers commented, "that ecologists recognize that microhabitat selection and usage is a process involving individuals, not species."

Pooling individuals is common because sample sizes are typically too small to test for selection by individual. However, the statistical tests usually used assume independence among sample units, which is often not the case in studies that consider each location a sample. Some techniques (Johnson 1980; Aebischer et al. 1993a, 1993b; Manly et al. 1993) consider animals as sample units, so lack of independence among locations within individuals is not problematic. However, these methods are still subject to difficulties with lack of independence if animals are gregarious (attracted to the same habitats because they are attracted to each other; e.g., bed sites of deer; Gilbert and Bateman 1983) or territorial (social exclusion precludes use of certain habitats), or if the study subjects are related (habitat preferences possibly affected by a common learning experience) or are from the same social group (group leaders dictate habitat use for all).

In an effort to alleviate the problem of a lack of independence among individuals, Neu et al. (1974) used groups of moose (*Alces alces*) and Schaefer and Messier (1995) used herds of muskoxen as their sample units, rather than individual animals. Similarly, although Gionfriddo and Krausman (1986) monitored habitat use of individual radiocollared mountain sheep (*Ovis canadensis*), they considered groups of sheep their sample unit. However, Millspaugh et al. (1998) contend that animals in a herd should be considered independent individuals if they congregate because of a resource rather than because of a biological dependence on each other. They provide a hypothetical example with elk, where 99 of 100 radiotagged animals congregated at a winter feeding area in one habitat and the remaining individual used a second habitat; at other times of the year the elk did not associate with each other. In this case, they argue that each radiotagged individual should be considered an independent sample. In contrast, predators that hunt together in a pack and are thus dependent on one another cannot be considered to use habitats independently. Millspaugh et al. (1998) recommend tests to evaluate independence of habitat use by seemingly associated individuals.

## MEASURING HABITAT AVAILABILITY

Measuring habitat availability is often more problematic than measuring use. Use–availability studies inherently assume that study animals have free and equal access to all habitats considered to be available. That is, at any given moment each study animal should be able to use any available habitat. This assumption may hold if use and availability are measured for each animal individually. However, the assumption may be violated when animals are pooled for analysis and the available habitat is considered to be the same for all, yet some individuals may not even have all habitat types within their home range.

Johnson (1980) suggested that the habitat composition of home ranges compared with the habitat composition of some broader area should indicate the level of selection animals exercise when establishing their home range. Often the broader available area is considered to be that encompassed by the composite of the home ranges of all study animals. However, there are several problems with this.

First, animals cannot really select their ideal mix of habitats to compose their home range. Animals can only choose home range borders that encompass the best mix of habitats from what exists on the landscape; they cannot alter the mix to suit their needs. By analogy, a person may pick a town to live in, among several available, based on the resources available. One may also choose where to live within the town, but one cannot alter the layout of the town or the array of features available.

Second, animals may not have free and equal access to all areas when establishing their home range. Home ranges may be established near the natal area just because of familiarity with resources or neighboring animals, not any choice related to habitat composition. Analogously, people might remain in their home state or country not because they consciously chose it among all others, but because they never had the opportunity to visit other places, or because moving elsewhere, even if it seemed desirable in some respects, had too many costs. Social constraints also may dictate choice of a home range by precluding access to certain areas. Extending the analogy with people, consider a house to be like a home range and a neighborhood a composite home range. The first few residents of a neighborhood might have selected where to live among houses that differed in various ways; however, as more people moved in, the choices narrowed, until no choice remained for the last resident. If all houses were used, regardless of their quality, one could not discern after the fact which houses were preferred unless the "colonization" process was observed. Fretwell and Lucas (1970) proposed a corresponding model for animal populations. In an expanding population, preferred habitats are settled first, but as these are taken, animals are forced to settle in poorer and poorer areas. However, unless they are strictly territorial, their ranges can overlap, so unlike the human example, they can choose to live in a preferred area even though another animal is already there. As animal density increases in the most preferred habitat, however, resources become less available to each individual, so the quality of the habitat from each resident's perspective diminishes. Thus unless individuals benefit from the presence of others (Smith and Peacock

1990), their home range selection is negatively influenced by conspecific density. Other competing species have a further interacting effect on habitat availability and hence selection (Ovadia and Abramsky 1995). Because competition changes each animal's perception of habitat availability, human measurement of availability, based on the assumption of free and equal access, is inevitably inaccurate. As a result, animals tend to be more uniformly distributed across patchy landscapes than predicted from studies of habitat selection (Kennedy and Gray 1993).

Another major problem in measuring habitat availability is the recognition and treatment of areas of nonhabitat that may exist within home ranges. Part of the difficulty arises simply because our concept of home range is too nebulous. Home range is typically defined as the area used by an animal for its normal activities (generally attributed to Burt 1943), but home range area is a human perception, not a biological entity. Humans may perceive the landscape as a mosaic of habitats that fit together like a jigsaw puzzle, on which are superimposed home ranges of animals. In contrast, animals may perceive the landscape as series of corridors or islands sprinkled in an ocean of nonhabitat. If we unwittingly define available habitat from our human perspective, and include large patches of nonhabitat that the animal does not really perceive as among its choices of places to live, a comparison of use to availability might demonstrate nothing more than avoidance of the nonhabitat. This would be grossly accurate, but not particularly insightful. An example was presented by Johnson (1980), where mallards (*Anas platyrhynchos*) rarely used open water areas far from shore, but the area of open water was large. Standard means of comparing use to availability, such as the chi-square test, might show open water to be avoided and all other habitats selected; however, a knowledgeable duck biologist would recognize this as a trivial result, and might elect to exclude this obvious nonhabitat from the analysis. Other cases may not be so clear-cut (figure 4.1). Manly et al. (1993:45–46) presented an example with California quail (*Callipepla californica*), taken from a study by Stinnett and Klebenow (1986). Bonferroni confidence limits, and hence perceptions of selection, depended on whether habitats that were not used as escape cover when the birds paired for mating were included or excluded from the analysis. In this case the habitats that were not used as escape cover during mating were not obvious nonhabitats because the birds used them in other circumstances and for other activities.

An advantage of Johnson's (1980) technique is that the results are rather robust to inclusion or exclusion of habitat types that are rarely used. A problem with Johnson's (1980) technique is that because it is based on rankings of

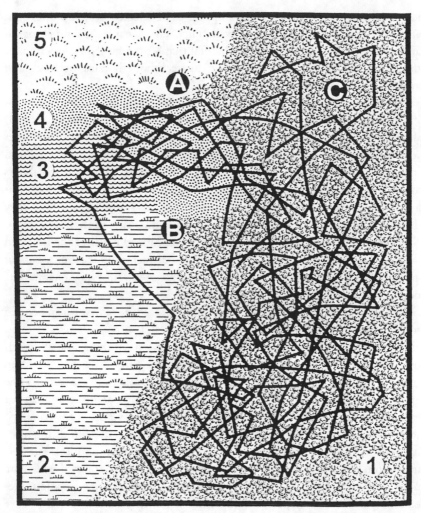

**Figure 4.1** Hypothetical movements of an animal overlaid on five (numbered) habitat types. Habitat selection is often assessed in terms of relative use compared to availability. In this example, habitats 1, 3, and 4 were used and thus also available. Habitat 2 (depicted as a swamp) appears to have been traversed, possibly just to get from habitat 1 to habitat 3; if it was used simply because of its location, not because of its habitat-related attributes, a question arises as to whether it should be considered in the analysis. Conversely, although habitat 5 was not used, it may or may not be considered available. Judged within the context of the home range boundaries, point A in habitat 5 appears to be unavailable, yet this point is closer to known locations of the animal than points B or C, which are both within the apparent home range. Habitat availability is a nebulous concept, and thus may be difficult to measure. Similarly, although the figure depicts a travel route, from which relative use of habitats might be deduced, most analyses deal with relative time, not distance, in each habitat (partly because telemetry data are generally comprised of point locations); it is unclear which is really a better measure of use.

use and availability, habitats will not appear to be selected if their proportionate use is ranked the same as their availability. Thus even if the animal spends an inordinate amount of time in the habitat that is most available, selection for this habitat will not be detected using this technique because both use and availability are ranked the same.

The Manly–Chesson index of habitat selection also does not fluctuate with inclusion or exclusion of seldom-used habitats, and Manly et al. (1993) showed that this index is much more versatile than Johnson's in many other respects. Recently, it was adapted by Arthur et al. (1996) to handle situations in which habitat availability changes. These authors recognized that habitats available to polar bears (*Ursus maritimus*) varied with changes in ice conditions and with movements of bears across their enormous home ranges. Thus they defined availability separately for each radiolocation, using the habitat composition of a circle with a radius (from the radiolocation) equal to the expected distance a bear would travel during the time between radiolocations; habitat availability within these circles was then compared with the type of habitat the bear actually used the next time it was located.

Another attribute of Manly et al.'s (1993) procedure is that it can be used to analyze data from site attribute studies as well as use–availability studies, although site attribute studies also face problems in assessing availability. If used sites are compared to random sites, the universe from which the random sites are drawn must be defined. As discussed earlier, that universe can be some arbitrarily defined study area, a composite home range of study animals, or each individual home range. Additional difficulties may arise if the comparison is between used and unused sites because errors may arise in distinguishing unused sites (i.e., nonobservation of use may not mean nonuse). Furthermore, unused sites may be vacant for a variety of reasons, some of which are unrelated to the physical habitat (e.g., human disturbance, exploitation, predation, parasites, interspecific competition). Some predictive models have fared poorly when they did not consider such variables (Diehl 1986; Laymon and Barrett 1986; O'Neil and Carey 1986). Geffen et al. (1992) found, unexpectedly, that Blanford's foxes (*Vulpes cana*) in desert environments were rarely observed near springs, where water and food were most abundant, probably because this habitat was favored by and provided cover for potential predators. In order to assess the criteria used by a species in selecting sites, investigators ideally should choose for comparison sites with both available resources and predators (or other confounding agents) present, as well as sites with only one or the other; however, such comparisons are unavailable in most field studies.

If a species is very selective in its choice of sites, differences between used and unused sites may be quite subtle; these subtleties would not be discernible in site attribute studies if the investigator chose unused or random sites that were very different from the used sites. The scale of comparison in this case would be too coarse. In an attempt to circumvent this difficulty, Capen et al. (1986) eliminated available sites in habitat types that were "radically different" from those that were used (analogous to eliminating nonhabitats in use–availability studies). Conversely, if in attempting to use a finer scale of comparison one picked random sites from too narrow a universe, such that they were all very similar to the used sites, habitat differences might not be detected if a large portion of the random sites were used. This points out the advantage of distinguishing unused sites instead of just random sites and of selecting unused sites that are similar in many respects to the used sites.

Use–availability studies do not distinguish unused areas and so may be especially prone to problems of too fine or too coarse a scale of comparison. The coarse-scale problem (used and available areas are too dissimilar to detect the true basis for selection) may occur when composition of home ranges or habitat use within home ranges is compared to some broader study area. The fine-scale problem (available area is too similar to the used area to detect differences) may occur when habitat use is compared to availability within home ranges. Thus these two scales of comparison may yield different results (Kilbride et al. 1992; Aebischer et al. 1993b; Boitani et al. 1994; Carroll et al. 1995; Paragi et al. 1996; MacCracken et al. 1997). McClean et al. (1998) examined the effects of varying the definition of available habitat, from the entire study area to progressively smaller-sized circles around individual radiolocations. They found that selection became increasingly difficult to detect as availability was defined by a smaller and smaller area. This result is not surprising because the radiolocation represents use, so habitat composition within smaller areas around each location more closely matches areas of actual use.

## ASSESSING HABITAT SELECTION: FATAL FLAW #1

Perceived habitat selection may vary with the technique chosen to compare use and availability or to compare attributes of used and unused (or available) sites. Some of this variation in perceived selection stems from the fact that different methods actually test different biological hypotheses (Alldredge and Ratti 1986, 1992; McClean 1998) and some is from the different assumptions inherent in these techniques and their sensitivity to violation of these assumptions (Thomas and Taylor 1990; Aebischer et al. 1993b; Manly et al. 1993).

Manly et al.'s (1993) technique can handle both use–availability and site attribute study designs. Moreover, it can be performed on an individual animal basis or with pooled data, it can be used to compare habitat selection among groups (e.g., species, sex–age classes, seasons, times of day, times within seasons), and it can incorporate both discrete and continuous variables. For these reasons, it has been heralded as a unified approach.

Manly et al.'s (1993) approach generates a resource selection probability function, giving the probability of a site being used as a function of various habitat variables. Each habitat variable can be tested to determine whether it contributes significantly to the probability of use. In the special case of only a single categorical habitat variable (i.e., habitat type), the function reduces to the Manly–Chesson selection index (Manly et al. 1972; Chesson 1978).

An advantage of this index, as discussed earlier, is that it is rather unaffected by the inclusion or exclusion of seldom-used habitats. In this sense, Chesson (1983:1297) suggested that the index is a measure of preference that "does not change with [resource] density unless [the animal's] behavior changes" and that it represents the expected use of the various resources if all were equally abundant. I think it is doubtful that this is true.

Consider first the simple example presented by Chesson (1978) to demonstrate the intuitiveness of the Manly–Chesson technique. The example deals with choice of foods, but I will adapt it for habitat selection. Suppose habitats A and B are equally available, and an animal spends 25 percent of its time in habitat A and 75 percent in habitat B (table 4.1). Because the Manly–Chesson selection index represents the expected use when resources are equally available, the index for each habitat in this case simply equals their proportional use (0.25 and 0.75 for A and B, respectively). Now suppose that the same animal is placed in an area composed of 80 percent habitat C and 20 percent habitat B, and it uses C 40 percent of the time and B 60 percent. The Manly–Chesson index would be 0.14 for habitat C and 0.86 for habitat B (table 4.1), suggesting that if habitats C and B had been equally available, they would have been used in these proportions. Because both A and C were compared against the same standard (habitat B), the results indicate that A would be preferred to C if those two types were offered together. However, given that the animal used A only 25 percent of the time but C 40 percent of the time, when in both cases the other choice was habitat B, the higher standardized selection index for A is not intuitive; these results are clearly a function of the higher availability of habitat C.

A *fatal flaw* of habitat selection studies in general, especially use–availability studies, is that they are based on the assumption that the more available a resource is, the more likely an animal should be to use it. This may not be true

**Table 4.1    Effect of Habitat Availability on Perceived Selection**

| Comparison<br>*Habitat* | % Available | % Used | Manly–Chesson<br>Selectivity<br>Index[a] | Manly–Chesson<br>Standardized<br>Index[b] |
|---|---|---|---|---|
| A vs B | | | | |
| A | 50 | 25 | 0.5 | 0.25 |
| B | 50 | 75 | 1.5 | 0.75 |
| C vs B | | | | |
| C | 80 | 40 | 0.5 | 0.14 |
| B | 20 | 60 | 3.0 | 0.86 |

Chesson (1978) used this comparison (with foods instead of habitats) to demonstrate the advantages of the Manly–Chesson index, but the lower standardized index for C than for A, despite C's greater use, is not intuitive.
[a]% Used/% available.
[b]Selectivity indices standardized so that they sum to 1 (selectivity index divided by sum of selectivity indices).

at all, may be true for only some resources, or may hold only within a narrow range of availabilities. Manly et al. (1993) made the explicit assumption, applicable for all models (except the previously discussed adaptation of Arthur et al. 1996) that availability remains constant for the period of study (if availability changes seasonally, data can be analyzed by season). This may seem like a benign assumption, but in reality it masks a fundamental weakness of the process. Of what value are measures of selection if they are specific to a single array of habitats? Measures of selection are supposed to be reflections of inherent preference—expected choices when availabilities of all habitat types are equal—so if selection appears to change as availability changes, then preference cannot be inferred from perceived selection when availabilities of habitats are unequal. In other words, if the goal is to assess habitat preferences for a population of animals based on habitat selection observed among a collection of individuals in that population, then something is amiss if selectivity appears to differ among these individuals simply because they have different habitat compositions available to them.

Consider a human analogy that demonstrates the effects of changes in availability on perceived selection. While at home a person spends 50 percent of the time sleeping and 20 percent preparing food and eating meals in the kitchen; the bedroom occupies 20 percent of the area of the house, and the kitchen 10 percent (table 4.2). Manly–Chesson selection indices for these rooms would be 0.51 and 0.40, respectively. Now suppose the person feels cramped in the kitchen and moves a wall, making it twice as big, at the expense of a room other than the bedroom. Afterwards the kitchen makes up 20 percent of the area of the house, the same as the bedroom, but use of the kitchen does not increase because it still takes the same amount of time to prepare and consume meals there. The selection index for the kitchen thus drops to 0.25, despite the fact that it is now more comfortable and better serves its purpose. Moreover, although no changes were made to the bedroom, its selection index improved to 0.63 as a result of the renovations to the kitchen. Superficially, it would appear that the expense for remodeling was not worth it.

Analogously, one might imagine a situation in which an animal used a habitat substantially more than its availability, but used it only for sleeping. If that habitat became more available, the animal would not be expected to sleep more, so its selection for it would appear to decline. A management agency that produced more of this habitat because results of a habitat selection study showed it to be used disproportionate to its availability would be disappointed to find that these efforts made the animal's selection for it drop.

**Table 4.2    Effect of Altered Availability (Floor Space) on Perceived Selection of Rooms in a House**

| Rooms | % Available | % Used | Manly–Chesson Selectivity Index[a] | Manly–Chesson Standardized Index[b] |
|---|---|---|---|---|
| Before renovation | | | | |
|   Kitchen | 10 | 20 | 2.00 | 0.40 |
|   Bedroom | 20 | 50 | 2.50 | 0.51 |
|   Others | 70 | 30 | 0.43 | 0.09 |
| After renovation | | | | |
|   Kitchen | 20 | 20 | 1.00 | 0.25 |
|   Bedroom | 20 | 50 | 2.50 | 0.63 |
|   Others | 60 | 30 | 0.50 | 0.12 |

This hypothetical example shows the nonintuitive result of diminished apparent selection for a kitchen after it was renovated to create more room. Neither use nor availability of the bedroom was changed, yet its standardized index increased after the kitchen was enlarged.
[a]% Used /% available.
[b]Selectivity indices standardized so that they sum to 1 (selectivity index divided by sum of selectivity indices).

These examples demonstrate cases in which the activity requires a fixed amount of time, so increasing availability of the preferred setting for that activity has no effect on how much time is spent there. This situation is just a special case demonstrating the point that use and availability are not inexorably linked. In the example of the house, the renovated kitchen might entice the person to spend more time there, but only up to a point (one certainly would not sleep there). Conversely, if the dining room had been remodeled at the expense of room in the kitchen, the person might not eat in the kitchen anymore, but, no matter how small it was, still prepare food there. Each room might thus have its own functional relationship between area and use. Similarly, if an animal prefers a certain habitat for resting because it offers protection from predators, it might spend more time resting in a larger patch of that habitat because it offers greater safety than a small patch. Enlarging a patch that offers virtually no predator protection to a size yielding some predator protection might thus cause significantly increased use of the patch; however, additional enlargements might have progressively lesser effects on use because they do not add much predator protection, and eventually further enlargements do nothing, or might even attract a different predator, thus deterring use. Various scenarios and corresponding relationships between patch size and use are plausible (figure 4.2). Considering that the relationship between patch size and use probably varies among habitat types and the mathematical relationship between use and availability also differs among the various selection indices (e.g., Manly–Chesson, Ivlev, and others; Lechowicz 1982), it seems doubtful that one could assess selection just by comparing relative use to the relative area of different habitats.

Mysterud and Ims (1998) proposed a logistic regression model to compare use:availability ratios among study subjects that had differing habitat compositions available to them. This model thus provides a test of the assumption that use increases with increased habitat availability. Their method is applicable to cases in which habitats can be categorized into two discrete types (e.g., forested vs. nonforested, oak vs. nonoak). They reexamined two data sets that Aebischer et al. (1993b) had analyzed using compositional analysis. In one, use increased with increased availability of a habitat for 9 of 12 ring-necked pheasants (*Phasianus colchicus*); however, three individuals did not fit this trend. In the second example, gray squirrels (*Sciurus carolinensis*) showed an inverse relationship between use and availability of open habitats within their home ranges (the same unexpected relationship therefore existed for the alternate, forested habitat). It was surmised that size and interspersion of habitat patches greatly affected the choices that these animals made, more so than just total

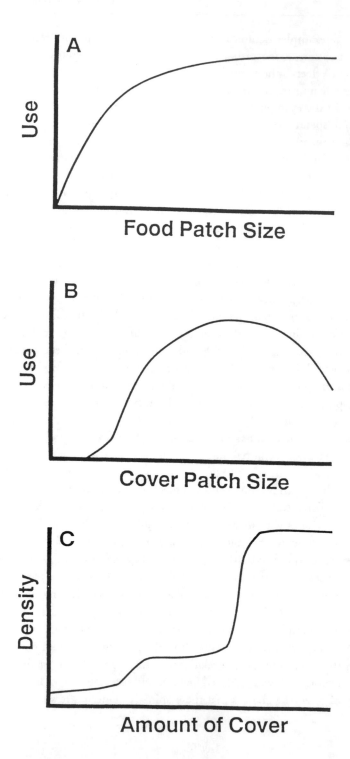

A

Use

Food Patch Size

B

Use

Cover Patch Size

C

Density

Amount of Cover

habitat area. Similarly, Mysterud and Ostbye (1995) found that although roe deer (*Capreolus capreolus*) in winter chose open canopy habitat for feeding and dense canopy for resting, they had to balance the advantages of being in each type of habitat against the energetic disadvantages of traveling between them, so patch size (distance between patches) affected habitat selection. Mysterud et al. (1999) suggested that for animals such as roe deer, which face tradeoffs in using different habitats, selection is not directly related to resource availability, so habitat rankings based simply on ratios of use to availability often are misleading. Bowyer et al. (1998) used a site attribute analysis to examine habitat selection related to various tradeoffs faced by black-tailed deer (*Odocoileus hemionus*).

Assessing selection can be extraordinarily complex because each habitat is not a single patch, but a series of patches of different sizes and shapes, each bordering other patches of different sizes, shapes, and habitat types. Otis (1997) offered a model that tests for the disproportionate use of habitat types as well as habitat patches, thereby providing a means of assessing things such as minimum patch size requirements. Data for this model (patch size distributions for each habitat type and locations of animals in specific patches) are available with modern geographic information system (GIS) coverages. This model still does not take into account habitat interspersion and juxtaposition, which probably have significant effects on selection for many species. For example, Porter and Church (1987) found that a standard use–availability analysis of habitat selection by wild turkeys indicated an avoidance of agricul-

Figure 4.2 (opposite page) Hypothetical relationships between area and use of habitat. Use–availability studies assume that habitat use increases linearly with area of available habitat. This is unlikely to be the case in many situations. (A) Relationship between use and size of a patch used mainly for foraging. A relationship like the one depicted might occur if different habitats offer different foods; the animal increases foraging time with increased availability of one habitat type, but this relationship asymptotes when the animal obtains enough of the food there and searches for alternative foods in other habitats. The same sort of relationship might occur for an animal that forages mainly near the edge of the patch, if size (*x*-axis) is in units of area but use increases with the perimeter. (B) Relationship between use and size of a patch used primarily for cover. In this case a very small patch offers virtually no benefit, so it is not used at all; use increases with increasing patch size, but then declines when the patch becomes large enough to attract another type of predator. (C) Relationship between density (a reflection of use) and cover (which in this case provides protection from predators, is used for food, and influences microclimatic conditions) that was shown (and partly hypothesized) for voles (*Microtus* spp.) (Birney at al. 1976). At low levels of cover, the area is occupied only by transients searching for a better place to live. The first threshold represents the point at which cover is adequate to attract residents. The second threshold represents a level of cover sufficient to enable the population to surge and eventually cycle. Although this second threshold was shown empirically, it is not well understood.

tural lands, when in reality turkeys used agricultural lands extensively, but only those near hardwood forests. In essence, the turkeys viewed the edge between field and forest as a separate habitat type. Similarly, Neu et al. (1974) posited that moose might feed preferentially in a recent burn, but not too far from the surrounding forest. Thus they defined four habitat types—the interior of the burn, the burn periphery, the forest edge adjoining the burn, and the remainder of the forest—and through a simple chi-square analysis showed selection for the edge (just inside or just outside the burn). Most situations probably are not this simple.

Many authors have admitted to the importance, but difficulty, of incorporating spatial aspects of habitats in use–availability analyses. Porter and Church (1987) proposed a method whereby the study area is gridded into cells and an assortment of habitat variables within those cells are examined through multivariate analyses to find those that best explain differential use of cells. Litvaitis et al. (1986) did just that in a study of bobcats (*Felis rufus*), which predated the paper by Porter and Church (1987). Litvaitis et al. (1986) looked for associations (using regression and DFA) between the number of radiolocations within 25-ha cells inside home ranges and measurements of several habitat variables sampled there; however, they found that these habitat variables poorly explained variation in frequency of use. Servheen and Lyon (1989) used a similar approach in assessing habitat selection by caribou (*Rangifer tarandus*). They measured habitat variables in 40-ha circles around telemetry locations and sought to find those that best differentiated the areas that the animals used seasonally. Although they had no real measure of juxtaposition or interspersion of habitats, their 40-ha circles contained habitats neighboring the one actually occupied, so the composition of these circles gave an indication of habitat combinations that corresponded with seasonal use. In another similar approach, Clark et al. (1993) used grid cells that could encompass several habitat types near the locations of radiocollared black bears. A suite of habitat characteristics (including the number of different habitat types) within each cell used by bears were combined to form what they called an ideal habitat profile. The habitat quality of each cell in the study area was then assessed by comparing it to this hypothetical ideal cell. Each of these studies looked at differential use, rather than use in terms of availability, and thus avoided the fatal flaw of habitat selection studies.

Site attribute studies are like the habitat use studies just discussed, except that instead of comparing cells with varying degrees of use, they categorize cells (sites) simply as used or unused; based on this, important habitat variables are identified. Interspersion and juxtaposition of habitats can thus be investigated. For example, Coker and Capen (1995) examined cowbird (*Molothrus ater*) selection for habitat patches of various size, shape, and location relative to other habitats by entering these variables in a logistic regression with use (used or not used) as the dependent variable. Similarly, Chapin et al. (1998) compared habitat variables (including an index of the extent of habitat edge) in grid cells of different sizes that were used (i.e., had at least one telemetry location) by American martens with those in cells not used by martens, and also compared characteristics of forest patches that were used and not used.

McLellan (1986) argued that observed use is a better indicator of habitat selection than use relative to availability. He reasoned that an animal familiar with its home range knows the availability and location of resources, so an animal's location at any given moment represents selection. He gave an example of a person at a buffet selecting a slice of beef from a 500-kg steer and an equal-sized slice of pork from a 100-kg pig; based on use alone, pork and beef were selected equally, but compared to availability, pork appears to be selected over beef, which is obviously absurd. However, had the steer and pig been cut up in equal-sized chunks and distributed over a large area, and after considerable searching the person still returned with an equal quantity of the two foods, active selection for pork would indeed seem apparent. The key difference is that in the latter case the person had to search for the food; selection was evidenced by the extra effort expended in finding the pork (and apparently bypassing chunks of beef). This searching for resources is really the basis for the development of use–availability comparisons and explains why it originated with studies of diet. In most cases animals do not know the location of all foods in their home range, so dietary selection based on availability may be appropriate. However, habitats are not spread around like chunks of pork and beef, but occur in large patches, the locations of which are known by the animals; thus habitats are probably more like McLellan's (1986) whole steer and whole pig than the cut up chunks of meat spread randomly around (figure 4.3).

Consider some actual examples of how observed use and use versus availability can lead to disparate interpretations of selection. Prayurasiddhi (1997) investigated use and selection among two large ungulates, gaur (*Bos gaurus*) and banteng (*B. javanicus*), in Thailand. He differentiated two general study area boundaries, one of which more closely matched the area that his radio-collared animals used most intensively. He also used actual home range boundaries as a third representation of the study area and hence the available habitat. He found that this variation in the area considered to be available habitat resulted in drastic differences in perceived habitat selection (table 4.3). One habitat that received 46 percent of use by gaur was deemed to be selected for,

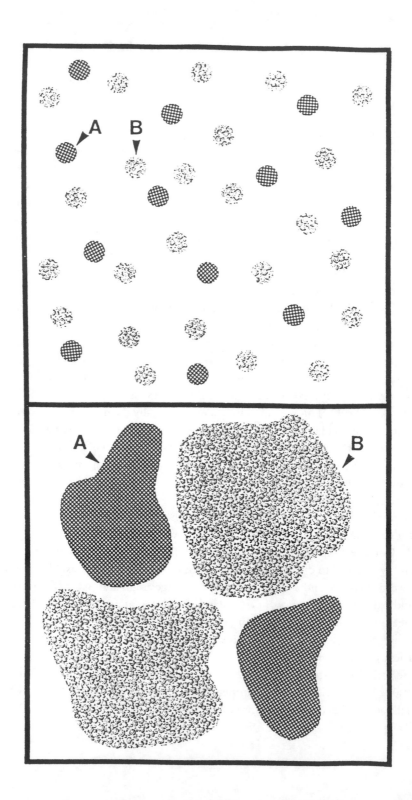

whereas another habitat that received 45 percent of use was seemingly "selected against." In one case banteng were judged to be unselective in their use of a habitat in which they spent 75 percent of their time. Prayurasiddhi (1997) recognized these difficulties and decided to evaluate seasonal changes in habitat selection and habitat-related differences among species based on use alone, rather than use compared to availability.

In another analogous situation, Macdonald and Courtenay (1996) found that crab-eating zorros (foxes, *Cerdocyon thous*) in Amazonian Brazil spent most of their time (64 percent) in wooded savanna and scrub habitats; however, because these two habitats were abundant within the home ranges of the study animals, they ranked the lowest in apparent preference (sixth and seventh among seven defined types) based on a comparison of use to availability. During the wet season, however, when lowland habitats flooded, the zorros used upland wooded savannas and scrub habitats even more, making the apparent preference ranking for these rise; in reality, the area of available (not flooded) lowland habitats diminished, but this reduction could not be measured and therefore was not taken into account in the use:availability calculations. The authors realized that the seasonal difference in apparent habitat preferences was thus an artifact of unmeasured changes in availability.

In another such case, Garrett et al. (1993) found that tidal flats were the principal foraging habitat for bald eagles (*Haliaeetus leucocephalus*) and observed that nearly one-fourth of their perch sites were within this habitat. That is, based on use alone, this area was clearly attractive to these birds. However, because this habitat was so widely available, especially at low tide, a comparison of use to availability suggested that the eagles avoided it. Apparent preference thus changed radically with tidal fluctuations.

**Figure 4.3** (opposite page)   The assumed linear relationship between use and availability of resources arises from a model in which the resources are scattered around in small bits (top panel), the locations of which are not known to the animal. In the case depicted, the dark-colored resource (A) is only half as available as the light-colored resource (B), so an animal that randomly encountered these would be expected to obtain (in the case of food) or use (in the case of habitat) resource A half as much as resource B. If resource A was used more than that, the animal must have bypassed B, thus demonstrating selection for A. In the lower panel, the two resources are still in the same proportions, but are clumped, thus representing a more realistic situation for habitats. An animal here would not wander around encountering and rejecting or accepting resources in its path, but would probably know the locations of habitat patches. Thus the time spent in each patch would be commensurate with the type of activity and attributes of that habitat, which may or may not include the area of the patch. If, in the case of the lower panel, an animal used (selected) the two habitats equally, it would be fallacious to assume that it was selecting habitat A over B simply because A was less available than B.

Table 4.3    Habitat Use, Availability, and Perceived Selection for Gaur and Banteng in Thailand During the Dry Season

| Species Habitat Type | Habitat Availability (%) | | | Habitat Use (%) and Selection | | |
|---|---|---|---|---|---|---|
| | Area 1 | Area 2 | Home Ranges | Area 1 | Area 2 | Home Ranges |
| Gaur | | | | | | |
| Evergreen | 31 | 40 | 27 | 28 n | 46 n | 46+ |
| Mixed deciduous | 56 | 49 | 61 | 70 + | 45 n | 45 – |
| Dry dipterocarp | 13 | 11 | 13 | 2 – | 9 n | 9 n |
| Banteng | | | | | | |
| Evergreen | 30 | 39 | 7 | 5 – | 8 – | 8 n |
| Mixed deciduous | 57 | 49 | 76 | 75 + | 75 + | 75 n |
| Dry dipterocarp | 13 | 12 | 17 | 20 n | 17 n | 17 n |

Values are rounded-off percentages from Prayurasiddhi (1997:230).
Use was compared with availability at three scales, denoted as area 1, area 2 (a somewhat smaller area with more intensive use by radiocollared animals), and home range boundaries. Selection was assessed with the methods of Neu et al. (1974) and Manly et al. (1993). Note changes in perceived selection (+ selected for, –selected against, n = no perceived selection) with changes in the area being considered.

A problem with assessing selection from use alone is that use often does change with availability. In the eagle study, tidal flats were rarely used during high tide because they were largely unavailable. The eagles may have preferred tidal flats to other habitats, but use of these areas depended on availability, up to a point. When availability exceeded some threshold they did not continue to increase their use, and when it declined below some threshold their use of the area ceased. Between these thresholds some relationship may exist between use and availability. However, relationships between use and availability probably vary among habitat types, and within each habitat may depend on what other habitats are available (figure 4.2a).

It might seem that site attribute studies avert these problems because selection is inferred not from differences between use and availability, but from differences between used sites and other sites. However, the comparative sites are taken proportionately from those available, so the comparison is still, in a sense, the same as in the use–availability design. Consider a study that locates 10 nest sites in each of two available habitat types and then compares the characteristics of these to 20 randomly chosen points. If one habitat type is much more common than the other, then the 20 randomly chosen points fall mostly within the more common habitat; this sampling could result in possible differences between the average characteristics of the randomly chosen sites and the real nests, suggesting that the species selected for nest sites in the less common habitat type. This logic falters, though, for the reason highlighted in McLellan's (1986) example of the buffet, in which a person chose equal portions of beef and pork. In that example, the equal choice of the two types of food was clear evidence of nonselection, or equal preference, regardless of availability. Accordingly, in the hypothetical case of the 20 nests, one should question why, if there was strong selection for one habitat type, did an equal number nest in each type? The conclusion that this represents evidence of preference for the less common type is (just as in use–availability comparisons) based on the assumption that if individuals had no preference, they would chose sites randomly, which is typically thought to mean in proportion to availability. However, suppose each individual, having no preference, just flipped a coin, so to speak, to decide which of the two types of habitats to nest in each year. The result would be as posed in the example, an equal number nesting in each type.

There are two other problems (really two aspects of the same problem) with detecting selection using a site attribute design. James and McCulloch (1990) showed that a species could be highly selective for a certain habitat component, as indicated by a low variance among values of this component at the sites it chose, but if the mean value for the component at selected sites approximately equaled the mean in the available habitat (where the variance was much higher than at selected sites), then selection might not be detected (type II error). Conversely, Rexstad et al. (1988) tested multivariate statistical procedures that are commonly used in site attribute studies and found that significant results, which in a real study would be indicative of selection for certain habitat components, were often derived from a collection of meaningless data (type I error). Taylor (1990) attributed Rexstad et al.'s spurious results to ill-defined hypotheses and inappropriate use of statistical procedures; however, Rexstad et al. (1990) showed that their study mimicked the majority of published applications of these techniques. North and Reynolds (1996) expanded on Rexstad et al.'s (1988, 1990) concerns that the statistical procedures used in site attribute studies may yield misleading results because assumptions are commonly violated.

## INFERRING HABITAT QUALITY: FATAL FLAW #2

If habitat selection studies were simply an attempt to better understand behavior and natural history, errors would merely be a setback in scientific inquiry. However, these studies are typically designed to provide management guidelines. Habitats are managed based on supposed importance in terms of fitness for members of the target population. High-quality habitats, by definition, produce animals with high reproduction and survival, and hence a population with a high growth rate (and high yield, for harvested species). However, the assumption that we can infer habitat quality or suitability from studies of habitat selection—that selection, even if accurately measured, is directly related to each habitat's potential contribution to individual fitness and hence the population's growth rate—represents what I consider the *second fatal flaw* in the process of habitat evaluation.

One of the best examples of a purported relationship between habitat selection and fitness was provided by a study of leaf-galling aphids (*Pemphigus betae*) that parasitize narrowleaf cottonwood trees (*Populus angustifolia*). These aphids overwinter in tree bark, then over a short period in the spring migrate up the tree and become entombed in expanding leaf tissue, where they reproduce. Whitham (1978, 1980) found that aphids apparently selected large leaves over small leaves (they colonized 100 percent of leaves over 15 cm but only 3 percent of leaves 5 cm or less), even though large leaves were less available (less than 2 percent of leaves were over 15 cm; more than 30 percent were 5 cm or less); moreover, those that colonized large leaves had better survival and reproduction than those colonizing small leaves. From this evidence it appeared that aphids had the ability to select leaves that offered the highest fitness; if not, the size distribution of colonized leaves would have matched the size distribution of leaves available on the tree (i.e., use would have equaled availability). However, Rhomberg (1984) observed that large leaves tended to be those near the tips of twigs (where they receive more light), and these opened a few days later than basal leaves. When the aphids begin their migration, basal leaves are already too mature to enable them to form a gall, so they must colonize the more distal leaves; thus, the basal leaves are in essence non-habitat. The migration appears to be timed so that the aphids inevitably colonize leaves that are destined to be favorable to their fitness, but individually, they do not select these among less favorable habitats. This is not to say that other animals do not select habitats favorable to their fitness; certainly they do. But if the comprehensive and seemingly compelling data on a simple system such as aphids on cottonwood trees can be misinterpreted, then clearly there is much room for misinterpreting relationships between habitat selection and fitness in more complicated systems.

The presumed link between selection and fitness is rarely tested, partly because such tests are difficult, but also because this relationship is viewed more often as fact than as assumption. However, there are many reasons why this relationship may not hold. Habitats that are used infrequently may be more important than suggested by the time spent there. Conversely, habitats used preferentially for activities that require a lot of time (e.g., resting) may be less important than indicated by their use; a resting habitat may be substitutable, with little effect on fitness, whereas a specific foraging habitat (or a water hole in an arid environment) may be more essential, even if not used very much. Differentiating activity-specific habitat use may help alleviate this difficulty (Forsman et al. 1984; Cavallini and Lovari 1991; Ternent 1995; Salas 1996; Tew et al. 2000), although this, too, is seldom done, and moreover, does not necessarily circumvent the problem. For example, Powell (1994) studied winter foraging of fishers, which preyed heavily on porcupines (*Erethizon dorsatum*) but spent a disproportionately small amount of time hunting in upland hardwood habitats where porcupines were common. The reason, he found, was that fishers rapidly located porcupines at known den sites, thus minimizing their search and chase times. In contrast, fishers had a harder time hunting snowshoe hares, so they spent a larger amount of time in lowland conifer habitats where hares were more common. Clearly, the relative importance of the two habitats to fishers was not reflected by their time spent hunting in each. For animals that feed on a variety of different foods, a mixture of different habitats may be more beneficial in many respects than a single, highly preferred type, and the time spent in each may be a poor indicator of importance of either the specific type or the overall mix. Omnivorous Egyptian mongooses (*Herpestes ichneumon*), for example, favor one specific habitat for resting, but use a mix of habitats when feeding (Palomares and Delibes 1992).

Fitness also may be affected by predation and interspecific or intraspecific competition. These factors can change the cost:benefit ratio of a habitat and hence alter an animal's habitat use (Douglass 1976; Holbrook and Schmitt 1988; Hughes et al. 1994; also see review and other citations in Lima and Dill 1990). Cowlishaw (1997) found that baboons (*Papio cynocephalus*) spent more time feeding in a habitat with poor food but a low risk of predation than in a food-rich habitat with a high risk of predation. Fitness was probably enhanced by this choice of habitats.

Fitness certainly was enhanced for a herd of caribou in Ontario, Canada, that spent most of the time on an island. The nearby mainland had higher-quality forage but also a high density of wolves. Ferguson et al. (1988) found that the island occupants sacrificed nutrition, which was reflected in smaller body and antler size as well as starvation of their calves, for overall higher survival, which enabled this population to persist while herds that remained on the mainland perished from predation. One could argue that the animals on the mainland made poor choices in habitat selection.

Animals may not always correctly perceive risks, especially in a novel or changing situation, so their choice of habitats may not necessarily be best in terms of their fitness (Wiens et al. 1986, Holt and Martin 1997). Pollution is one such risk that is rarely evaluated in habitat studies. Consider Mallory et al.'s (1994) study of the effects of acid rain on habitat quality for common goldeneyes (*Bucephala clangula*). Acid rain killed fish in lakes, which enabled proliferation of invertebrates, thus providing more food and enhanced reproduction for goldeneyes. However, these ducks may inevitably suffer other consequences of the acidified habitat. In some situations the choices made by animals may maximize fitness over the long term (e.g., if the changes are ephemeral); in other cases animals may require time to adapt.

Suppose that an animal uses alternate habitats more often when the normally preferred habitat becomes less profitable because of increased density and hence competition for resources. If the alternative habitats are more available than the preferred habitat, then use of these alternative habitats can end up higher than that of the otherwise preferred habitat. Situations like this seemingly justify considering habitat use in terms of availability. However, Hobbs and Hanley (1990) showed both intuitively and with computer modeling that intraspecific competition may alter use:availability ratios such that they may not reflect the true quality of a habitat in terms of its potential contribution to population growth. Certainly interspecific competition and predation are integral components of habitat quality, and managers concerned about a given species should not evaluate habitats independent of these factors. However, perceptions of habitat quality that are affected by density of the targeted species are an artifact of the method. Recall, as Peek (1986) noted, that habitat preference should be innate, and intricately tied to fitness; if our perception of preference declines as the preferred habitat becomes more densely populated, then our perception is clearly mistaken. Hobbs and Hanley (1990: 520) concluded that "use/availability data inherently reflect differences in animal density among habitats" (independent of social behavior) and therefore reveal little about habitat quality. Consequently, they recommended abandonment of the use–availability design in favor of an approach that enables a direct investigation of the link between habitats and fitness.

Site attribute studies often focus on sites of biological importance (as reviewed earlier) and therefore may provide more direct insights into habitat variables that affect fitness. However, as pointed out by North and Reynolds (1996), conclusions from studies of this sort are based on two fundamental, but questionable, underlying assumptions: that used sites are in suitable habitat and that unused sites are in unsuitable habitat. Consider the study by Meyer et al. (1998), who observed significant habitat-related differences among sites that were occupied versus unoccupied by spotted owls (*Strix occidentalis*). They found, as expected, that these owls tended to select sites having a large proportion and large patches of old-growth forest, but suggested that the owls could tolerate certain amounts of clearcut and young forest (based on the fact that used sites contained such areas). Their reproductive data were inconclusive about what the owls could tolerate from a population standpoint, but Meyer et al (1998:47) nonetheless asserted "that resource selection probability functions [which they generated using Manly et al.'s 1993 approach] . . . are a reliable tool for assessing capability of landscapes to support northern spotted owls." Hence they concluded (abstract, p. 5) that their habitat selection model "can be used to predict the probability that a given landscape mosaic will be a *suitable* spotted owl site" (emphasis mine). This is exactly what I refer to as *fatal flaw #2*.

To avoid assumptions related to suitability in site attribute studies, North and Reynolds (1996) proposed that instead of comparing habitat characteristics of used sites with unused or random sites, used sites be categorized by their intensity of use (e.g., 1–2 percent of locations = low use, 3–10 percent medium use, more than 10 percent = high use), and habitat characteristics compared among such categories (i.e., excluding unused sites) with a form of logistic regression. Thus this is not really a site attribute design (according to my definition), nor is it even applicable to the types of situations in which site attribute studies are commonly used (i.e., in which an animal remains at a single site for a fairly long period of time). Rather, this procedure represents just another means of investigating habitat characteristics associated with differing degrees of use. I previously discussed other studies of this sort (Litvaitis et al. 1986; Porter and Church 1987; Servheen and Lyon 1989; Clark et al. 1993) as a special case of the use–availability design because, although lacking an assessment of habitat availability, they are based on proportional use. Whereas these methods may avoid many of the statistical and general assumptions of site attribute studies, they digress further from the issue of habitat-related fitness because the studied sites, instead of being of some special biological significance, are simply points where the animal was located some number of times. As North and Reynolds (1996) recognized, intensity of use may not be a good indicator of habitat quality.

## ■ Advantages and Problems of the Demographic Response Design

The best measure of habitat quality would be a test of its effects on demographic parameters such as population growth and carrying capacity. Such tests are extraordinarily difficult in most situations, as evidenced by the scant published studies of this nature. Most demographic response studies I reviewed examined potential relationships between habitat and animal density. Because habitat-specific density is actually a reflection of differential habitat use, investigations of habitat-related density suffer the same drawbacks as studies of habitat use.

Density tends to be an ineffective measure of habitat quality because it may fluctuate widely, is subject to sizable errors in estimation, and may be largely influenced by social factors. Van Horne (1983) gave several examples of situations in which juveniles were restricted from settling in the best habitats and thus accumulated in large numbers in poorer quality habitats. She indicated that such circumstances are likely to be common among generalist species with high reproductive rates and a social hierarchy. For these species in particular, then, habitat-specific density would probably be a poor indicator of habitat quality unless the population is well below carrying capacity. A good example was provided by Messier et al. (1990), who showed that density of muskrats (*Ondatra zibethicus*) during a general population increase swelled 30- to 90-fold in low-quality habitats but much less in high-quality habitats.

Considering that, in general, animals in poor-quality habitats should be trying to leave and ones in high-quality habitats trying to stay (and keep competitors out), Winker et al. (1995) posited that turnover rate would be a better index of habitat quality than density. They measured turnover rates for wood thrushes (*Catharus mustelinus*) by examining recapture rates and telemetry movement data; low-quality habitat was defined as that in which recapture rates were low and many radiotagged birds were transient visitors. They found density and habitat quality, assessed in terms of turnover rates, to be inversely related. Notably, Winker et al.'s (1995) turnover rate model may not be applicable to other species, even other territorial, noncolonial songbirds, some of which are preferentially attracted to habitats occupied by conspecifics, which they use as a cue to habitat quality (Muller et al. 1997). Conspecific attraction tends to perpetuate use of the same areas across generations, so even if habitat quality deteriorates, high densities may be maintained through tradition.

Other competing species or unidentified confounding variables also may weaken the linkage between habitat quality and density. Maurer (1986) measured density and various habitat characteristics for five species of grassland birds; the habitat models developed to explain species-specific density in one study area were inexplicably poor predictors of density in a nearby area with similar habitat. Kellner et al. (1992) found that density was positively related to reproductive success in only 7 of 17 bird studies that they reviewed; more studies showed a negative relationship. Sherry and Holmes (1996) felt that density should be relied on as an indicator of habitat quality only if it is corroborated by other data, as was the case in their study, in which population density and weight loss of wintering migrant birds were correlated (high weight loss in areas with low densities) and both were related to habitat type.

Reproduction and survival data may be more apt to reflect real influences of habitat on demographics. However, reproduction and survival are probably also tied to habitat in a complex manner. For example, a number of studies observed a direct relationship between cover and the survival (and thus density) of voles (*Microtus* spp.), but a lower threshold exists below which reductions in cover have little effect on vole density; above the threshold, survival and density increase but eventually reach an upper asymptote (Birney et al. 1976; Adler and Wilson 1989; Peles and Barrett 1996; figure 4.2C). The vole studies found that cover provides food as well as protection from predators, and also may affect microclimate, activity patterns, and interactions among conspecifics, all of which affect the cover–density relationship. Moreover, male and female voles have different responses to varying cover (Ostfeld et al. 1985; Ostfeld and Klosterman 1986), and cover–demographic relationships tend to be different for other small grassland rodents (Kotler et al. 1988). Each of the various habitat components that relate to an animal's fitness probably has thresholds, asymptotes, and inflection points, and these limits may vary with the mix, shape, size, and juxtaposition of habitat components available; however, few attempts have been made to assess any of these factors individually (Harper et al. 1993; Whitcomb et al. 1996), let alone in combination.

Several studies also found that density-dependent effects may reduce reproduction or survival independent of habitat quality (Kaminski and Gluesing 1987; Clark and Kroeker 1993; Clark 1994) or may even result in higher fitness in low-quality, less crowded habitats (Pierotti 1982; Fernandez 1999). Zimmerman (1982) found that nesting success of dickcissels (*Spiza americana*) did not differ between habitats and was unrelated to density, but females nested preferentially in the habitat preferred by males; males chose this more heterogeneous habitat because they could sequester more nest sites and thus mate with more females. In other studies, reproduction and survival were found to be unrelated to measured habitat variables, despite evidence of habitat selection, possibly because of confounding effects of weather, human disturbance, measurement error, and other factors (McEwan and Hirth 1979; Hines 1987; Rumble and Hodoroff 1993; Bruggink et al. 1994; Gilbert et al. 1996; Maxson and Riggs 1996). Even experimental demographic response studies have been plagued with unexpected variations in confounding variables (Taitt et al. 1981; Harper et al. 1993). Nevertheless, some carefully designed studies observed links between habitat and reproduction or survival and through further investigation discovered the underlying causes (Chasko and Gates 1982; Brown and Litvaitis 1995; Greenwood et al. 1995; Loegering and Fraser 1995).

A final major problem, discussed previously in relation to use–availability and site attribute designs, is scale. Levin (1992) showed clearly that there is no single correct scale for studying ecological relationships. Animals view and react to their environment at various scales. Human perceptions of ecological systems are inescapably biased or incomplete because they are filtered by the observational scale chosen for the investigation. Some demographic response studies have recognized this and have adopted a multiscaled approach. For example, Orians and Wittenberger (1991) found that densities of yellow-headed blackbirds (*Xanthocephalus xanthocephalus*) were higher on marshes with higher food (insect) abundance but that density of blackbird territories within these marshes was related to vegetational structure, not food. The authors postulated that these birds may not select food-rich territories because they often hunt outside their territories. Also, they establish their territories before the full emergence of insects and hence may not be able to predict future food abundance on a scale smaller than the marsh level. Pedlar et al. (1997) developed habitat models on two different scales to explain variation in raccoon (*Procyon lotor*) density. One model was fit to macrohabitat features and the other to microhabitat variables, after which the two were combined to form a more comprehensive model relating density to habitat at both scales. Morris (1984, 1987, 1992) observed both macrohabitat and microhabitat differences among several species of small mammals, suggesting habitat selection on both scales, but found that variation in density was much more evident at the macrohabitat level. On a larger scale, Dooley and Bowers (1998) discovered, counter to their expectations, that densities and population growth rates of voles were higher in patches within a fragmented landscape than in an unfragmented landscape, whereas total population size was higher in the unfragmented landscape (because more total habitat was available). Landscape fragmentation caused overall habitat loss but, on a finer scale, enhanced reproduction within individual habitat fragments. Similarly, Brown and Litvaitis

(1995) noted that some mammalian predators that hunt preferentially in forests nonetheless often exist at lower densities in homogeneous forests than in forests interspersed with disturbed areas. Each of these examples represents cases in which a single-scale investigation would have failed to detect habitat variables affecting population demography.

The previous examples all concerned spatial scale. Time scale may be equally important. The demographic value of a habitat may become evident only in the long term, after a population has been subjected to the stresses of a periodic drought, severe winter, or failed food crop (Beyer et al. 1996; Pelton and van Manen 1996).

## ■ Applications and Recommendations

A great deal of effort continues to be invested in habitat-related studies of wildlife. In the United States, federal land management agencies in particular have focused on developing formalized procedures for evaluating habitat for wildlife (Morrison et al. 1998). The procedures that have been adopted rely on models derived by species experts, who in constructing these models tend to rely more on experience than on empirical data (Schamberger and Krohn 1982; Thomas 1982). Therefore, the models are really hypotheses in need of testing. However, because these models hypothesize explicit relationships between habitat attributes and animal populations (so-called habitat suitability indices), they cannot be rejected or accepted in normal scientific fashion; that is, none of the relationships are likely to be exactly correct. Thus it seems inappropriate to suggest, as is common parlance, that they should (or even could) be "verified" or "validated."

Brooks (1997) and Morrison et al. (1998) proposed steps for verifying or validating habitat suitability models. Unfortunately, these authors and most others writing on this subject have misused these terms. *Verification* means establishment of truth and *validation* technically refers to establishment of legitimacy (i.e., in the case of a model, showing that there are no logical or mathematical flaws; Oreskes et al. 1994). The general misapplication of these terms in reference to model testing is not merely a semantic issue, but rather a real misrepresentation of accomplishment. Because of the complexity of natural systems, habitat models invariably exclude some relevant parameters and presume relationships that are not exactly or not at all correct. There is simply no way to perfectly model these sorts of open systems (i.e., systems in which all variables and relationships are not known). In a well-reasoned discussion of the subject, Oreskes et al. (1994:643) argued, "A match between predicted and obtained output does not verify an open model. . . . If a model fails to reproduce observed data, then we know that the model is faulty in some way, but the reverse is never the case." Assuming so is a logical fallacy called affirming the consequent. "Numerical models are a form of highly complex scientific hypothesis [unlike simple null models that we are accustomed to testing]; . . . verification is impossible." The utility of models is to guide further study or help make predictions and decisions regarding complicated systems; thus they warrant testing, but that testing should be viewed as a never-ending process of refinement, properly called benchmarking or calibration. Given the basis of habitat suitability models and the complexity of their many interacting variables, it is likely that any such model could be improved through rigorous testing.

Several attempts have been made to test such models. Often this process is circular, involving just another panel of experts making qualitative assessments (O'Neil et al. 1988). In other cases, models have been tested using results of a study on habitat use (Lancia et al. 1982) or use relative to availability (Thomasma et al. 1991; Powell et al. 1997), with the inevitable associated shortcomings discussed in detail in this chapter. In some instances, habitat management prescriptions based on "common knowledge" or expert opinion have, through collection of better data, been proven faulty (Brown and Batzli 1984; Bart 1995; Beyer et al. 1996). I found one case in which model-derived habitat scores for individual home ranges were compared with reproduction, juvenile growth rates, and home range size, but no significant relationships were observed (Hirsch and Haufler 1993).

Often, models have been tested by comparing habitat-specific densities to model predictions. However, even if a model explains a significant portion of the variation in density (Cook and Irwin 1985), the data collected to test (or purportedly validate) the model could be better used to modify it or build new one (Roseberry and Woolf 1998). In most cases, habitat models have proved to be poor predictors of animal density, indicating either defectiveness of the model, lack of a clear habitat–density relationship, or effects of other confounding factors, such as hunting pressure, which are also habitat-related (Bart et al. 1984; Laymon and Barrett 1986; Robel et al. 1993; Rempel et al. 1997). Bender et al. (1996) found that if the variance around the estimated values of the model inputs were taken into account (a process that is not commonly done), suitability scores for a variety of habitats that appeared very disparate were not significantly different; that is, the parameter estimates were not precise enough to even enable the model to be tested.

A major inherent but generally unstated (maybe unrecognized) assumption of habitat suitability models is that high-quality habitats (i.e., habitats that confer high fitness) are in fact suitable (i.e., able to sustain a population; Kellner et al. 1992). Explicit tests of this assumption are rarely conducted, yet counterexamples exist. In one case, a habitat suitability model for Florida scrub-jays (*Aphelocoma coerulescens*) correlated well with demographic performance (reproduction, survival, and density), but most of the area was found to be a population sink where mortality exceeded reproduction (Breininger et al. 1998). Apparently, as birds competed to occupy the best habitats, they were less alert to predators, and thus suffered high mortality. In another example, Kirsch (1996) found that interior least terns (*Sterna antillarum,* least terns nesting in noncoastal areas) selected high-quality nesting habitats, but possibly because of disturbance, their productivity was not sufficient to maintain population size (i.e., the nesting habitats were unsuitable). Lomolino and Channell (1995, 1998) observed that remnant populations of endangered mammals often occur near the periphery of their former ranges; because the periphery of the range represents the edge of suitable habitat, studies of habitat suitability of endangered mammals based on habitat use in existing populations are likely to be misleading. These are not gratifying results.

It has not been for lack of effort, expense, or analytical developments that relationships between habitats and population growth often elude detection. It is simply the complexity of the interactions between animals and their environment that make such relationships deceptively difficult to understand. A case in point is the spotted owl, which has undergone intense scrutiny because of its threatened status and apparent proclivity for old-growth forest in a region where the economy is tied largely to timber; despite a plethora of habitat studies, significant debate persists among ecologists as to the critical habitat requirements of this species, and why it prefers old-growth forest (Forsman et al. 1984; Carey et al. 1992; Rosenberg et al. 1994; Carey 1995).

I am not suggesting that ecologists have not been creative in their efforts. However, substantive flaws in the most commonly used techniques for studying wildlife–habitat relationships apparently have not been widely recognized. I believe it is time to reconsider the ways these techniques are used in evaluating habitat quality.

Studies of the use of habitat have merit, but also many limitations. Habitats in which an animal spends a large proportion of its time are clearly selected among others available. Even if widely available, frequently used habitats are certainly not "selected against" or "avoided" (except maybe among life forms that have no memory of where they were). However, frequent use suggests nothing about habitat importance or substitutability in terms of fitness. Correspondingly, infrequent use may not be indicative of lack of suitability. A habitat may be used infrequently because it serves little value, because its value can be extracted in a short amount of time, because it is not readily available, or because access is constrained by threats (social pressures, competition, predation) or physical barriers. A high use:availability ratio might suggest (correctly) that such a habitat is more important than indicated by its infrequent use.

Studies of habitat use would benefit greatly from replication. Significant insights might be gained from comparisons of habitat use and use:availability among individuals, among groups of individuals in different portions of a study area, among study areas, among time periods, and so on. If individuals are disparate in their use or apparent selection of habitats (Holbrook et al. 1987; Ehlinger 1990; Donázar et al. 1993; Boitani et al. 1994; Macdonald and Courtenay 1996), inferences regarding habitat quality become more equivocal. In contrast, if the data are partitioned and the subsets show consistent patterns (e.g., use:availability ratios for each habitat are similar among individuals despite large differences in availability within individual home ranges) or if curvilinear relationships between use and availability can be ascertained (figure 4.4), inferences are strengthened; even so, studies of this sort provide only a superficial understanding of the effects of habitat on population dynamics. Certainly no strong prescriptions for habitat manipulation are warranted from interpretations of selection based solely on observed patterns of habitat use.

Site attribute studies tend to provide stronger inferences about habitat selection. However, in identifying myriad habitat characteristics that are apparently preferred, even for activities that impinge on the animal's survival and reproduction, these studies still cannot assume that population growth would be significantly higher in more "ideal" settings. Controlled experiments can help sort out the factors important in the animal's selection processes (Danell et al. 1991; Parrish 1995) and can thus provide a better understanding of its behavior, but without corresponding demographic measurements, the importance of various habitat components in terms of their contribution to the animal's fitness cannot be appraised. I agree with Kirsch (1996:37–38): "Unfortunately, proximate habitat features may not indicate habitat suitability, nor do they reveal the possible selective pressures that influence habitat selection in a system. One must measure components of fitness, determine factors that influence fitness, and relate fitness and factors influencing fitness to habitats or habitat features."

Demographic response studies are the only means of truly evaluating the relative importance and suitability of habitats for supporting animal popula-

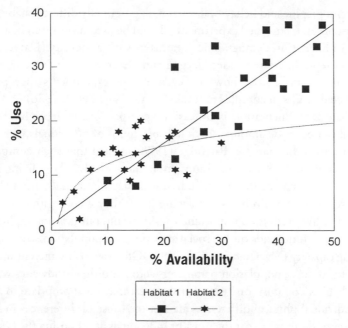

**Figure 4.4**    The assumed linear relationship between use and availability of habitats can be tested, to an extent, by plotting use and area of each habitat in individual home ranges. The figure depicts such a plot for 20 hypothetical home ranges (each point represents habitat use and availability in one home range) and two (of several) types of habitat (sums for both use and availability for all habitat types for each individual would total 100%). The data suggest that increased availability of habitat 1 prompted a corresponding increase in use. In habitat 2, however, the animals showed an asymptotic relationship between use and availability (as in figure 4.2a); 10% availability seemed to represent an approximate threshold, above which use no longer increased. That is, if at least 10% of their home range was composed of habitat 2, the animals could obtain whatever resources they needed from this habitat by spending 10–20% of their time there. A similar threshold may exist for habitat 1 at availabilities >50%; assumptions should not be made beyond the data.

tions. These should be given greater emphasis. Oddly, Hansson (1996) suggested that habitat-specific survival might confuse perceptions of habitat selection. In his small mammal study, perceived habitat selection appeared to be a consequence rather than a cause of differential survival. It seems to me that such knowledge of habitat-specific survival is exactly the desired objective; habitat selection studies are just an indirect approach toward this end.

For many species, habitat-specific densities may be easier to measure than habitat-specific reproduction or survival, but density studies may yield uncertain or misleading results because density is the end result of various processes, both demographic and behavioral, each with potentially different habitat-specific responses. Controlled experiments should be used more often in assessing effects of habitat on demographic parameters such as density (Darveau et al. 1995) or reproduction (Vander Haegen and DeGraaf 1996; Siikamäki 1995; Holt and Martin 1997); however, even elegant experiments may produce unexpected results in complex natural systems (Wiens et al. 1986).

Where experimentation is unfeasible, comparisons of reproduction and survival can be made among individuals or groups of individuals with different patterns of habitat use. Partridge (1978) warned that such comparisons may be confounded by competition among individuals: Those living in preferred habitats may be dominant individuals that are naturally fitter. That is, they live where they do because they are fitter; they are not fitter because of where they live. This is an important consideration, but it is not applicable to all situations; it depends on the social structure and population size relative to carrying capacity (MacCracken et al. 1997). One way to circumvent the problem is to compare population parameters across multiple study sites with differing habitat compositions. Often, though, it may be appropriate to assume that individual physical differences are not the cause of differences in habitat use, or even if they are, that the most fit individuals are choosing the best habitats, thereby justifying comparisons within a single study site.

For studies in a single study site, an effective design would be to monitor habitat use of various individuals (e.g., using radiotelemetry), and then compare their frequency of use of different habitats or the habitat composition of their home ranges to their eventual reproduction and survival. Many studies of birds have used this sort of approach to examine relationships between habitat characteristics of nesting sites and nesting success. However, few attempts have been made to relate reproduction or survival to variation in habitat use within home ranges. In one example, Aanes and Andersen (1996) observed a relationship between survival of roe deer (*Capreolus capreolus*) fawns and the habitat types that they used. Similarly, Aebischer et al. (1993a) found that habitat use by individual pheasants was related to their survival. In another example, Fuller (1989) observed that wolf (*Canis lupus*) territories were limited to areas with low road density (a habitat component that relates to human access) and then looked for (but did not find) a relationship between survival of radio-collared wolves and road density within individual territories. In another study of the same species, Massolo and Meriggi (1998) found relationships between reproduction (documented presence of pups) and various natural and anthropic habitat features, including road density. In studies in which data on survival and reproduction are not obtainable or too scant for making comparisons with habitat use, surrogates (e.g., body size, home range size) may, in some cases, be appropriate and could add measurably to assessments of habitat quality.

Surprisingly, even intensive, long-term studies of radiocollared animals with detailed data on both habitat use and demographics have typically not attempted to relate the two. This should be a goal for the future.

### Acknowledgments

I thank K. V. Noyce and T. K. Fuller for many thoughtful discussions about this subject, and participants of the workshop for their input following my presentation. K. V. Noyce, P. L. Coy, M. S. Boyce, and W. F. Porter provided helpful suggestions on the manuscript. The Minnesota Department of Natural Resources supported my study of black bears, which stimulated my thinking about habitat evaluation and prompted me to use a study design like that described at the end of this chapter.

### Literature Cited

Aanes, R. and R. Andersen. 1996. The effects of sex, time of birth, and habitat on the vulnerability of roe deer fawns to red fox predation. *Canadian Journal of Zoology* 74: 1857–1865.

Adler, G. H. and M. L. Wilson. 1989. Demography of the meadow vole along a simple habitat gradient. *Canadian Journal of Zoology* 67: 772–774.

Aebischer, N. J., V. Marcström, R. E. Kenward, and M. Karlbom. 1993a. Survival and habitat utilisation: A case for compositional analysis. In J. D. Lebreton and P. M. North, eds., *Marked individuals in the study of bird populations*, 343–353. Basel, Switzerland: Birkhauser Verlag.

Aebischer, N. J., P. A. Robertson, and R. E. Kenward. 1993b. Compositional analysis of habitat use from animal radio-tracking data. *Ecology* 74: 1313–1325.

Alldredge, J. R. and J. T. Ratti. 1986. Comparison of some statistical techniques for analysis of resource selection. *Journal of Wildlife Management* 50: 157–165.

Alldredge, J. R. and J. T. Ratti. 1992. Further comparison of some statistical techniques for analysis of resources selection. *Journal of Wildlife Management* 56: 1–9.

Andersen, R., O. Hjeljord, and B.-E. Sæther. 1992. Moose defecation rates in relation to habitat quality. *Alces* 28: 95–100.

Armleder, H. M., M. J. Waterhouse, D. G. Keisker, and R. J. Dawson. 1994. Winter habitat use by mule deer in the central interior of British Columbia. *Canadian Journal of Zoology* 72: 1721–1725.

Arthur, S. M., B. F. J. Manly, L. L. McDonald, and G. W. Garner. 1996. Assessing habitat selection when availability changes. *Ecology* 77: 215–227.

Arthur, S. M. and C. C. Schwartz. 1999. Effects of sample size on accuracy and precision of brown bear home range models. *Ursus* 11. In press.

Bart, J. 1995. Amount of suitable habitat and viability of northern spotted owls. *Conservation Biology* 9: 943–946.

Bart, J., D. R. Petit, and G. Linscombe. 1984. Field evaluation of two models developed following the habitat evaluation procedures. *Transactions of the North American Wildlife and Natural Resources Conference* 49: 489–499.

Beier, P. and R. H. Barrett. 1987. Beaver habitat use and impact in Truckee River Basin, California. *Journal of Wildlife Management* 51: 794–799.

Belk, M. C., H. D. Smith, and J. Lawson. 1988. Use and partitioning of montane habitat by small mammals. *Journal of Mammalogy* 69: 688–695.

Bender, L. C., G. J. Roloff, and J. B. Haufler. 1996. Evaluating confidence intervals for habitat suitability models. *Wildlife Society Bulletin* 24: 347–352.

Beyer, D. E., Jr., R. Costa, R. G. Hooper, and C. A. Hess. 1996. Habitat quality and reproduction of red-cockaded woodpecker groups in Florida. *Journal of Wildlife Management* 60: 826–835.

Beyer, D. E., Jr. and J. B. Haufler. 1994. Diurnal versus 24-hour sampling of habitat use. *Journal of Wildlife Management* 58: 178–180.

Birney, E. C., W. E. Grant, and D. D. Baird. 1976. Importance of vegetative cover to cycles of *Microtus* populations. *Ecology* 57: 1043–1051.

Boitani, L., L. Mattei, D. Nonis, and F. Corsi. 1994. Spatial and activity patterns of wild boars in Tuscany, Italy. *Journal of Mammalogy* 75: 600–612.

Bowers, M. A. 1995. Use of space and habitats by the eastern chipmunk, *Tamias striatus*. *Journal of Mammalogy* 76: 12–21.

Bowyer, R. T., J. G. Kie, and V. Van Ballenberghe. 1998. Habitat selection by neonatal black-tailed deer: Climate, forage, or risk of predation. *Journal of Mammalogy* 79: 415–425.

Breininger, D. R., V. L. Larson, B. W. Duncan, and R. B. Smith. 1998. Linking habitat suitability to demographic success in Florida scrub-jays. *Wildlife Society Bulletin* 26: 118–128.

Brennan, L. A., W. M. Block, and R. J. Gutiérrez. 1986. The use of multivariate statistics for developing habitat suitability index models. In J. Verner, M. L. Morrison, and C. J. Ralph, eds., *Modeling habitat relationships of terrestrial vertebrates,* 177–182. Madison: University of Wisconsin Press.

Brooks, R. P. 1997. Improving habitat suitability index models. *Wildlife Society Bulletin* 25:163–167.

Brown, A. L. and J. A. Litvaitis. 1995. Habitat features associated with predation of New England cottontails: What scale is appropriate? *Canadian Journal of Zoology* 73: 1005–1011.

Brown, B. W. and G. O. Batzli. 1984. Habitat selection by fox and gray squirrels: A multivariate analysis. *Journal of Wildlife Management* 48: 616–621.

Bruggink, J. G., T. C. Tacha, J. C. Davies, and K. F. Abraham. 1994. Nesting and brood-rearing ecology of Mississippi Valley population Canada geese. *Wildlife Monographs* 126.

Burt, W. H. 1943. Territoriality and home range concepts as applied to mammals. *Journal of Mammalogy* 24: 346–352.

Byers, C. R., R. K. Steinhorst, and P. R. Krausman. 1984. Clarification of a technique for the analysis of utilization-availability data. *Journal of Wildlife Management* 48: 1050–1053.

Capen, D. E., J. W. Fenwick, D. B. Inkley, and A. C. Boynton. 1986. Multivariate models of songbird habitat in New England forests. In J. Verner, M. L. Morrison, and C. J. Ralph, eds., *Modeling habitat relationships of terrestrial vertebrates,* 171–175. Madison: University of Wisconsin Press.

Carey, A. B. 1995. Sciurids in Pacific Northwest managed and old-growth forests. *Ecological Applications* 5: 648–661.

Carey, A. B., S. P. Horton, and B. L. Biswell. 1992. Northern spotted owls: influence of prey base and landscape character. *Ecological Monographs* 62: 223–250.

Carey, A. B., J. A. Reid, and S. P. Horton. 1990. Spotted owl home range and habitat use in southern Oregon coast ranges. *Journal of Wildlife Management* 54: 11–17.

Carroll, J. P., R. D. Crawford, and J. W. Schulz. 1995. Gray partridge winter home range and use of habitat in North Dakota. *Journal of Wildlife Management* 59: 98–103.

Cavallini, P. and S. Lovari. 1991. Environmental factors influencing the use of habitat in the red fox, *Vulpes vulpes. Journal of Zoology* 223: 323–339.

Chapin, T. G., D. J. Harrison, and D. D. Katnik. 1998. Influence of landscape pattern on habitat use by American marten in an industrial forest. *Conservation Biology* 12: 1327–1337.

Chasko, G. G. and J. E. Gates. 1982. Avian habitat suitability along a transmission-line corridor in an oak–hickory forest region. *Wildlife Monographs* 82.

Cherry, S. 1996. A comparison of confidence interval methods for habitat use–availability studies. *Journal of Wildlife Management* 60: 653–658.

Chesson, J. 1978. Measuring preference in selective predation. *Ecology* 59: 211–215.

Chesson, J. 1983. The estimation and analysis of preference and its relationship to foraging models. *Ecology* 64: 1297–1304.

Clark, J. D., J. E. Dunn, and K. G. Smith. 1993. A multivariate model of female black bear habitat use for a geographic information system. *Journal of Wildlife Management* 57: 519–526.

Clark, W. R. 1994. Habitat selection by muskrats in experimental marshes undergoing succession. *Canadian Journal of Zoology* 72: 675–680.

Clark, W. R. and D. W. Kroeker. 1993. Population dynamics of muskrats in experimental marshes at Delta, Manitoba. *Canadian Journal of Zoology* 71: 1620–1628.

Coker, D. R. and D. E. Capen. 1995. Landscape-level habitat use by brown-headed cowbirds in Vermont. *Journal of Wildlife Management* 59: 631–637.

Collins, W. B. and P. J. Urness. 1981. Habitat preferences of mule deer as rated by pellet-group distributions. *Journal of Wildlife Management* 45: 969–972.

Collins, W. B. and P. J. Urness. 1984. The pellet-group census technique as an indicator of relative habitat use: response to Leopold et al. *Wildlife Society Bulletin* 12: 327.

Cook, J. G. and L. L. Irwin. 1985. Validation and modification of a habitat suitability model for pronghorns. *Wildlife Society Bulletin* 13: 440–448.

Cowlishaw. 1997. Trade-offs between foraging and predation risk determine habitat use in a desert baboon population. *Animal Behaviour* 53: 667–686.

Danell, K., L. Edenius, and P. Lundberg. 1991. Herbivory and tree stand composition: Moose patch use in winter. *Ecology* 72: 1350–1357.

Darveau, M., P. Beauchesne, L. Bélanger, J. Huot, and P. Larue. 1995. Riparian forest strips as habitat for breeding birds in boreal forest. *Journal of Wildlife Management* 59: 67–78.

Diehl, B. 1986. Factors confounding predictions of bird abundance from habitat data. In J. Verner, M. L. Morrison, and C. J. Ralph, eds., *Modeling habitat relationships of terrestrial vertebrates,* 229–233. Madison: University of Wisconsin Press.

Donázar, J. A., J. J. Negro, and F. Hiraldo. 1993. Foraging habitat selection, land-use changes and population decline in the lesser kestrel *Falco naumanni. Journal of Applied Ecology* 30: 515–522.

Dooley, J. L. Jr. and M. A. Bowers. 1998. Demographic responses to habitat fragmentation: Experimental tests at the landscape and patch scale. *Ecology* 79: 969–980.

Douglass, R. J. 1976. Spatial interactions and microhabitat selections of two locally sympatric voles, *Microtus montanus* and *Microtus pennsylvanicus. Ecology* 57: 346–352.

Douglass, R. J. 1989. The use of radio-telemetry to evaluate microhabitat selection by deer mice. *Journal of Mammalogy* 70: 648–652.

Dunn, P. O. and C. E. Braun. 1986. Summer habitat use by adult female and juvenile sage grouse. *Journal of Wildlife Management* 50: 228–235.

Edge, W. D., C. L. Marcum, and S. L. Olson-Edge. 1987. Summer habitat selection by elk in western Montana: A multivariate approach. *Journal of Wildlife Management* 51: 844–851.

Edge, W. D., C. L. Marcum, and S. L. Olson-Edge. 1988. Summer forage and feeding site selection by elk. *Journal of Wildlife Management* 52: 573–577.

Ehlinger, T. J. 1990. Habitat choice and phenotype-limited feeding efficiency in bluegill: Individual differences and trophic polymorphism. *Ecology* 71: 886–896.

Elston, D. A., A. W. Illius, and I. J. Gordon. 1996. Assessment of preference among a range of options using log ratio analysis. *Ecology* 77: 2538–2548.

Ferguson, S. H., A. T. Bergerud, and R. Ferguson. 1988. Predation risk and habitat selection in the persistence of a remnant caribou population. *Oecologia* 76: 236–245.

Fernandez, F. A. S., N. Dunstone, and P. E. Evans. 1999. Density-dependence in habitat utilisation by wood mice in a sitka spruce successional mosaic: The roles of immigration, emigration, and variation among local demographic parameters. *Canadian Journal of Zoology* 77: 397–405.

Flores, R. E. and W. R. Eddleman. 1995. California black rail use of habitat in southwestern Arizona. *Journal of Wildlife Management* 59: 357–363.

Folk, M. J. and T. C. Tacha. 1990. Sandhill crane roost site characteristics in the North Platte River Valley. *Journal of Wildlife Management* 54: 480–486.

Forsman, E. D., E. C. Meslow, and H. M. Wight. 1984. Distribution and biology of the spotted owl in Oregon. *Wildlife Monographs* 87.

Fretwell, S. D. and H. L. Lucas, Jr. 1970. On territorial behavior and other factors influencing habitat distribution in birds. I. Theoretical development. *Acta Biotheoretica* 19: 16–36.

Fuller, T. K. 1989. Population dynamics of wolves in north-central Minnesota. *Wildlife Monographs* 105.

Gaines, E. P. and M. R. Ryan. 1988. Piping plover habitat use and reproductive success in North Dakota. *Journal of Wildlife Management* 52: 266–273.

Garrett, M. G., J. W. Watson, and R. G. Anthony. 1993. Bald eagle home range and habitat use in the Columbia River estuary. *Journal of Wildlife Management* 57: 19–27.

Geffen, E., R. Hefner, D. W. Macdonald, and M. Ucko. 1992. Habitat selection and home range in the Blanford's fox, *Vulpes cana:* Compatibility with the resource dispersion hypothesis. *Oecologia* 91: 75–81.

Gese, E. M., O. J. Rongstad, and W. R. Mytton. 1988. Home range and habitat use of coyotes in southeastern Colorado. *Journal of Wildlife Management* 52: 640–646.

Gilbert, D. W., D. R. Anderson, J. K. Ringelman, and M. R. Szymczak. 1996. Response of nesting ducks to habitat management on the Monte Vista National Wildlife Refuge, Colorado. *Wildlife Monographs* 131.

Gilbert, F. F. and M. C. Bateman. 1983. Some effects of winter shelter conditions on white-tailed deer, *Odocoileus virginianus,* fawns. *Canadian Field-Naturalist* 97: 391–400.

Gionfriddo, J. P. and P. R. Krausman. 1986. Summer habitat use by mountain sheep. *Journal of Wildlife Management* 50: 331–336.

Greenwood, R. J., A. B. Sargeant, D. H. Johnson, L. M. Cowardin, and T. L. Shaffer. 1995. Factors associated with duck nest success in the prairie pothole region of Canada. *Wildlife Monographs* 128.

Hacker, A. L. and B. E. Coblentz. 1993. Habitat selection by mountain beavers recolonizing Oregon coast range clearcuts. *Journal of Wildlife Management* 57: 847–853.

Hall, L. S., P. R. Krausman, and M. L. Morrison. 1997. The habitat concept and a plea for standard terminology. *Wildlife Society Bulletin* 25: 173–182.

Hansson, L. 1996. Habitat selection or habitat-dependent survival: On isodar theory for spatial dynamics of small mammals. *Oikos* 75: 539–542.

Harper, S. J., E. K. Bollinger, and G. W. Barrett. 1993. Effects of habitat patch shape on population dynamics of meadow voles (*Microtus pennsylvanicus*). *Journal of Mammalogy* 74: 1045–1055.

Heisey, D. M. 1985. Analyzing selection experiments with log-linear models. *Ecology* 66: 1744–1748.

Hines, J. E. 1987. Winter habitat relationships of blue grouse on Hardwicke Island, British Columbia. *Journal of Wildlife Management* 51: 426–435.

Hirsch, J. G. and J. B. Haufler. 1993. Evaluation of a forest habitat model for black bear. *Proceedings of the International Union of Game Biologists* 21(1): 330–337.

Hobbs, N. T. and T. A. Hanley. 1990. Habitat evaluation: Do use/availability data reflect carrying capacity? *Journal of Wildlife Management* 54: 515–522.

Holbrook, S. J. and R. J. Schmitt. 1988. The combined effects of predation risk and food reward on patch selection. *Ecology* 69:125–134.

Holbrook, H. T., M. R. Vaughan, and P. T. Bromley. 1987. Wild turkey habitat preferences and recruitment in intensively managed piedmont forests. *Journal of Wildlife Management* 51: 182–187.

Holt, R. F. and K. Martin. 1997. Landscape modification and patch selection: The demography of two secondary cavity nesters colonizing clearcuts. *Auk* 114: 443–455.

Huegel, C. N., R. B. Dahlgren, and H. L. Gladfelter. 1986. Bedsite selection by white-tailed deer fawns in Iowa. *Journal of Wildlife Management* 50: 474–480.

Hughes, J. J., D. Ward, and M. R. Perrin. 1994. Predation risk and competition affect habitat selection and activity of Namib Desert gerbils. *Ecology* 75: 1397–1405.

Hurlbert, S. H. 1984. Pseudoreplication and the design of ecological field experiments. *Ecological Monographs* 54: 187–211.

Ivlev, V. S. 1961. *Experimental ecology of the feeding of fishes*. New Haven, Conn.: Yale University Press.

James, F. C. and C. E. McCulloch. 1990. Multivariate analysis in ecology and systematics: Panacea or Pandora's box? *Annual Review of Ecology and Systematics* 21: 129–166.

Johnson, D. H. 1980. The comparison of usage and availability measurements for evaluating resource preference. *Ecology* 61: 65–71.

Kaminski, R. M. and E. A. Gluesing. 1987. Density- and habitat-related recruitment in mallards. *Journal of Wildlife Management* 51: 141–148.

Kellner, C. J., J. D. Brawn, and J. R. Karr. 1992. What is habitat suitability? And how should it be measured? In D. R. McCullough and R. H. Barrett, eds., *Wildlife 2001: Populations,* 476–488. London: Elsevier.

Kennedy, M. and R. D. Gray. 1993. Can ecological theory predict the distribution of foraging animals? A critical analysis of experiments on the Ideal Free Distribution. *Oikos* 68: 158–166.

Kilbride, K. M., J. A. Crawford, K. L. Blakely, and B. A. Williams. 1992. Habitat use by breeding female California quail in western Oregon. *Journal of Wildlife Management* 56: 85–90.

Kincaid, W. B. and E. H. Bryant. 1983. A geometric method for evaluating the null hypothesis of random habitat utilization. *Ecology* 64: 1463–1470.

Kincaid, W. B., G. N. Cameron, and E. H. Bryant. 1983. Patterns of habitat utilization in sympatric rodents on the south Texas coast prairie. *Ecology* 64: 1471–1480.

Kirsch, E. M. 1996. Habitat selection and productivity of least terns on the lower Platte River, Nebraska. *Wildlife Monographs* 132.

Klinger, R. C., and M. J. Kutilek, and H. S. Shellhammer. 1989. Population responses of black-tailed deer to prescribed burning. *Journal of Wildlife Management* 53: 863–871.

Knight, T. W. and D. W. Morris. 1996. How many habitats do landscapes contain? *Ecology* 77: 1756–1764.

Kotler, B. P., M. S. Gaines, and B. J. Danielson. 1988. The effects of vegetative cover on the community structure of prairie rodents. *Acta Theriologica* 33: 379–391.

Krausman, P. R. and B. D. Leopold. 1986. Habitat components for desert bighorn sheep in the Harquahala Mountains, Arizona. *Journal of Wildlife Management* 50: 504–508.

Kufeld, R. C., D. C. Bowden, and J. M. Siperek, Jr. 1987. Evaluation of a telemetry system for measuring habitat usage in mountainous terrain. *Northwest Science* 61: 249–256.

Lacki, M. J., M. D. Adam, and L. G. Shoemaker. 1993. Characteristics of feeding roosts of Virginia big-eared bats in Daniel Boone National Forest. *Journal of Wildlife Management* 57: 539–543.

Lancia, R. A., S. D. Miller, D. A. Adams, and D. W. Hazel. 1982. Validating habitat quality assessment: an example. *Transactions of the North American Wildlife and Natural Resources Conference* 47: 96–110.

Laymon, S. A. and R. H. Barrett. 1986. Developing and testing habitat-capability models: Pitfalls and recommendations. In J. Verner, M. L. Morrison, and C. J. Ralph, eds., *Modeling habitat relationships of terrestrial vertebrates,* 87–91. Madison: University of Wisconsin Press.

Lechowicz, M. 1982. The sampling characteristics of electivity indices. *Oecologia* 52: 22–30.

Lehmkuhl, J. F. and M. G. Raphael. 1993. Habitat pattern around northern spotted owl locations on the Olympic Peninsula, Washington. *Journal of Wildlife Management* 57: 302–315.

Leopold, A. 1933. *Game management.* New York: Scribner.

Levin, S. A. 1992. The problem of pattern and scale in ecology. *Ecology* 73: 1943–1967.

Lima, S. L. and L. M. Dill. 1990. Behavioral decisions made under the risk of predation: A review and prospectus. *Canadian Journal of Zoology* 68: 619–640.

Litvaitis, J. A., J. A. Sherburne, and J. A. Bissonette. 1986. Bobcat habitat use and home range size in relation to prey density. *Journal of Wildlife Management* 50: 110–117.

Loeb, S. C. 1993. Use and selection of red-cockaded woodpecker cavities by southern flying squirrels. *Journal of Wildlife Management* 57: 329–335.

Loegering, J. P. and J. D. Fraser. 1995. Factors affecting piping plover chick survival in different brood-rearing habitats. *Journal of Wildlife Management* 59: 646–655.

Lomolino, M. V. and R. Channell. 1995. Splendid isolation: Patterns of geographic range collapse in endangered mammals. *Journal of Mammalogy* 76: 335–347.

Lomolino, M. V. and R. Channell. 1998. Range collapse, re-introductions, and biogeographic guidelines for conservation. *Conservation Biology* 12: 481–484.

MacCracken, J. G., V. Van Ballenberghe, and J. M. Peek. 1997. Habitat relationships of moose on the Copper River Delta in coastal south-central Alaska. *Wildlife Monographs* 136.

Macdonald, D. W. and O. Courtenay. 1996. Enduring social relationships in a population of crab-eating zorros, *Cerdocyon thous,* in Amazonian Brazil. *Journal of Zoology* 239: 329–355.

Mallory, M. L., D. K. McNicol, and P. J. Weatherhead. 1994. Habitat quality and reproductive effort of common goldeneyes nesting near Sudbury, Canada. *Journal of Wildlife Management* 58: 552–560.

Manly, B. F. J., L. L. McDonald, and D. L. Thomas. 1993. *Resource selection by animals. Statistical design and analysis for field studies.* London: Chapman & Hall.

Manly, B. F. J., P. Miller, and L. M. Cook. 1972. Analysis of a selective predation experiment. *American Naturalist* 106: 719–736.

Marcum, C. L. and D. O. Loftsgaarden. 1980. A nonmapping technique for studying habitat preferences. *Journal of Wildlife Management* 44: 963–968.

Massold, A. and A. Meriggi. 1998. Factors affecting habitat occupancy by wolves in Northern Apennines (Northern Italy): A model of habitat suitability. *Ecography* 21: 97–107.

Maurer, B. A. 1986. Predicting habitat quality for grassland birds using density-habitat correlations. *Journal of Wildlife Management* 50: 556–566.

Maxson, S. J. and M. R. Riggs. 1996. Habitat use and nest success of overwater nesting ducks in westcentral Minnesota. *Journal of Wildlife Management* 60: 108–119.

McClean, S. A., M. A. Rumble, R. M. King, and W. L. Baker. 1998. Evaluation of resource selection methods with different definitions of availability. *Journal of Wildlife Management* 62: 793–801.

McEwan, L. C. and D. H. Hirth. 1979. Southern bald eagle productivity and nest site selection. *Journal of Wildlife Management* 43: 585–594.

McLellan, B. N. 1986. Use–availability analysis and timber selection by grizzly bears. In G. P. Contreras and K. E. Evans, eds., *Proceedings—Grizzly bear habitat symposium,* 163–166. U.S. Forest Service General Technical Report INT-207.

Messier, F., J. A. Virgl, and L. Marinelli. 1990. Density-dependent habitat selection in muskrats: A test of the ideal free distribution model. *Oecologia* 84: 380–385.

Meyer, J. S., L. L. Irwin, and M. S. Boyce. 1998. Influence of habitat abundance and fragmentation on northern spotted owls in western Oregon. *Wildlife Monographs* 139.

Millspaugh, J. J., J. R. Skalski, B. J. Kernohan, K. J. Raedeke, G. C. Brundige, and A. B. Cooper. 1998. Some comments on spatial independence in studies of resource selection. *Wildlife Society Bulletin* 26: 232–236.

Moen, R., J. Pastor, Y. Cohen, and C. C. Schwartz. 1996. Effects of moose movement and habitat use on GPS collar performance. *Journal of Wildlife Management* 60: 659–668.

Morris, D. W. 1984. Patterns and scale of habitat use in two temperate-zone, small mammal faunas. *Canadian Journal of Zoology* 62: 1540–1547.

Morris, D. W. 1987. Ecological scale and habitat use. *Ecology* 68: 362–369.

Morris, D. W. 1992. Scales and costs of habitat selection in heterogeneous landscapes. *Evolutionary Ecology* 6: 412–432.

Morrison, M. L., B. G. Marcot, and R. W. Mannan. 1998 (2d ed.). *Wildlife–habitat relationships: Concepts and applications.* Madison: University of Wisconsin Press.

Muller, K. L., J. A. Stamps, V. V. Krishnan, and N. H. Willits. 1997. The effects of conspecific attraction and habitat quality on habitat selection in territorial birds (*Troglodytes aedon*). *American Naturalist* 150: 650–661.

Mysterud, A. and R. A. Ims. 1998. Functional responses in habitat use: Availability influences relative use in trade-off situations. *Ecology* 79: 1435–1441.

Mysterud, A., P. K. Larsen, R. A. Ims, and E. Østbye. 1999. Habitat selection by roe deer and sheep: Does habitat ranking reflect resource availability? *Canadian Journal of Zoology* 77: 776–783.

Mysterud, A. and E. Østbye. 1995. Bed-site selection by European roe deer (*Capreolus capreolus*) in southern Norway during winter. *Canadian Journal of Zoology* 73: 924–932.

Nadeau, S., R. Décarie, D. Lambert, and M. St-Georges. 1995. Nonlinear modeling of muskrat use of habitat. *Journal of Wildlife Management* 59: 110–117.

Nams, V. O. 1989. Effects of radiotelemetry error on sample size and bias when testing for habitat selection. *Canadian Journal of Zoology* 67: 1631–1636.

Neu, C. W., C. R. Byers, and J. M. Peek. 1974. A technique for analysis of utilization–availability data. *Journal of Wildlife Management* 38: 541–545.

Nixon, C. M., L. P. Hansen, and P. A. Brewer. 1988. Characteristics of winter habitats used by deer in Illinois. *Journal of Wildlife Management* 52: 552–555.

North, M. P. and J. H. Reynolds. 1996. Microhabitat analysis using radiotelemetry locations and polytomous logistic regression. *Journal of Wildlife Management* 60: 639–653.

Ockenfels, R. A. and D. E. Brooks. 1994. Summer diurnal bed sites of Coues white-tailed deer. *Journal of Wildlife Management* 58: 70–75.

O'Neil, L. J. and A. B. Carey. 1986. Introduction: When habitats fail as predictors. In J. Verner, M. L. Morrison, and C. J. Ralph, eds., *Modeling habitat relationships of terrestrial vertebrates*, 207–208. Madison: University of Wisconsin Press.

O'Neil, L. J., T. H. Roberts, J. S. Wakeley, and J. W. Teaford. 1988. A procedure to modify habitat suitability index models. *Wildlife Society Bulletin* 16: 33–36.

Oreskes, N., K. Shrader-Frechette, and K. Belitz. 1994. Verification, validation, and confirmation of numerical models in the earth sciences. *Science* 263: 641–646.

Orians, G. H. and J. F. Wittenberger. 1991. Spatial and temporal scales in habitat selection. *American Naturalist* (suppl.) 137: S29–S49.

Ostfeld, R. S. and L. L. Klosterman. 1986. Demographic substructure in a California vole population inhabiting a patchy environment. *Journal of Mammalogy* 67: 693–704.

Ostfeld, R. S., W. Z. Lidicker, Jr., and E. J. Heske. 1985. The relationship between habitat heterogeneity, space use, and demography in a population of California voles. *Oikos* 45: 433–442.

Otis, D. L. 1997. Analysis of habitat selection studies with multiple patches within cover types. *Journal of Wildlife Management* 61: 1016–1022.

Ovadia, O. and Z. Abramsky. 1995. Density-dependent habitat selection: Evaluation of the isodar method. *Oikos* 73: 86–94.

Palomares, F. and M. Delibes. 1992. Data analysis design and potential bias in radio-tracking studies of animal habitat use. *Acta Oecologia* 13: 221–226.

Paragi, T. F., W. N. Johnson, D. D. Katnik, and A. J. Magoun. 1996. Marten selection of postfire seres in the Alaskan taiga. *Canadian Journal of Zoology* 74: 2226–2237.

Parrish, J. D. 1995. Effects of needle architecture on warbler habitat selection in a coastal spruce forest. *Ecology* 76: 1813–1820.

Partridge, L. 1978. Habitat selection. In J. R. Krebs and N. B. Davies, eds., *Behavioural ecology. An evolutionary approach*, 351–376. Sunderland, Mass.: Sinauer.

Pedlar, J. H., L. Fahrig, and H. G. Merriam. 1997. Raccoon habitat use at 2 spatial scales. *Journal of Wildlife Management* 61: 102–112.

Peek, J. M. 1986. *A review of wildlife management.* Englewood Cliffs, N.J.: Prentice Hall.

Peles, J. D. and G. W. Barrett. 1996. Effects of vegetative cover on the population dynamics of meadow voles. *Journal of Mammalogy* 77: 857–869.

Pelton, M. R. and F. T. van Manen, 1996. Benefits and pitfalls of long-term research: A case study of black bears in Great Smoky Mountains National Park. *Wildlife Society Bulletin* 24: 443–450.

Pierotti, R. 1982. Habitat selection and its effect on reproductive output in the herring gull in Newfoundland. *Ecology* 63: 854–868.

Porter, W. F. and K. E. Church. 1987. Effects of environmental pattern on habitat preference analysis. *Journal of Wildlife Management* 51: 681–685.

Powell, R. A. 1994. Effects of scale on habitat selection and foraging behavior of fishers in winter. *Journal of Mammalogy* 75: 349–356.

Powell, R. A., J. W. Zimmerman, and D. E. Seaman. 1997. *Ecology and behaviour of North American black bears: Home ranges, habitat and social organization.* London: Chapman & Hall.

Prayurasiddhi, T. 1997. *The ecological separation of gaur (Bos gaurus) and banteng (Bos javanicus) in Huai Kha Khaeng Wildlife Sanctuary, Thailand.* Ph.D. thesis. Minneapolis: University of Minnesota.

Prescott, D. R. C. and D. M. Collister. 1993. Characteristics of occupied and unoccupied loggerhead shrike territories in southeastern Alberta. *Journal of Wildlife Management* 57: 346–352.

Rempel, R. S., P. C. Elkie, A. R. Rodgers, and M. J. Gluck. 1997. Timber management and natural disturbance effects on moose habitat: Landscape evaluation. *Journal of Wildlife Management* 61: 517–524.

Rexstad, E. A., D. D. Miller, C. H. Flather, E. M. Anderson, J. W. Hupp, and D. R. Anderson. 1988. Questionable multivariate statistical inference in wildlife habitat and community studies. *Journal of Wildlife Management* 52: 794–798.

Rexstad, E. A., D. D. Miller, C. H. Flather, E. M. Anderson, J. W. Hupp, and D. R. Anderson. 1990. Questionable multivariate statistical inference in wildlife habitat and community studies: A reply. *Journal of Wildlife Management* 54: 189–193.

Rhomberg, L. 1984. Inferring habitat selection by aphids from the dispersion of their galls over the tree. *American Naturalist* 124: 751–756.

Robel, R. J., L. B. Fox, and K. E. Kemp. 1993. Relationship between habitat suitability index values and ground counts of beaver colonies in Kansas. *Wildlife Society Bulletin* 21: 415–421.

Roseberry, J. L. and A. Woolf. 1998. Habitat–population density relationships for white-tailed deer in Illinois. *Wildlife Society Bulletin* 26: 252–258.

Rosenberg, D. K., C. J. Zabel, B. R. Noon, and E. C. Meslow. 1994. Northern spotted owls: Influence of prey base—A comment. *Ecology* 75: 1512–1515.

Rosenzweig, M. L. and Z. Abramsky. 1986. Centrifugal community organization. *Oikos* 46: 339–348.

Rumble, M. A. and R. A. Hodoroff. 1993. Nesting ecology of Merriam's turkeys in the black hills, South Dakota. *Journal of Wildlife Management* 57: 789–801.

Salas, L. A. 1996. Habitat use by lowland tapirs (*Tapirus terrestris* L.) in the Tabaro River valley, southern Venezuela. *Canadian Journal of Zoology* 74: 1452–1458.

Schaefer, J. A. and F. Messier. 1995. Habitat selection as a hierarchy: The spatial scales of winter foraging by muskoxen. *Ecography* 18: 333–344.

Schamberger, M. and W. B. Krohn. 1982. Status of the habitat evaluation procedures. *Transactions of the North American Wildlife and Natural Resources Conference* 47: 154–164.

Schooley, R. L. 1994. Annual variation in habitat selection: Patterns concealed by pooled data. *Journal of Wildlife Management* 58: 367–374.

Schooley, R. L. and C. R. McLaughlin. 1992. Observer variability in classifying forested habitat from aircraft. *Northeast Wildlife* 49: 10–16.

Servheen, G. and L. J. Lyon. 1989. Habitat use by woodland caribou in the Selkirk Mountains. *Journal of Wildlife Management* 53: 230–237.

Sherry, T. W. and R. T. Holmes. 1996. Winter habitat quality, population limitation, and conservation of Neotropical–Nearctic migrant birds. *Ecology* 77: 36–48.

Siikamäki, P. 1995. Habitat quality and reproductive traits in the pied flycatcher–an experiment. *Ecology* 76: 308–312.

Smith, A. A. and R. W. Mannan. 1994. Distinguishing characteristics of Mount Graham red squirrel midden sites. *Journal of Wildlife Management* 58: 437–445.

Smith, A. T. and M. M. Peacock. 1990. Conspecific attraction and determination of metapopulation colonization rates. *Conservation Biology* 4: 320–323.

Stauffer, D. F. and S. R. Peterson. 1985. Seasonal micro-habitat relationships of ruffed grouse in southeastern Idaho. *Journal of Wildlife Management* 49: 605–610.

Stinnett, D. P. and D. A. Klebenow. 1986. Habitat use of irrigated lands by California quail in Nevada. *Journal of Wildlife Management* 50: 368–372.

Taitt, M. J., J. H. W. Gipps, C. J. Krebs, and Z. Dundjerski. 1981. The effects of extra food and cover on declining populations of *Microtus townsendii*. *Canadian Journal of Zoology* 59: 1593–1599.

Taylor, J. 1990. Questionable multivariate statistical inference in wildlife habitat and community studies: A comment. *Journal of Wildlife Management* 54: 186–189.

Ternent, M. A. 1995. *Management and ecology of black bears in Camp Ripley Military Reservation, Minnesota.* M.S. thesis. Minneapolis: University of Minnesota.

Tew, T. E., I. A. Todd, and D. W. Macdonald. 2000. Arable habitat use by wood mice (*Apodemus sylvaticus*). 2. Microhabitat. *Journal of Zoology.* In press.

Thomas, D. L., and E. J. Taylor. 1990. Study designs and tests for comparing resource use and availability. *Journal of Wildlife Management* 54: 322–330.

Thomas, J. W. 1982. Needs for and approaches to wildlife habitat assessment. *Transactions of the North American Wildlife and Natural Resources Conference* 47: 35–46.

Thomasma, L. E., T. D. Drummer, and R. O. Peterson. 1991. Testing the habitat suitability index model for the fisher. *Wildlife Society Bulletin* 19: 291–297.

Thompson, F. R. III, D. A. Freiling, and E. K. Fritzell. 1987. Drumming, nesting, and brood habitats of ruffed grouse in an oak-hickory forest. *Journal of Wildlife Management* 51: 568–575.

Todd, I. A., T. E. Tew, and D. W. Macdonald. 2000. Habitat use of the arable ecosystem by wood mice (*Apodemus sylvaticus*). 1. Macrohabitat. *Journal of Zoology*. In press.

Vander Haegen, W. M. and R. DeGraaf. 1996. Predation on artificial nests in forested riparian buffer strips. *Journal of Wildlife Management* 60: 542–550.

Van Horne. 1983. Density as a misleading indicator of habitat quality. *Journal of Wildlife Management* 47: 893–901.

Whitcomb, S. D., F. A. Servello, and A. F. O'Connell, Jr. 1996. Patch occupancy and dispersal of spruce grouse on the edge of its range in Maine. *Canadian Journal of Zoology* 74: 1951–1955.

White, G. C. and R. A. Garrott. 1986. Effects of biotelemetry triangulation error on detecting habitat selection. *Journal of Wildlife Management* 50: 509–513.

Whitham, T. G. 1978. Habitat selection by *Pemphigus* aphids in response to resource limitation and competition. *Ecology* 59: 1164–1176.

Whitham, T. G. 1980. The theory of habitat selection: Examined and extended using *Pemphigus* aphids. *American Naturalist* 115: 449–466.

Wiens, J. A., J. T. Rotenberry, and B. Van Horne. 1986. A lesson in the limitations of field experiments: Shrubsteppe birds and habitat alteration. *Ecology* 67: 365–376.

Winker, K., J. H. Rappole, and M. A. Ramos. 1995. The use of movement data as an assay of habitat quality. *Oecologia* 101: 211–216.

Zimmerman, J. L. 1982. Nesting success of dickcissels (*Spiza americana*) in preferred and less preferred habitats. *Auk* 99: 292–298.

## Chapter 5

# Investigating Food Habits of Terrestrial Vertebrates

JOHN A. LITVAITIS

Why study food habits? Probably one of the most fundamental questions that ecologists attempt to answer is, "What resources does a particular species require to exist?" Indeed, the first principle among wildlife ecologists is to have a thorough understanding of the food, cover, and water requirements of an animal before initiating any effort to alter the factors that may be limiting it. Information on food habits is therefore an important introduction to the natural history of any species. This has been a justification for many studies of food habits of vertebrates (Martin et al. 1961) and is still a valid reason to investigate the diet of any species when little information is available (Salas and Fuller 1996). Food habits have been investigated for a variety of other reasons. Such information is essential in understanding the potential competitive interactions among sympatric species (Jaksic et al. 1992; Wiens 1993) or in determining how the foraging patterns of individuals affect community composition. For example, how does grazing by wildebeests (*Connochaetes taurinus*) affect the diversity of grasses and forbs? Does predation by lions (*Panthera leo*) limit that same wildebeest population? A simple list of foods used by wildebeests or lions will not answer these questions. However, determining the biomass consumed and abundance of alternative forage or prey is an important first step in understanding how these two species influence community composition.

In human-dominated landscapes, information on the food habits of common terrestrial vertebrates has been useful in understanding the "economic food niche" of many species. Losses of livestock, agricultural crops, or game populations are serious economic concerns. Limiting these losses is a major charge of government wildlife management agencies. Historically, efforts to control depredating wildlife have included indiscriminate attempts to reduce populations of the offending species. Well-known examples of such an approach include the efforts to reduce large carnivores in North America. Detailed investigations have revealed that the actual nuisance individuals may be only a portion of the population and depredation may be restricted to a limited time period (Till and Knowlton 1983). As a result, control efforts can be more exact and cause less ecological damage.

A recent motivation to examine animal food is the emergence of environmental assessments during large-scale habitat alterations, such as road construction and commercial timber harvesting. Biologists must identify the important food and cover resources in an affected area if they are to mitigate the effects of these activities.

Regardless of the specific question being addressed, nearly all investigations of food habits can be distilled to two basic questions. What is the importance of a specific food to the fitness of an organism (especially survival and reproductive success)? How does the feeding niche of an organism affect community composition? In this chapter I compare the relative effectiveness of methods commonly used to investigate food habits and how appropriate each method is at answering these questions. I also consider several recent innovations and how potential improvements may improve our ability to understand the significance of food usage. Readers interested in learning more about the actual procedures should consult references by Korschegen (1980), Cooperrider (1986), Reynolds and Aebischer (1991), and Litvaitis et al. (1994).

## ■ Conventional Approaches and Their Limitations

### DIRECT OBSERVATION

Direct observations have been widely applied to document the forage or prey used by a variety of species. Individual animals or groups are observed through binoculars as they graze or feed on an animal carcass. Observations at bird nests also have provided information on foods brought to juveniles (Errington 1932; Marti 1987; Bielefedt et al. 1992). The basic approach is simple and relies on limited equipment. For researchers studying herbivores, bite counts or feeding minutes by plant species are recorded. These values can then be translated into relative occurrence in the diet by comparing total bites or minutes of foraging and the contribution of each species to the total observed. Biomass consumed can be approximated by estimating the average mass per bite for each species incorporated in the diet (Smith and Hubbard 1954). Additionally, direct observations are useful in identifying differences in foraging among sexes or age classes (Illius and Gordon 1987). Unfortunately, this technique is hampered by several limitations (table 5.1). Observations are usually limited to species that occupy open habitat (grassland, savanna, or tundra) and forage during daylight periods. Consistently identifying the grasses and forbs consumed by herbivores may limit the application of this technique to habitats where forage diversity is limited or species are easily differentiated. For wide-ranging animals, the observer must follow subject animals and thus may affect their foraging behavior or prey availability (Mills 1992). Among carnivores, observations are usually limited to identifying large prey that are not eaten whole (Schaller 1972; Mills 1992).

## LEAD ANIMALS

Concern about the suitability of direct observation has prompted some investigators to use tame, hand-reared animals to investigate food use. Subject animals are followed closely by the researcher, allowing accurate identification of the foods consumed and those avoided (Gill et al. 1983). Lead animals provide an opportunity to observe the consumption of items that are consumed whole (e.g., fruits and small forbs) or where there is little evidence of consumption. Although this approach may eliminate some of difficulties associated with direct observations, the diet summaries reported for lead animals have been criticized as being artificial (Wallmo et al. 1973). Physiological condition, hunger, presence of conspecifics, and previous foraging experience may affect food use by tame animals and result in erroneous conclusions. Although this approach has had only limited application among carnivores, it may be useful under unique circumstances (such as a falconer using a trained raptor) to investigate differential vulnerability of prey if captured prey are compared to a random sample of the target population (Temple 1987).

## FEEDING SITE SURVEYS

Feeding site surveys are among the earliest approaches used to investigate food habits and were initially developed to determine the foods of livestock. This method relies on an inventory of plants consumed or identification of prey remains.

### Survey of food remains

Because of the limitations imposed by cover or nocturnal activity, the food habits of many species cannot be observed directly. As a result, some investigators have relied on an examination of feeding sites. This technique obviously depends on the ability of the investigator to locate sites where feeding has

**Table 5.1 Evaluation of Methods Used to Investigate Vertebrate Food Habits**

| Method | Advantages | Disadvantages |
|---|---|---|
| Direct observation | Inexpensive<br>Sample age or sex differences if these can be determined | Limited to diurnal periods and open habits<br>Presence of observer in the field affects activity of consumer and potential prey |
| Lead animals | Provide very precise information<br>Selection can be investigated by sampling available foods | Expensive<br>Limited sample size<br>Results considered artificial unless subject acclimated before data collection |
| Feeding site surveys | | |
| Food remains | Can provide summary of major foods consumed<br>Can estimate biomass consumed | Small or completely consumed foods are not surveyed<br>Cannot examine age or sex differences in food use |
| Before–after comparisons | Can estimate biomass consumed | Must be able to differentiate consumers |
| Exclosures | Useful in evaluating long-term effects of herbivory on plant community dynamics | Provides information only on major plants consumed<br>No information on diets of age or sex categories<br>Must be able to distinguish foraging by sympatric herbivores |
| Postingestion samples | | |
| Pellets or feces | Able to sample large segments of population throughout year<br>Inexpensive | Cannot differentiate samples by age or sex categories<br>Differential digestibility limits evaluation of importance of various foods |
| Gastrointestinal tracts | Able to examine age and sex differences<br>Can examine other parameters (e.g., physical condition and reproductive rates) from other carcass samples | Samples usually limited to legal harvest by hunters or trappers |

occurred. If more than one species is known to forage in the area, than there must be a way to differentiate consumers (e.g., foraging by lagomorphs versus foraging by ungulates). Among herbivores, grazed grasses and forbs or browsed twigs are counted and converted into percentages of the total forage consumed. For woody plants, it is possible to estimate the biomass consumed by measuring the diameter of the twig at the point it was clipped and using regression tables specific for each species (Basile and Hutchings 1966; Telfer 1969). Such surveys may be biased by "invisible consumption" when an entire plant is consumed and there is no residual indicating the occurrence of that plant (McInnis et al. 1983). For carnivores, this approach has been used to identify the age, sex, and physical condition of large prey (Mech 1966; Sinclair and Arcese 1995), but provides no information on prey that are eaten whole or cached. A major disadvantage of this approach is the lack of opportunity to investigate sex or age differences in food use (table 5.1).

*Difference comparisons*

This approach is restricted to herbivores and relies on comparisons between used and unused plots or sites surveyed before and after foragers have passed through an area. Species consumed (based on inventory) or biomass removed can be estimated by clipping used and unused plots (Bobek et al. 1975; Cook and Stubbendieck 1986). An obvious limitation of used and unused sites is the ability of the investigator to locate comparable sites. Also, unless the investigator is able to verify nearly exclusive use of a site by the herbivore under investigation, before and after comparisons may be clouded by a variety of herbivores foraging at the same site (table 5.1). In general, difference methods do not detect small differences in use. Therefore, they should not be considered unless use is expected to be more than 50 percent on the forage species (Cooperrider 1986).

A more accurate but labor-intensive technique for assessing use has been applied in the western United States (Nelson 1930; Smith and Urness 1962). Stems of potential browse species are tagged and total length of annual twig growth is measured during autumn. In spring, the investigator returns and measures the length of annual growth that remains. Although the technique does not directly measure biomass removed, the length of the twig removed is highly correlated with the amount of forage removed (Smith and Urness 1962). A limitation of this approach is that it requires some knowledge of the plants that are likely to be consumed (table 5.1).

## EXCLOSURES

A slight variation on difference comparisons is the comparison of used sites and sites where access has been restricted by an exclosure. Exclosures are usually wire fences or cages that limit access by herbivores but do not hinder plant recruitment or growth. Exclosures can reveal information on general food habits of herbivores when based on short-term differences between paired (fenced and open) plots (Bobek et al. 1975). This technique should not be applied to estimates of use during the growing season because protected plants grow at different rates than grazed plants. Probably the more common use of exclosures is to reveal the effects of herbivory on plant community composition. Here, investigators usually erect a smaller number of large exclosures that are monitored for 1–2 years (Huntly 1987) or longer (Alverson et al. 1988; Brander et al. 1990; Brown and Heske 1990; McInnes et al. 1992). Periodic inventories are conducted to compare species composition and growth rates between exclosures and open sites. Exclosure studies have revealed that herbivores can affect the abundance, biomass, and diversity of plants (review by Huntly 1991). For example, moose (*Alces alces*) in boreal forests affect the recruitment of forage and nonforage species (Brander et al. 1990; McInnes et al. 1992). Long-term exclosure studies also have revealed that this herbivore not only alters the species composition of forest stands, but also can directly affect the structure of herb, shrub, and canopy layers (McInnes et al. 1992). Although exclosures can help identify major forage plants and responses in plant communities, they are not an appropriate method to gain information on infrequently used forage (table 5.1).

## POSTINGESTION SAMPLES

The most common technique for analyzing food habits of terrestrial vertebrates involves sampling either during or after the digestive process. Samples may be collected from various stages of digestion for use in identifying food habits of herbivores and carnivores. All postmastication sampling requires identification of materials that may not be easily recognized.

### Feces or pellets

Examining the content of feces or regurgitated pellets has become widely used because this approach is nondestructive and large samples can be collected. This technique has been applied with equal success to waterfowl (Owen

1975), upland birds (Eastman and Jenkins 1970), and raptors (Craighead and Craighead 1956) as well as small (Maser et al. 1985) and large mammals (Green et al. 1986). To distinguish fecal samples, field guides are available that rely on size, shape, and color to identify the source (Webb 1943; Murie 1974). If a less ambiguous approach is needed, especially where sympatric species have feces or pellets that are similar, fecal pH or bile acid differences can be used to distinguish samples (Howard 1967, Johnson et al. 1984). A newer approach uses information on mitochondrial DNA from epithelial cells shed from the intestines of the animal that defecated or regurgitated the pellet (Höss et al. 1992; Litvaitis and Litvaitis 1996; Paxinos et al. 1997).

Among herbivores, partially digested seeds and fruits can be identified macroscopically. However, most of the analysis of herbivorous materials relies on microhistological techniques to identify characteristic cells and structures of foods consumed. A reference collection of potential food items is crucial. Microscope slides must be prepared from all potential food plants in the same manner as sample materials. In addition, a collection of local seeds and fruits should also be made for reference. Generally, less than half of what appears on a typical slide will be identifiable plant fragments. The cellular characteristics used to identify plant fragments are those that survive the mastication and digestive process and are generally composed of epidermal tissue (Storr 1961). These include cuticle, stomata, cell walls, aperites, glands, trichomes, silica cells, druses, crystals, starch grains, and silica–suberose couples as well as general cellular configurations, size, and other structural characteristics.

Because the ratio of identifiable to nonidentifiable fragments changes during digestion and sample preparation (Havstad and Donart 1978; Holechek 1982) and because certain browse species have a low proportion of epidermal material in relation to their biomass (Westoby et al. 1976), correction factors may be developed to improve the approximation of diet composition (Dearden et al. 1975). Several researchers have recommended that hand-mixed diets be used to test the assumption that the actual diet matches the diet estimated from microhistological analysis (Westoby et al. 1976; Vavra and Holechek 1980; Holechek et al. 1982). Others have suggested that the differences are too small to justify developing correction factors or that correction factors do not consistently improve the estimation of diet composition, particularly when the diets contain a variety of grasses, forbs, and browse (Hansson 1970; Gill et al. 1983).

Investigators of herbivore diets that relied on fecal analysis have been criticized for identifying fewer species than can be found in rumen samples. Generally, easily digested forbs are underestimated and the less digestible items are overestimated (Anthony and Smith 1974; Vavra et al. 1978; Smith and Shandruk 1979; McInnis et al. 1983). Inaccuracy of the technique, particularly when applied to diets of mixed (grasses, forbs, and browse combined) feeders, has led some researchers to question its usefulness for herbivores other than grazers (Gill et al. 1983). Lack of experience and training in identifying plant fragments are often cited as the most important sources of error in using the microhistological technique (Holechek et al. 1982).

In contrast to herbivore samples, fecal material and pellets from carnivores require minimal preparation. Depending on the size of prey and the level of mastication, much of the material in a sample from a carnivore may be identified using field guides to birds, mammals, and insects for comparison. Other items may be identified through bones, teeth, hair, feathers, or scale patterns. Therefore, reference materials may include complete skeletons of vertebrates, samples of hair, feathers, scales of fish and reptiles, and exoskeletons of insects. A collection of dorsal guard hairs of the mammals likely to be encountered is particularly useful because of the characteristic features of color banding, medullary pigment patterns, and the morphology of cuticular scales (Adorjan and Kolenosky 1969).

Recently, several investigators have used postingestion samples to explore selectivity of prey consumption or differential vulnerability. These studies were based on skeletal remains in feces that could be distinguished to different sex or age categories (Dickman et al. 1991; Koivunen et al. 1996; Zalewski 1996). Composition of prey remains was then compared to trapped samples. Although this approach has increased the information obtained using postingested samples, it is important to recognize that for larger prey it is not possible to determine whether the prey was killed or scavenged.

Probably the most contentious aspect of using postingested samples is the lack of an unambiguous method to quantify the contribution of specific foods to the total biomass consumed. A variety of approaches have been used, including frequency of occurrence of specific prey among samples (Murie 1944; Litvaitis and Shaw 1980; Ackerman et al. 1984), measured weights of remains recovered in fecal samples (Johnson and Hansen 1979; Corbett 1989), relative volume of prey remains (Hellgren and Vaughan 1988), and estimated biomass consumed calculated with digestibility coefficients (Lockie 1959; Floyd et al. 1978; Greenwood 1979; Johnson and Hansen 1979; Weaver 1993; Hewitt and Robbins 1996). Evaluations of these approaches relative to ranking prey importance indicated that the most common inconsistencies occurred among small prey (Corbett 1989; Ciucci et al. 1996; figure 5.1).

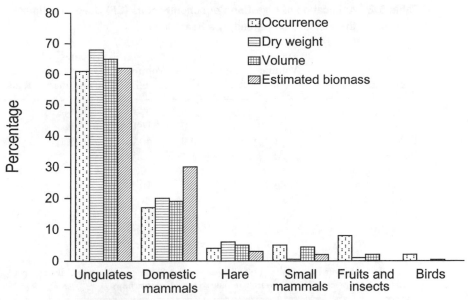

**Figure 5.1** Comparison of four methods used to investigate prey use by wolves. Fecal samples obtained from free-ranging wolves in Italy. *Occurrence* was the percentage of each prey relative to total occurrences of all prey. *Dry weight* was a percentage of all weight of all prey remains recovered in fecal samples. *Volume* was estimated using a reference grid and each prey was expressed as a percentage of the total volume of the feces. *Estimated biomass* was estimated using correction factors based on feeding trials conducted by Weaver (1993). Redrawn from Ciucci et al. (1996).

Development of digestibility coefficients or conversion factors seems to have the greatest potential for estimating actual biomass consumed. Correction factors are generated by feeding known amounts of forage or prey, collecting the resulting fecal material, and weighing the contents. These results can then be used with volumetric estimates or weights of prey remains to create a diet profile (table 5.2). Ideally, an investigator should use consumer-specific correction factors. Hewitt and Robbins (1996) also recommended developing correction factors for specific foods. Until such correction factors are derived, however, it may be useful for investigators to apply existing correction factors developed for general food or prey categories (e.g., foliage, fruits, nuts, small vertebrates) as long as the feeding trails used to generate the correction factors and their subsequent application are trophic-level specific (i.e., correction factors developed for carnivores are applied to carnivores and not to herbivores). Although there may be interspecific differences in digestion efficiencies within a trophic level (compare Litvaitis and Mautz 1980 with Powers et

**Table 5.2    Application of Digestion Correction Factors (CF) Used to Estimate the Biomass Consumed by a Bear**

| Diet Item | Fecal Volume (%) | Rank | CF | Volume × CF | Percentage of Dry Matter Consumed | Rank |
|---|---|---|---|---|---|---|
| Ungulates | 31 | 2 | 3.0 | 93.0 | 54 | 1 |
| Rodents | 4 | 4 | 4.0 | 16.0 | 9 | 3 |
| Fish | 1 | 7 | 40.0 | 40.0 | 23 | 2 |
| Insects | 3 | 5 | 1.1 | 3.3 | 2 | 6 |
| Coarse vegetation | 3 | 5 | 0.16 | 0.5 | <1 | 8 |
| Graminoid foliage | 45 | 1 | 0.24 | 10.3 | 6 | 4 |
| Forbs | 5 | 3 | 0.26 | 1.3 | 1 | 7 |
| Roots | 5 | 3 | 1.0 | 5.0 | 3 | 5 |
| Fleshy fruit | 2 | 6 | 1.2 | 2.4 | 1 | 7 |
| Seeds | 1 | 7 | 1.5 | 1.5 | 1 | 7 |

After table 2 of Hewitt and Robbins (1996). CF = correction factor: grams dry matter ingested per milliliter residue in feces, determined from feeding trials with captive bears.

al. 1989), the biases associated with the subsequent estimates of biomass consumed will probably be smaller than those based on percentage occurrence or percentage volume.

## Gastrointestinal samples

Contents of alimentary tracts are generally collected only from wild animals with large populations because they usually involve sacrificing the animal. Investigators use this approach often on collections during legal hunting and trapping seasons. These harvests usually span only a short portion of the year, limiting the application of this technique. For carnivores, harvest samples also may be biased by foods used as bait. On the other hand, information about the sex, age, and body condition of the sampled animal and volume of prey consumed is an advantage of this method (table 5.1).

Crop contents of granivores are a very different sample because only limited digestion has occurred. Often the investigator can separate and identify the majority of the sample. The volume of each seed type also can be estimated with a graduated cylinder or by the displacement of a known quantity of water in a burette (Inglis and Barstow 1960). Seeds and fruit can be identified by comparison with a local reference collection or through the use of reference books (Musil 1963).

Among herbivores, esophageal- and rumen-fistulated animals have been used instead of sacrificed animals. Fistulating involves installing a permanent device in the digestive tract of a living animal, allowing samples to be taken as food passes that point in the digestive process (Torell 1954; Short 1962; McManus 1981). Fistulation has been used extensively in describing the diets of domestic animals (Vavra et al. 1978), but rarely has this approach been applied to wild ruminants (Rice 1970). Taming and hand-rearing fistulated animals are the only ways a researcher can approach them to collect samples. But as with using domesticated "wild" herbivores for bite count data collection, the time and money required are generally prohibitive and tame animals may not reflect the true food habits of their wild counterparts.

Emetics, flushing tubes, and manual expression of the gullet have been also been used, primarily on birds, to purge the upper portion of the digestive tract without harming the animal (Errington 1932; Vogtman 1945). More recently, nonlethal sampling of stomach contents has been successfully done on small vertebrates, especially reptiles (Shine 1986; Henle 1989).

## ■ Evaluating the Importance of Specific Foods and Prey

### USE, SELECTION, OR PREFERENCE?

*Use, selection,* and *preference* have been applied interchangeably when discussing food use patterns, resulting in some confusion. Use simply indicates consumption of a specific food. *Selection* implies that an animal is choosing among alternative foods that are available. Use is selective if foods are consumed disproportionally to their availability in the environment (Johnson 1980). Preference is independent of availability. For example, animals can be provided different foods on an equal basis (cafeteria experiment) to determine preference among the foods provided.

### AVAILABILITY VERSUS ABUNDANCE

As I have just indicated, any evaluation of the selection of foods can be accomplished only with information on food availability. Unfortunately, the availability of forage or prey species can be difficult to estimate. Physical access to forage plants may be constrained by the reach of a herbivore or snow coverage

(Keith et al. 1984). Snow (Halpin and Bissonette 1988; Fuller 1991; Brown and Litvaitis 1995) and vegetation structure (Beier and Drennan 1997) also may influence the availability or vulnerability of prey to carnivores. Likewise, the presence of predators can affect the distribution of foraging activity by herbivores (Lima and Dill 1990; Barbour and Litvaitis 1993). As a result, many investigators have relied on an estimate of food abundance as a surrogate to availability. Is this an appropriate compromise? Estimates of relative abundance are often in units that do not necessarily correlate with density or biomass, such as captures per 100 trapnights or individuals observed per kilometer of transect surveyed (Windberg and Mitchell 1990). Obviously, large differences between consumption and relative abundance are pertinent to understanding forage or prey selection. However, studies that incorporate estimates of food abundance without considering the limitations of estimates of availability should be viewed as inferential.

## CAFETERIA EXPERIMENTS

The ability of an investigator to provide equal access to all foods as a method of identifying preference is obviously limited to captive situations. Rather than simply identifying preferred foods, most researchers who have used cafeteria experiments have attempted to identify the components that affect diet composition (Rodgers 1990) and therefore complement field observations (Topping and Kruuk 1996). For example, Klein (1977) observed that snowshoe hares (*Lepus americanus*) consumed twigs from sprout growth disproportionately less than twigs from older trees. Nutritional analysis (including protein content) did not explain this differential consumption. Klein speculated that hares might have been responding to antiherbivore chemicals in juvenile plants. Bryant and colleagues (Bryant 1981; Bryant et al. 1994) later examined this relationship by extracting resins from juvenile plants and then compared consumption rates of resin-coated and uncoated twigs by captive hares, providing experimental evidence that these compounds do exist and may indeed explain foraging preferences by snowshoe hares (but see Sinclair and Smith 1984).

## ■ Innovations

### IMPROVEMENTS ON LEAD ANIMAL STUDIES

Recent improvements in the use of lead animals may cause this approach to become more commonly used if very detailed information on food use is needed. Hand-reared animals can be equipped with radiotransmitters and released into the wild several months before data collection. Lead animals are often visited by their handler to remain acclimated to humans, but should no longer be naive about natural foods (Heim 1988). Use and availability can be accessed by walking with the animal as it feeds, recording use, and returning to measure availability along a marked trail (Heim 1988).

## USE OF ISOTOPE RATIOS

Biogeochemists have demonstrated the utility of comparing the relative concentration of various isotopes of carbon, nitrogen, and sulfur to reveal elemental cycles (Petersen and Fry 1987). This approach has been applied to examine current (Hobson and Clark 1992; Hilderbrand et al. 1996) and historic (Chisholm and Schwarcz 1982; Tieszen et al. 1989) food use patterns of some vertebrates. Essentially, the analysis of carbon isotopes can be used to determine the relative contributions of marine and terrestrial sources to an individual's carbon pool (Ramsey and Hobson 1991; Hobson and Welch 1992). In marine ecosystems, carbon enters as a bicarbonate and $^{13}C$ is enriched relative to terrestrial ecosystems (Petersen and Fry 1987; figure 5.2). This information can then be used to compare the importance of marine and terrestrial sources of forage. For example, in the northwestern United States, diets of historic populations of brown bears (*Ursus arctos*) were assessed by examining isotopic signatures from hair and bone collagen from specimens collected 140 years ago (Hilderbrand et al. 1996). The relative abundance $^{13}C$ also can be used to examine consumption of $C_4$ (many grasses) versus $C_3$ (trees and shrubs) plants (Tieszen et al. 1989). Nitrogen-15 is enriched at each step of the food chain (DeNiro and Epstein 1980), probably because the preferential excretion for lighter $^{14}N$ in urine (Peterson and Fry 1987; figure 5.3).

## EXPERIMENTAL MANIPULATIONS

Although it is widely accepted that populations of vertebrates are ultimately limited by the abundance of food, there has been substantial controversy among ecologists on how food abundance affects population dynamics and community structure (Lack 1954; Hairston et al. 1960; Wiens 1977; White 1978). This has led a number of investigators to experimentally manipulate food availability. Most of these studies were concerned with understanding how food supply affected such characteristics as reproduction and density, not with food selection or preference. In his review of more than 130 studies,

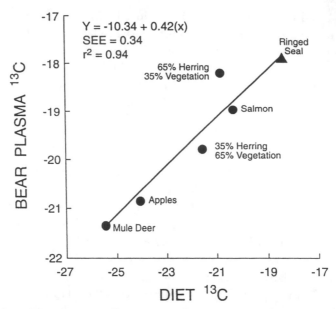

**Figure 5.2** Relationship between $^{13}C$ signatures of the diet of equilibrated plasma in black bears (*Ursus americanus*, (●) and polar bears (*U. maritimus*, (▲). Copyright 1996 Hilderbrand et al. Reprinted by permission of the *Canadian Journal of Zoology*.

Boutin (1990) found that most studies increased food abundance. Among individuals with access to supplemental foods, there were some clear tendencies: reduced home ranges, increased body weights, and advanced age and timing of reproduction. Populations typically doubled to tripled in density where food was essentially unlimited. However, food addition did not prevent declines in fluctuating populations (Boutin 1990).

The ability of herbivores to shape the composition and structure of vegetation via their foraging patterns has generated considerable interest in a variety of ecosystems (Bazley and Jefferies 1986; Brown and Heske 1990; Huntly 1991; Jones et al. 1994, Johnston 1995). This has also been the subject of several experimental manipulations. For instance, Ostfeld and Canham (1993) planted seedlings and monitored herbivory by meadow voles (*Microtus pennsylvanicus*) to reveal that this small mammal can effectively delay the invasion of trees into old-field communities. Spatial variation in foraging by voles also may create a mosaic of woody and nonwoody vegetation in these sites (Ostfeld and Canham 1993).

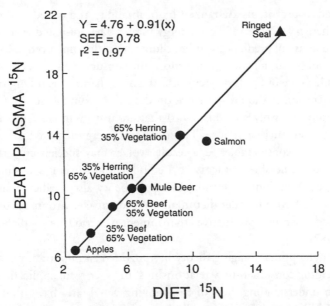

**Figure 5.3** Relationship between $^{15}$N signatures of the diet of equilibrated plasma in black bears (*Ursus americanus*, (•) and polar bears (*U. maritimus*, (▲). Copyright 1996 Hilderbrand et al. Reprinted by permission of the *Canadian Journal of Zoology*.

## THE ROLE OF FORAGING THEORY IN UNDERSTANDING FOOD HABITS

Most animals exploit a narrower range of food items than they are capable of consuming. In an effort to understand the constraints that may determine diet width, ecologists have organized their evaluations of food selection into a body of theory called optimal foraging theory (Schoener 1986; Begon et al. 1996; Perry and Pianka 1997). Two distinct approaches have developed to address this issue. The first considers that an animal selects among various food or prey items that are distributed in some fashion (e.g., clumped) throughout a generally suitable habitat. The second approach examines how animals discriminate among various patches of habitat that vary in productivity and suitability (Morrison et al. 1992) and can be viewed as an evaluation of habitat selection.

Early efforts to understand diet width relied on evaluating potential food items in terms of cost (search and handling time) and benefit (energy) (MacArthur and Pianka 1966; Charnov 1976). According to the basic assumptions of foraging theory, an animal should have a diet that maximizes energy intake and minimizes time to obtain nourishment (Schoener 1971). As such, food or prey items are ranked by profitability and added to the diet as long as there is an increase in net energy intake. The optimal diet model provides several useful predictions. If handling times (the time needed to pursue, capture, and consume) are typically short, the consumer should be a generalist (use a wide range of foods or prey). On the other hand, if handling times are long, the consumer should specialize on the most profitable foods. Consider prey selection by wolves (*Canis lupus*) that are usually in close proximity to large ungulates, such as moose. The time and energy required to capture a moose may be considerable. As a result, wolves may specialize on the most profitable or vulnerable segments of the population (juveniles and older animals in poor condition). Optimal foraging theory also predicts that a consumer should have a broader diet in an unproductive environment or during lean periods than in a productive environment or periods of food abundance (Gray 1987).

Although optimal foraging theory has provided an important platform for understanding consumer–prey relationships, the successful application of this theory to understanding diets of free-ranging vertebrates have been limited (Perry and Pianka 1997). The predictions of this theory are based on a series of assumptions that may not be justified (Pierce and Ollason 1987; Perry and Pianka 1997). The first is that the foraging behavior exhibited by present-day animals was favored by natural selection and continues to enhance the fitness of animals; the second is that high fitness is achieved by a high rate of net energy intake (Begon et al. 1996). Numerous field investigations in the last decade have revealed that diet selection is probably the consequence of fairly complex interactions of external, internal, and phylogenetic factors (figure 5.4).

External factors may include prey availability, risk of predation (Lima and Dill 1990) and social interactions (e.g., competition) (Perry and Pianka 1997). Internal factors include animal condition or hunger (McNamara and Houston 1984), learned experiences, age, sex and reproductive state, macro- and micronutrient requirements, and concentration of toxins or distasteful compounds. Phylogenetic factors include morphological constraints (e.g., mouth shape), sensory limitations, and physiological limitations. With such a complex array factors now known to affect foraging decisions, hindsight is quite clear: General models will probably fall short in contributing to our understanding of foraging patterns. However, recent innovations (e.g., using phylogenetic comparative methods) and continued use of manipulative experiments will undoubtedly advance our ability to identify parameters that are influential in complex environments.

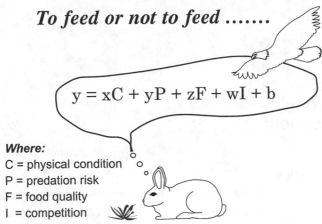

*To feed or not to feed .......*

$$y = xC + yP + zF + wI + b$$

**Where:**
C = physical condition
P = predation risk
F = food quality
I = competition

**Figure 5.4**   Internal and external factors affecting foraging decisions by a lagomorph (drawing courtesy of D. F. Smith).

## ■ Lessons

### SAMPLE RESOLUTION AND INFORMATION OBTAINED

As we have seen, the many methods used to investigate vertebrate food habits do not provide comparable information. The approaches summarized in table 5.1 differ according to the sample unit and resolution of information obtained. On one hand, exclosures provide information on relative use only by the population that is restricted by the fencing. On the other hand, samples obtained from gastrointestinal tracts can provide information on the actual biomass consumed by a specific individual of known sex and age (figure 5.5). Studies using the latter approach are therefore much more revealing to biologists concerned with the effects of foraging patterns.

Obviously, selection of a method to investigate food use is related to the application of the subsequent results to the investigator's objectives. For example, to understand community relationships, it may be essential to partition food use patterns by age and sex of the organism being studied. Greater sample resolution also may be needed among sexually dimorphic taxa, where a polygamous breeding system may result in spatial segregation, or among taxa that exhibit behavioral hierarchies that result in resource partition and habitat segregation. An example using indices of diet overlap may illustrate this point. A variety of indices have been used to investigate similarity of diets or other niche parameters and often vary from 0 (no overlap or similarity) to approxi-

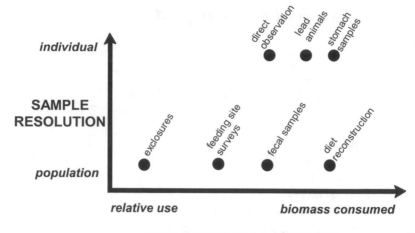

**INFORMATION CONTENT**

**Figure 5.5**   Information content (ranging from relative use to estimated biomass consumed) and sample resolution (individual animal or population) of common methods used to investigate vertebrate food habits.

mately 1 (identical diet; Krebs 1989; Litvaitis 1992). During an investigation of carnivore interactions (Litvaitis 1992), the overlap between coyotes (*Canis latrans*) and bobcats (*Felis rufus*) was 0.83. Because adult male bobcats are 50–100 percent larger than adult females, male bobcats are able to exploit larger prey and thus may be more similar to coyotes, which are substantially larger than female bobcats. To examine this possibility, samples from bobcats were separated into two classes based on body mass and diet; overlap between each size class of bobcats and coyotes was calculated. As suspected, overlap was substantially greater between adult male bobcats and coyotes (0.95) than between coyotes and female and juvenile bobcats (0.78) (Litvaitis unpublished data). This example suggests the benefits of obtaining greater sample resolution and not averaging samples from a population where resource segregation has occurred.

### IMPROVING SAMPLE RESOLUTION AND INFORMATION CONTENT

There may be ways to enhance the information obtained from conventional approaches to examining food habits. Fecal samples are still the most convenient, nonintrusive method to examine food habits of vertebrates. Methods are currently available and others are being developed that may increase the information obtained from such samples. Steroid concentrations (especially estrogen) have been used to examine pregnancy rates among free-ranging mammals (Kirkpatrick et al. 1990). This technique could be modified to distinguish male- and female-derived fecal samples. Even greater sample resolution is possible by using emerging molecular techniques. As indicated earlier, fecal samples contain epithelial cells shed from the intestine walls of the animal depositing the sample. DNA extracted from these cells has been used to identify the species that deposited the sample. Recently, several investigators have used this approach to identify sex and individual genetic markers (Kohn and Wayne 1997; Reed et al. 1997). Therefore, it is possible to substantially increase the resolution of fecal samples so that researchers can track the diet of identified free-ranging individuals. The information obtained from fecal samples could be enhanced even more by using digestibility correction factors that estimate biomass consumed. The resulting data set would probably prove very useful in evaluating diet selection and effects of consumption patterns on the forage or prey community.

As should be apparent by now, substantial information on food use patterns of vertebrates has been collected. Yet the ability of biologists to apply this information to understand factors that affect an organism's fitness or role in community structure has been limited. Perhaps the most needed change is to ensure that future investigations have a more complete context associated with them. Rather than partitioning studies into separate efforts to examine food and habitat use, these investigations should occur (and be reported) simultaneously.

Recent advances in molecular biology will enable vertebrate ecologists to generate a more complete picture of food use patterns by specific segments of a population. Such detailed information will enhance our ability to understand community relationships and spatialemporal patterns of vertebrate abundance. Rather than addressing general questions on the natural history of a specific species or population, clearly defined investigations of animal food habits may enhance our ability to answer the important how and why questions of vertebrate ecology.

### Literature Cited

Ackerman, B. B., F. G. Lindzey, and T. P. Hemker. 1984. Cougar food habits in southern Utah. *Journal of Wildlife Management* 48: 147–155.

Adorjan, A. A. and G. B. Kolenosky. 1969. *A manual for the identification of hairs of selected Ontario mammals.* Ontario Department of Lands and Forests, Research Report (Wildlife) no. 90.

Alverson, W. S., D. M. Waller, and S. L. Solheim. 1988. Forests too deer: Effects in northern Wisconsin. *Conservation Biology* 2: 348–358.

Anthony, R. G. and N. S. Smith. 1974. Comparison of rumen and fecal analysis to describe deer diets. *Journal of Wildlife Management* 38: 535–540.

Barbour, M. S. and J. A. Litvaitis. 1993. Niche dimensions of New England cottontails in relation to habitat patch size. *Oecologia* 95: 321–327.

Basile, J. V. and S. S. Hutchings. 1966. Twig diameter–length–weight relationships of bitterbrush. *Journal of Range Management* 19: 34–38.

Bazley, D. R. and R. L. Jefferies. 1986. Changes in composition and standing crop of salt marsh communities in response to removal of a grazer. *Journal of Ecology* 74: 693–706.

Begon, M., J. L. Harper, and C. R. Townsend. 1996. *Ecology*. Cambridge, Mass.: Blackwell.

Beier, P. and J. E. Drennan. 1997. Forest structure and prey abundance in foraging areas of northern goshawks. *Ecological Applications* 7: 564–571.

Bielefeldt, J., R. N. Rosenfield, and J. M. Papp. 1992. Unfounded assumptions about the diet of the Cooper's hawk. *Condor* 94: 427–436.

Bobek, B., S. Borowski, and R. Dzieciolowski. 1975. Browse supply in various forest ecosystems. *Polish Ecological Studies* 1: 17–32.

Boutin, S. 1990. Food supplementation experiments with terrestrial vertebrates: patterns, problems, and the future. *Canadian Journal of Zoology* 69: 203–220.

Brander, T. A., R. O. Peterson, and K. L. Risenhoover. 1990. Balsam fir on Isle Royale: Effects of moose herbivory and population density. *Ecology* 71: 155–164.

Brown, A. L. and J. A. Litvaitis. 1995. Habitat features associated with predation of New England cottontails: What scale is appropriate? *Canadian Journal of Zoology* 73: 1005–1011.

Brown, J. H. and E. J. Heske. 1990. Control of a desert–grassland transition by a keystone rodent guild. *Science* 250: 1705–1707.

Bryant, J. P. 1981. Phytochemical deterrence of snowshoe hare browsing by adventitious shoots of four Alaskan trees. *Science* 213: 889–890.

Bryant, J. P., R. K. Swihart, P. B. Reichardt, and L. Newton. 1994. Biogeography of woody plant chemical defense against snowshoe hare browsing: Comparison of Alaska and eastern North America. *Oikos* 70: 385–395.

Charnov, E. 1976. Optimal foraging, the marginal value theorem. *Theoretical Population Biology* 9: 129–136.

Chisholm, B. S. and H. P. Schwarcz. 1982. Stable-carbon isotope ratios as a measure of marine versus terrestrial protein in ancient diets. *Science* 216: 1131–1132.

Ciucci, P., L. Boitani, E. Raganella Pelliccioni, M. Rocco, and I. Guy. 1996. A comparison of scat-analysis methods to assess the diet of the wolf *Canis lupus*. *Wildlife Biology* 2: 37–48.

Cook, C. W. and J. Stubbendieck. 1986. Methods of measuring herbage and browse utilization. In C. W. Cook and J. Stubbendieck, eds., *Range research: Basic problems and techniques,* 120–121. Denver, Colo.: Society of Range Management.

Cooperrider, A. Y. 1986. Food habits. In A. Y. Cooperrider, R. J. Boyd, and H. R. Stuart, eds., *Inventory and monitoring of wildlife habitat,* 699–710. Denver, Colo.: U.S. Bureau of Land Management.

Corbett, L. K. 1989. Assessing the diet of dingoes from feces: A comparison of 3 methods. *Journal of Wildlife Management* 53: 343–346.

Craighead, J. J. and F. C. Craighead. 1956. *Hawks, owls, and wildlife.* Harrisburg, Pa.: Stackpole.

Dearden, B. L., R. E. Pegau, and R. M. Hansen. 1975. Precision of microhistological estimate of ruminant food habits. *Journal of Wildlife Management* 39: 402–407.

DeNiro, M. J. and S. Epstein. 1980. Influence of diet on the distribution of nitrogen isotopes in animals. *Geochimica et Cosmochimica Acta* 45: 341–351.

Dickman, C. R., M. Predavec, and A. J. Lyman. 1991. Differential predation of size and sex classes of mice by the barn owl, *Tyto alba. Oikos* 62: 67–76.

Eastman, D. S. and D. Jenkins. 1970. Comparative food habits of red grouse in northeast Scotland using fecal analysis. *Journal of Wildlife Management* 34: 612–620.

Errington, P. L. 1932. Techniques of raptor food habits study. *Condor* 34: 75–86.

Floyd, T. J., L. D. Mech, and P. A. Jordan. 1978. Relating wolf scat content to prey consumed. *Journal of Wildlife Management* 42: 528–532.

Fuller, T. K. 1991. Effect of snow depth on wolf activity and prey selection in north central Minnesota. *Canadian Journal of Zoology* 69: 283–287.

Gill, R. B., L. H. Carpenter, R. M. Bartmann, D. L. Baker, and G. G. Schoonveld. 1983. Fecal analysis to estimate mule deer diets. *Journal of Wildlife Management* 47: 902–915.

Gray, R. D. 1987. Faith and foraging: A critique of the "paradigm argument from design." In A. C. Kamil, J. R. Krebs, and H. R. Pulliam, eds., *Foraging behavior,* 69–140. New York: Plenum.

Green, G. A., G. W. Witmer, and D. S. DeCalesta. 1986. NaOH preparation of mammalian predator scats for dietary analysis. *Journal of Mammalogy* 67: 742.

Greenwood, R. J. 1979. Relating residue in raccoon feces to food consumed. *American Midland Naturalist* 102: 191–193.

Hairston, N. G., F. E. Smith, and L. B. Slobodkin. 1960. Community structure, population control and competition. *American Naturalist* 94: 421–425.

Halpin, M. A. and J. A. Bissonette. 1988. Influence of snow depth on prey availability and habitat use by red fox. *Canadian Journal of Zoology* 66: 587–592.

Hansson, L. 1970. Methods of morphological diet analysis in rodents. *Oikos* 21: 255–266.

Havstad, K. M. and G. B. Donart. 1978. The microhistological technique: Testing two central assumptions in south central New Mexico. *Journal of Range Management* 31: 469–470.

Heim, S. J. 1988. *Late winter and spring food habits of tame free-ranging white-tailed deer in southern New Hampshire.* M.S. thesis. Durham: University of New Hampshire.

Hellgren, E. C. and M. R. Vaughan. 1988. Seasonal food habits of black bears in Great Dismal Swamp, Virginia–North Carolina. *Proceedings of the Annual Conference of Southeast Fish and Wildlife Agencies* 42: 295–305.

Henle, K. 1989. Ecological segregation in a subterranean reptile assemblage in arid Australia. *Amphibia–Reptilia* 10: 277–295.

Hewitt, D. G. and C. T. Robbins. 1996. Estimating grizzly bear food habits from fecal analysis. *Wildlife Society Bulletin* 24: 547–550.

Hilderbrand, G. V., S. D. Farley, C. T. Robbins, T. A. Hanley, K. Titus, and C. Servheen. 1996. Use of stable isotopes to determine diets of living and extinct bears. *Canadian Journal of Zoology* 74: 2080–2088.

Hobson, K. A. and R. G. Clark. 1992. Assessing avian diets using stable isotopes I: Turnover of $^{13}$C in tissues. *Condor* 94: 181–188.

Hobson, K. A. and H. E. Welch. 1992. Determination of trophic relationships within a high arctic marine food web using $^{13}$C and $^{15}$N analysis. *Marine Ecology Progress Series* 84: 9–18.

Holechek, J. L. 1982. Sample preparation techniques for microhistological analysis. *Journal of Range Management* 35: 267–268.

Holechek, J. L., M. Vavra, S. Mady Dado, and T. Stephenson. 1982. Effect of sample preparation, growth stage, and observer on microhistological analysis. *Journal of Wildlife Management* 46: 502–505.

Höss, M., S. Pääbo, F. Knauer, and W. Schroder. 1992. Excrement analysis by PCR. *Nature* 359: 199.

Howard, V. W., Jr. 1967. Identifying fecal groups by pH analysis. *Journal of Wildlife Management* 31: 190–191.

Huntly, N. 1987. Effects of refuging consumers (pikas: *Ochotona princeps*) on subalpine vegetation. *Ecology* 68: 274–283.

Huntly, N. 1991. Herbivores and the dynamics of communities and ecosystems. *Annual Review of Ecology and Systematics* 22: 477–503.

Illius, A. W. and I. J. Gordon. 1987. The allometry of food intake in grazing ruminants. *Journal of Animal Ecology* 56: 989–999.

Inglis, J. M. and C. J. Barstow. 1960. A device for measuring the volume of seeds. *Journal of Wildlife Management* 24: 221–222.

Jaksic, F. M., P. Feinsinger, and J. E. Jimenez. 1992. A long-term study on the dynamics of guild structure among predatory vertebrates at a semi-arid Neotropical site. *Oikos* 67: 87–96.

Johnson, D. H. 1980. The comparison of usage and availability measurements for evaluating resource preference. *Ecology* 61: 65–71.

Johnson, M. K., R. C. Belden, and D. R. Aldred. 1984. Differentiating mountain lion and bobcat scats. *Journal of Wildlife Management* 48: 239–244.

Johnson, M. K. and R. M. Hansen. 1979. Estimating coyote food intake from undigested residues in scat. *American Midland Naturalist* 102: 362–367.

Johnston, C. A. 1995. Effects of animal species on physical landscape pattern. In L. Hannson, L. Fahrig, and G. Merriam, eds., *Mosaic landscapes and ecological processes*. New York: Chapman & Hall.

Jones, C. G., J. H. Lawton, and M. Shachak. 1994. Organisms as ecosystem engineers. *Oikos* 69: 373–386.

Keith, L. B., J. R. Cary, O. J. Rongstad, and M. C. Brittingham. 1984. Demography and ecology of a declining snowshoe hare population. *Wildlife Monographs* 90: 1–43.

Kirkpatrick, J. F., S. E. Shideler, J. W. Turner, Jr. 1990. Pregnancy determination in uncaptured feral horses based on steroid metabolites in urine-soaked snow and free steroids in feces. *Canadian Journal of Zoology* 68: 2576–2579.

Klein, D. R. 1977. Winter food preferences of snowshoe hares (*Lepus americanus*) in Alaska. *Proceedings of the International Congress of Game Biologists* 13: 266–275.

Kohn, M. H. and R. K. Wayne. 1997. Facts from feces revisited. *Trends in Ecology and Evolution* 12: 223–227.

Koivunen, V., E. Korpimaki, H. Hakkarainen, and K. Norrdahl. 1996. Prey choice of Tengmalm's owls (*Aegolius funerus funerus*): Preference for substandard individuals? *Canadian Journal of Zoology* 74: 816–823.

Korschegen, L. J. 1980 (4th ed.). Procedures for food habits analyses. In S. D. Schemnitz, ed., *Wildlife management techniques manual,* 113–127. Washington, D.C.: The Wildlife Society.

Krebs, C. J. 1989. *Ecological methodology.* New York: Harper & Row.

Lack, D. 1954. *The natural regulation of animal numbers.* New York: Oxford University Press.

Lima, S. L. and L. M. Dill. 1990. Behavioral decisions made under the risk of predation: A review and prospectus. *Canadian Journal of Zoology* 68: 619–640.

Litvaitis, J. A. 1992. Niche relations between coyotes and sympatric Carnivora. In A. H. Boer, ed., *Ecology and management of the eastern coyote,* 73–86. Fredericton: Wildlife Research Unit, University of New Brunswick.

Litvaitis, J. A. and W. W. Mautz. 1980. Food and energy utilization by captive coyotes. *Journal of Wildlife Management* 44: 56–61.

Litvaitis, J. A. and J. H. Shaw. 1980. Coyote movements, habitat use, and food habits in southwestern Oklahoma. *Journal of Wildlife Management* 44: 62–68.

Litvaitis, J. A., K. Titus, and E. Anderson. 1994. Measuring vertebrate use of terrestrial habitats and foods. In T. Bookhout, ed., *Research and management techniques for wildlife and habitats,* 254–274. Washington, D.C.: The Wildlife Society.

Litvaitis, M. K. and J. A. Litvaitis. 1996. Using mitochondrial DNA to inventory the distribution of remnant populations of New England cottontails. *Wildlife Society Bulletin* 24: 725–730.

Lockie, J. D. 1959. The estimation of the food of foxes. *Journal of Wildlife Management* 23: 224–227.

MacArthur, R. H. and E. R. Pianka. 1966. On optimal use of patchy environments. *American Naturalist* 100: 603–609.

Marti, C. D. 1987. Raptor food habits studies. In B. A. Giron Pendleton, B. A. Milsap, K. W. Cline, and D. M. Bird, eds., *Raptor management techniques manual,* 67–80. Washington, D.C.: National Wildlife Federation.

Martin, A. C., H. S. Zim, and A. L. Nelson. 1961. *American wildlife and plants: A guide to wildlife food habits.* New York: Dover.

Maser, Z., C. Maser, and J. M. Trappe. 1985. Food habits of the northern flying squirrel (*Glaucomys sabrinus*) in Oregon. *Canadian Journal of Zoology* 63: 1084–1088.

McInnes, P. F., R. J. Naiman, J. Pastor, and Y. Cohen. 1992. Effects of moose browsing on vegetation and litter of the boreal forest, Isle Royale, Michigan, USA. *Ecology* 73: 2059–2075.

McInnis, M. L., M. Vavra, and W. C. Krueger. 1983. A comparison of four methods used to determine the diets of large herbivores. *Journal of Range Management* 36: 302–307.

McManus, W. R. 1981. Oesophageal fistulation technique as an aid to diet evaluation of the grazing ruminant. In J. L. Wheeler and R. D. Mochrie, eds., *Forage evaluation: Concepts and techniques,* 249–260, Lexington, Ky.: American Forage and Grassland Council.

McNamara, J. M. and A. I. Houston. 1984. Starvation and predation as factors limiting population size. *Ecology* 68: 1515–1519.

Mech, L. D. 1966. *The wolves of Isle Royale.* U.S. National Park Service Fauna Series 7.

Mills, M. G. L. 1992. A comparison of methods used to study food habits of large African carnivores. In D. R. McCullough and R. H. Barrett, eds., *Wildlife 2001: Populations,* 1112–1124. New York: Elsevier.

Morrison, M. L., B. G. Marcot, and R. W. Mannan. 1992. *Wildlife–habitat relationships: Concepts and applications.* Madison: University of Wisconsin Press.

Murie, A. 1944. *The wolves of Mount McKinley.* U.S. Park Service, Fauna Series 5.

Murie, O. J. 1974. *Animal tracks.* Boston: Houghton Mifflin.

Musil, A. F. 1963. *Identification of crop and weed seeds.* U.S. Department of Agriculture Handbook 219.

Nelson, E. W. 1930. Methods of studying shrubby plants in relation to grazing. *Ecology* 11: 764–769.

Ostfeld, R. S. and C. D. Canham. 1993. Effects of meadow vole population density on tree seedling survival in old fields. *Ecology* 74: 1792–1801.

Owen, M. 1975. An assessment of fecal analysis technique in waterfowl feeding studies. *Journal of Wildlife Management* 39: 271–279.

Paxinos, E., C. McIntosh, K. Ralls, and R. Fleischer. 1997. A noninvasive method for distinguishing among canid species: Amplification and enzyme restriction of DNA from dung. *Molecular Ecology* 6: 483–486.

Perry, G. and E. R. Pianka. 1997. Animal foraging: Past, present and future. *Trends in Ecology and Evolution* 12: 360–364.

Petersen, B. J. and B. Fry. 1987. Stable isotopes in ecosystem studies. *Annual Review of Ecology and Systematics* 18: 293–320.

Pierce, G. J. and J. G. Ollason. 1987. Eight reasons why optimal foraging theory is a complete waste of time. *Oikos* 49: 111–118.

Powers, J. G., W. W. Mautz, and P. J. Pekins. 1989. Nutrient and energy assimilation of prey. *Journal of Wildlife Management* 53: 1004–1008.

Ramsey, M. A. and K. A. Hobson. 1991. Polar bears make little use of terrestrial food webs: Evidence from stable-carbon isotope analysis. *Oecologia* 86: 598–600.

Reed, J. Z., D. J. Tollit, P. M. Thompson, and W. Amos. 1997. Molecular scatology: The use of molecular genetic analysis to assign species, sex and individual identity to seal fae-ces. *Molecular Ecology* 6: 225–234.

Reynolds, J. C. and N. J. Aebischer. 1991. Comparison and qualification of carnivore diet by faecal analysis: A critique, with recommendations based on a study of the fox *Vulpes vulpes*. *Mammal Review* 21: 97–122.

Rice, R. W. 1970. Stomach content analyses: A comparison of the rumen vs. esophageal techniques. In *Range and wildlife habitat evaluation: A research symposium,* 127–132. U.S. Forest Service Miscellaneous Publication 1147.

Rodgers, A. R. 1990. Evaluating preference in laboratory studies of diet selection. *Canadian Journal of Zoology* 68: 188–190.

Salas, L. A. and T. K. Fuller. 1996. Diet of the lowland tapir (*Tapirus terrestris* L.) in the Tabaro River valley, southern Venezuela. *Canadian Journal of Zoology* 74: 1444–1451.

Schaller, G. B. 1972. *The Serengeti lion*. Chicago: University of Chicago Press.

Schoener, T. 1971. Theory of feeding strategies. *Annual Review of Ecology and Systematics* 2: 369–404.

Schoener, T. 1986. A brief history of optimal foraging theory. In A. C. Kamil, J. R. Krebs, and H. R. Pulliam, eds., *Foraging behavior,* 5–68. New York: Plenum.

Shine, R. 1986. Food, habits, habitats and reproductive biology of four sympatric species of varanid lizards in tropical Australia. *Herpetologica* 42: 346–360.

Short, H. L. 1962. The use of a rumen fistula in a white-tailed deer. *Journal of Wildlife Management* 20: 21–25.

Sinclair, A. R. E. and P. Arcese. 1995. Population consequences of predation-sensitive for-aging: The Serengeti wildebeest. *Ecology* 76: 882–891.

Sinclair, A. R. E. and J. N. M. Smith. 1984. Do plant secondary compounds determine feeding preferences of snowshoe hares? *Oecologia* 61: 403–410.

Smith, A. D., and R. L. Hubbard. 1954. Preference ratings for winter deer forages from northern Utah ranges based on browsing time and forage consumed. *Journal of Range Management* 7: 262–265.

Smith, A. D. and J. L. Shandruk. 1979. Comparison of fecal, rumen, and utilization meth-ods for ascertaining pronghorn diets. *Journal of Range Management* 32: 275–279.

Smith, A. D. and P. J. Urness. 1962. *Analysis of the twig-length method of determining uti-lization of browse*. Utah State Department of Fish and Game Publication 69-9.

Storr, G. M. 1961. Microscopic analysis of faeces, a technique for ascertaining the diet of herbivorous mammals. *Australian Journal of Biology* 14: 157–164.

Telfer, E. S. 1969. Twig weight–diameter relationships for browse species. *Journal of Wildlife Management* 33: 917–921.

Temple, S. A. 1987. Do predators always capture substandard individuals disproportion-ately from prey populations? *Ecology* 68: 669–674.

Tieszen, L. L., T. W. Boutton, W. K. Ottichilo, D. E. Nelson, and D. H. Brandt. 1989. An assessment of long-term food habits of Tsavo elephants based on stable carbon and nitrogen ratios of bone collagen. *African Journal of Ecology* 27: 219–226.

Till, J. A. and F. F. Knowlton. 1983. Efficacy of denning in alleviating coyote depredations upon domestic sheep. *Journal of Wildlife Management* 47: 1018–1025.

Topping, M. and H. Kruuk. 1996. Size selection of prey by otters, *Lutra lutra* L.: An experimental approach. *Zeitschrift für Saügetierkunde* 61: 1–3.

Torell, D. T. 1954. An esophageal fistula for animal nutrition studies. *Journal of Animal Science* 13: 878–882.

Vavra, M., and J. L. Holechek. 1980. Factors influencing microhistological analyses of herbivore diets. *Journal of Range Management* 33: 371–374.

Vavra, M., R. W. Rice, and R. M. Hansen. 1978. A comparison of esophageal fistula and fecal material to determine steer diets. *Journal of Range Management* 31: 11–13.

Vogtman, D. B. 1945. Flushing tube for determining food of gamebirds. *Journal of Wildlife Management* 9: 255–257.

Wallmo, O. C., R. B. Gill, L. H. Carpenter, and D. W. Reichert. 1973. Accuracy of field estimates of deer food habits. *Journal of Wildlife Management* 37: 556–562.

Weaver, J. L. 1993. Refining the equation for interpreting prey occurrence in gray wolf scat. *Journal of Wildlife Management* 57: 534–538.

Webb, J. 1943. Identification of rodents and rabbits by their fecal pellets. *Transactions of the Kansas Academy of Science* 43: 479–481.

Westoby, M., G. R. Rost, and J. A. Weis. 1976. Problems with estimating herbivore diets by microscopically identifying plant fragments from stomachs. *Journal of Mammalogy* 57: 167–172.

White, T. 1978. The importance of a relative shortage of food in animal ecology. *Oecologia* 33: 71–86.

Wiens, J. A. 1977. On competition and variable environments. *American Scientist* 65: 590–597.

Wiens, J. A. 1993. Fat times, lean times and competition among predators. *Trends in Ecology and Evolution* 8: 348–349.

Windberg, L. A. and C. D. Mitchell. 1990. Winter diets of coyotes in relation to prey in southern Texas. *Journal of Mammalogy* 71: 639–647.

Zalewski, A. 1996. Choice of age classes of bank voles *Clethrionomys glareolus* by pine marten *Martes martes* and tawny owl *Strix aluco* in Bialowieza National Park. *Acta Oecologia* 17: 233–244.

## Chapter 6

# Detecting Stability and Causes of Change in Population Density

JOSEPH S. ELKINTON

Other chapters in this volume focus on various methods for quantifying density or other population qualities. Here I focus on the techniques ecologists use to extract the dynamics of population systems from such data. Population ecologists seek to explain why some animals are rare whereas others are common, as well as what accounts for observed changes in density. They have focused on two analytical questions: Are populations stabilized by negative feedback mechanisms, and what are the causes of density change? Here I examine some of the techniques that have been developed to answer these questions.

The concept of a balance of nature goes back to the very early days of ecology. It is obvious that unlimited capacity of all animals to increase in population size or density is inevitably checked by competition for resources or the action of natural enemies. If any of these factors cause systematic changes in survival or fecundity of a population as the density increases, they are said to be density dependent. If fecundity or survival decreases sufficiently as the population increases, then the per capita birth rate will decline to a value equal to or less than the per capita death rate and population growth will stop. In this manner, density-dependent processes constitute negative feedbacks on population growth that can maintain densities at or near an equilibrium value indefinitely.

For more than 50 years ecologists have debated whether population densities of most species are stabilized by such density-dependent factors. Howard and Fiske (1911) were the first to articulate the idea that populations cannot long persist unless they contain at least one density-dependent factor that causes the average fecundity to balance the average mortality. Other early proponents of this idea were Nicholson (1933, 1957) and Lack (1954). In contrast, Andrewartha and Birch (1954) argued that most populations are not held at equilibrium density. Rather, densities merely fluctuate. In their view most species avoid extinction because they comprise what we now call metapopulations (Levins 1969). These consist of a series of subpopulations whose densities fluctuate independently of one another, but are linked by dispersal. Extinction of subpopulations occurs quite often, but these are recolonized by individuals dispersing from other subpopulations, allowing the species to persist indefinitely over the entire region. This process has been called spreading of risk (den Boer 1968; Reddingius 1971). In recent years metapopulation dynamics have been explored by way of simulations (Hanski 1989) that have revealed that such systems eventually go extinct in the absence of density dependence.

The debate about the ubiquity of density-dependent processes has persisted to the present day despite the efforts of various ecologists to terminate the discussion either because it was bankrupt (Krebs 1991; Wolda 1995) or because they deemed that prevalence of density dependence was too obvious to deny (Royama 1977, 1992). Turchin (1995) provides a comprehensive review of the current status of the debate. Nevertheless, many ecologists have insisted that no conclusion regarding the existence of density dependence in a population system can be made unless their action can be demonstrated in data collected from the populations. This has proved difficult to achieve. Until recently, adequate methods for detecting density dependence in population systems have been lacking and earlier methods have been shown to be statistically invalid. Several new methods have been proposed over the last decade, most of which involve a variety of computer-based resampling procedures. I review the most promising or widely used of these tests here and discuss their limitations.

Ecologists used to assume that populations governed by density-dependent processes had simple dynamics. They supposed that densities either remained close to equilibrium or exhibited regular oscillations about the equilibrium value. The pioneering work on deterministic chaos by Robert May (1974, 1976) taught us otherwise. May studied the behavior of the discrete logistic, arguably the simplest possible density-dependent population model, and showed that densities would fluctuate erratically and unpredictably if the reproductive rates were sufficiently high. Before this ecologists had assumed that the erratic fluctuations characteristic of most natural populations were caused by random influences such as weather conditions that disturbed the system away from equilibrium. There ensued an effort to determine whether natural populations were indeed chaotic. Early studies (Hassell et al. 1976) concluded that most populations were not chaotic. These studies were based on attempts to fit natural populations to simple models and then to see whether the values of model parameters representing, for example, density dependence, time delays, or reproductive rate were such that chaotic behavior would be expected. The problem was that the conclusion depended on the particular model used, which was always a simplistic abstraction of the inevitably complicated dynamics of real populations. Subsequent investigators offered techniques that were more general and did not assume particular population models (Schaffer and Kot 1985; Turchin and Taylor 1992). Applications of these techniques have indicated that some but not all population systems are chaotic. A more detailed discussion of these techniques can be found in chapter 8. Here I focus on techniques to demonstrate density dependence and the causes of change in density.

## ■ Detection of Density Dependence

### ANALYSIS OF TIME SERIES OF DENSITY

For data that consist of a time series, that is, a sequence of periodic estimates of density from a population, a variety of tests have been proposed to detect the existence of density-dependent processes. If we define $R$ as change in population density on a log scale,

$$R = \log N_{t+1}/N_t = X_{t+1} - X_t \qquad (6.1)$$

where $N_t$ is density and $X_t$ is log density at time $t$, then if density-dependent processes are at work, population change should be correlated with density.

$$R = \alpha + \beta \log N_t + \varepsilon_t \qquad (6.2)$$

where $\alpha$ and $\beta$ are coefficients representing density-independent and density-dependent processes, respectively, and $\varepsilon_t$ is any source of random fluctuation. The processes involved may affect fecundity, mortality, or both. If there is no density dependence, $\beta = 0$. If $\beta = 0$ it means there is no general upward or downward trend in density over time, in the absence of density dependence. Alternatively, equation 6.2 can be written as

$$X_{t+1} = \alpha + (\beta + 1) X_t + \varepsilon_t \tag{6.3}$$

If both $\beta = 0$ and $\alpha = 0$, the model is known as a random walk:

$$X_{t+1} = X_t + \varepsilon_t \tag{6.4}$$

Thus to look for density dependence we could plot either $R$ or $X_{t+1}$ against $X_t$. Under the null hypothesis of density independence we could test whether $\beta = 0$ or $\beta + 1 = 1$.

Morris (1959) was the first to use plots of $X_{t+1}$ against $X_t$ to search for density dependence in population systems, in his case spruce budworm populations. Smith (1961) used this technique to show density dependence in populations of the flower thrips, which Andrewartha and Birch (1954) had used as an example of an insect governed by density independent processes. Equivalently one can plot $R$ against $X_t$, which I illustrated with my own data on gypsy moth density (Elkinton et al. 1996) collected from eight populations over a 10-year period (figure 6.1).

The striking downward trend evident in figure 6.1 seems to indicate a strong density dependence: Populations decline ($R < 0$) when densities are high and increase ($R > 0$) when densities are low. Standard regression procedures indicate a significant negative slope (solid line in figure 6.1). However, there is a statistical problem. The axes are not independent and this produces a negative bias in regression estimates of slope (Watt 1964; St. Amant 1970; Eberhardt 1970; Reddingius 1971). Data generated from a random walk process (no density dependence) will show strong negative slope. To illustrate this I fit a random walk model to the data given in figure 6.1 and selected 100 time series of length $n = 10$ (years) based on values of $\varepsilon_t$ chosen at random from a normal distribution with a variance that matches that of the data. The resulting average slope (dotted line in figure 6.1) is strongly negative, implying that population growth declines with density, even though there is no density dependence and hence no stability in this model of the population system. This example illustrates that usual methods of statistical inference based on regression of $X_{t+1}$ or $R$ on $X_t$ are fundamentally flawed as a way of detecting density dependence in a population time series.

Various investigators have suggested solutions to the problem just illustrated. Varley and Gradwell (1968) advocated plotting $X_{t+1}$ against $X_t$, interchanging them as independent and dependent variables. Only if both regressions were significantly different from the slope of the null model ($\beta + 1 = 1$)

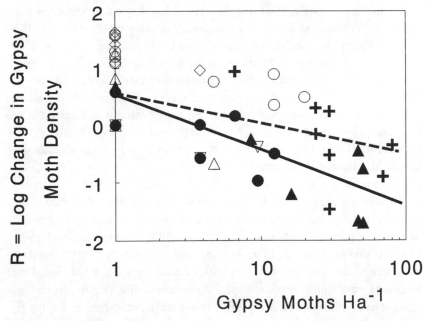

**Figure 6.1**   Change in gypsy moth density ($R = \log_{10}[N_{t+1}/N_t]$).plotted against gypsy moth density ($N_t$) compared to slope (dotted line) expected with no density dependence (Elkinton et al. 1996). Solid line is regression equation fit to the data. Values of $R < 0$ indicate declining densities; values of $R > 0$ indicate increasing densities. Densities are estimated egg masses per hectare.

in the same direction would density dependence be confirmed. Their approach has been subsequently shown to be extremely conservative, such that very few data sets would pass the test (Holyoak 1993).

Pollard et al. (1987) proposed a so-called randomization test that has emerged in several comparisons as the most accurate and powerful of extant methods (Holyoak 1993). They calculated a null distribution of slopes (i.e., those expected if there were no density dependence) to compare against the slope obtained from the actual data. This was done by generating a large number (e.g., 1,000) of permutations of the list of successive annual census data values. Because the data values were scrambled in this way, the dependence of each value on the preceding value was obliterated, yet the overall variance of the densities was retained. For each permutation they regressed $R$ versus $X_t$ and then compared the slope from the actual data with the distribution of slopes obtained from the 1,000 permutations. Reddingius (1971) proposed a very similar test.

Dennis and Taper (1994) offered a procedure known as the parametric bootstrap likelihood ratio test that was similar to that of Pollard et al. (1987). Instead of permuting the data they generated 1,000 simulated data sets under the null model (equation 6.2) of density independence with randomly chosen error terms. As Pollard et al. (1987) had done, they then calculated a distribution of slopes with which they compared the actual slope generated from the data by ordinary linear regression. After applying both of these techniques to the data given in figure 6.1, I found only equivocal support for density dependence (Elkinton et al. 1996). Three of eight population time series were identified as density dependent (at $p < 0.05$) with the test of Dennis and Taper (1994) and two out of eight with the test of Pollard et al. (1987).

However, neither of these techniques is without problems; first they all lack statistical power. One needs 20-30 generations of data to reliably find density dependence when it exits (Solow and Steele 1990; Dennis and Taper 1994). Data sets that long are rare in ecology. Second, Shenk et al. (1998) simulated the effect of measurement error on both of these tests. They concluded that both tests were highly prone to type I error (concluding density dependence when it did not exist) and hence of little use when measurement error is significant, as it usually is in most data sets. Dennis and Taper (1994) conducted analogous simulations and found that their test was robust against measurement error. The difference in the two studies was in how measurement error was modeled and it is too early to tell which view will prevail. Finally, correlations between error terms from one year to the next can also lead to spurious conclusions of density dependence when it does not exist (Solow 1990; Reddingius 1990). It would not be very surprising to have autocorrelated errors because they include the effects of all other variables on population growth other than density. Suppose a population was determined largely by the action of a generalist predator whose density was not linked to that of its prey and that might or might not cause density-dependent mortality. Variation in predation rates caused by fluctuation in predator density would be embodied in the error term and would probably be influenced by predator densities at previous time steps. The error term would thus be autocorrelated (Williams and Liebhold 1995).

## ANALYSIS OF DATA ON MORTALITY OR SURVIVAL

Many investigators collect data not just on density but on mortality or fraction surviving in particular age categories or life stages. They may be interested in particular agents of mortality and want to know whether these are capable of stabilizing the population. Plots of some measure of percentage mortality versus density of the population reveal density dependence. For example, Varley and Gradwell (1968) presented data on the winter moth, a defoliator of oak trees in Great Britain. They collected data on sources of mortality on successive instars or life stages over a period of 15 generations (years) at one site. They expressed mortality as $k$-values ($k = -\log_{10}$(proportion surviving)) and plotted $k$-values for each cause of mortality against the log density of the individuals present at the beginning of the stage or age category on which the agent of mortality acted. They used standard linear regression to determine whether mortality increased or decreased with density. The strong negative bias described earlier for analysis of density time series was not present here because the measurements of mortality differed from the measures of density. However, a number of statistical problems involve violations of the usual assumptions of linear regression. The regression may be nonlinear, measurement error may affect the estimates of both density and mortality, the variance of the $k$-values may vary systematically with density, and the error terms may not be independent because the data are obtained from time series. Solutions have been proposed for several of these problems (for example, see Hassell et al. 1987). However, Vickery (1991) analyzed the various extant methods and found them all either biased or lacking in statistical power. He advocated using a randomization test identical to that of Pollard et al. (1987), but applied to data on stage-specific mortality instead of time series of density.

Use of these techniques rests on the assumption that the data collected for mortality and density are accurate and unbiased. This may not be easy to achieve, particularly when ages or life stages overlap temporally. Various techniques may be used to convert densities or numbers present in periodic samples to estimates of numbers entering particular life stages or age categories (see reviews in Bellows et al. 1992). Similar techniques exist to convert mortalities or rates of infection obtained from periodic samples to the stage-specific mortality that best represents the overall impact of the agent of mortality (van Driesche et al. 1991). Additional techniques may be required when two or more agents of mortality act contemporaneously (Elkinton et al. 1992). It is beyond the scope of this review to describe these techniques here.

Density dependence is often reported in studies in which data on mortality or survival are obtained simultaneously from several different populations that vary in density. It is important to realize that the processes that give rise to density dependence in such studies may not be the same as those producing density dependence in studies wherein mortality and density are shown to vary over time from one or more populations. For example, Gould et al. (1990)

showed that mortality of gypsy moths caused by a parasitic fly, *Compsilura concinnata,* increased dramatically with gypsy moth density on a series of experimental populations created with different densities at several locations in the same year (figure 6.2).

The density-dependent response was evidently a behavioral one by the fly. It was not at all clear the extent to which such responses would occur in studies in which density varied temporally instead of spatially. Only in the latter studies would the reproductive response of the fly to changes in gypsy moth density be measured. Indeed, a 10-year study of parasitism in naturally occurring populations of gypsy moth (Williams et al. 1992) revealed no evidence of temporal density dependence and far lower levels of parasitism by *C. concinnata* (figure 6.2). The ability of this fly to regulate low densities of gypsy moth is thus questionable.

Several investigators have surveyed the published literature on life tables in particular taxa in order to ascertain how often density dependence has been

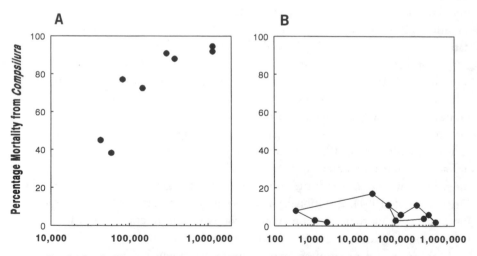

## Initial Density of Gypsy Moth Larvae

**Figure 6.2**    (A) Percentage mortality of gypsy moth caused by the parasitic fly *Compsilura concinnata* to a series of experimental populations created with different densities in the same year (Gould et al. 1990). (B) Time series of percentage mortality caused by *C. concinnata* in a 10-year study of gypsy moth in naturally occurring populations (Williams et al. 1992). Solid line in (B) connects consecutive generations.

detected. A typical conclusion is that evidence of density dependence is rare (Dempster 1983; Stiling 1987). Part of the reason for this is that most life table studies are of short duration and the problems of statistical power indicated earlier limit our ability to detect density dependence when it exists. In addition, data from most studies contain a considerable degree of measurement error. Such error can also obscure density-dependent relationships (Hassell 1985). Finally, the action of many density-dependent mortalities may lag behind those of their hosts. The methods described earlier will not detect their action. I discuss this in the next section.

## ■ Detection of Delayed Density Dependence

Time lags in density-dependent responses are common in population systems. For example, it is typical for a predator or parasitoid to respond numerically to changes in density of its host, but this response typically lags behind that of its host by at least one generation. The result is that peak predator density and hence peak mortality of the host occurs after the host has declined dramatically from peak density. Plots of mortality against density may reveal no positive relationship between the two, even if it is clear that the predator is regulating its host. Such responses are known as delayed density dependence. Different techniques have been developed to detect it.

The first of these techniques were graphical in nature (Hassell and Huffaker 1969; Varley et al. 1973). If one plots mortality against density and connects successive years, a counterclockwise spiral is evident (figure 6.3). If the data consist of census data rather than mortality, that is, successive generations of density counts, then connection of successive years on a graph of $R$ plotted against $N_t$ or $X_t$ yields a clockwise spiral (figure 6.3).

A major advance in detection of delayed density dependence was developed by Turchin (1990). He applied time series analyses (Box and Jenkins 1976) that have had wide application in econometrics and the physical sciences. The methods involve fitting a model similar to equation 6.2 but with terms representing the effects of density in generations before the last one:

$$X_{t+1} = \alpha + \beta X_t + \gamma X_{t-1} \ldots + \varepsilon_t \qquad (6.5)$$

Partial autocorrelation analysis tells you whether there is significant delayed density dependence.

To illustrate this method, I give two examples. The first one is undoubtedly

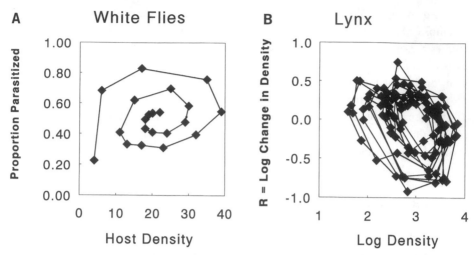

**Figure 6.3** Graphic detection of delayed density dependence wherein points representing consecutive generations are connected. (A) Percentage mortality caused by the parasitoid *Encarsia formosa* on the greenhouse white fly (*Trialeuroides vaporariorium*). (Redrawn from Varley et al. 1973, Fig. 4.5; data from Burnett 1958). (B) Change in density on a log scale (*R*) vs. density of the Canadian lynx (redrawn from Royama 1977; data from Elton and Nicholson 1942).

the best known time series in ecology, the snowshoe hare and lynx oscillation in Canada based on pelts delivered to the Hudson Bay Company over a time period exceeding 100 years (Elton and Nicholson 1942). These data have been analyzed by Royama (1992). Figure 6.4 shows fluctuations in density of the lynx populations, along with the autocorrelation function (ACF) and the partial autocorrelation function (PACF). The ACF expresses the correlation between each measure of density and the densities 1, 2, . . . *n* generations back (the lag time). For time series that display pronounced cycles, as in figure 6.4, the ACF reaches a peak value at a lag time corresponding to cycle period.

The ACF can be used to determine whether time series are truly cyclic. Although the cycles are obvious in figure 6.4, in many time series the fluctuations are more irregular and detecting the difference between cyclic behavior and random fluctuations is not at all obvious. Delayed density dependence can be detected by looking at the PACF. The PACF represents the correlation that remains at each lag time (between $X_t$ and $X_{t-n}$), with the correlations due to smaller lag times removed. For example, in most population time series there is a fairly high correlation between $X_t$ and $X_{t-1}$ (figure 6.4). For the same reason $X_{t-1}$ is correlated with $X_{t-2}$, and, consequently, a positive correlation exists between $X_t$ and $X_{t-2}$, albeit a weaker one (figure 6.4). The PACF removes these effects due to shorter lags and measures the correlation that remains. For most population systems, delayed density dependence is manifest in a significantly negative PACF at lag 2 (figure 6.4), where significance ($p < 0.05$) is indicated by a values that cross the dashed, horizontal cutoff lines. In figure 6.4 I give an example from Turchin (1990) of the pine spinner moth, *Dendrolimus pini*, in Germany, where fluctuations are erratic and there is no evidence of cyclicity in the ACF and yet significant delayed density dependence in the PACF (figure 6.4). Turchin (1990) used these procedures to analyze data from 14 species of forest lepidoptera to show that nearly all of them had significant lag 2 or higher effects. In contrast, Hanski and Woiwod (1991) analyzed 5,715 annual time series of moths and aphids captured in survey traps. They found a high incidence (67-91 percent) of density dependence but less than the 5 percent incidence of delayed density dependence they would have expected by chance alone. Holyoak (1994b) suggests that part of the unexpectedly low incidence of delayed density dependence may be caused by the multiple generations that elapsed between the annual samples for many of these species in contrast to the forest lepidoptera analyzed by Turchin (1990).

Several of the same limitations of tests for direct density dependence also apply to tests for delayed density dependence. The techniques have little or no statistical power for the short time series that are typical of most ecological data (Holyoak 1994a). Furthermore, autocorrelations in the error terms can lead to spurious conclusions of delayed density dependence (Williams and Liebhold 1995). Berryman and Turchin (1997) argue that Williams and Liebhold's conclusion is overly pessimistic and based on simulations using unrealistic parameter values. Williams and Liebhold (1997) reply that the parameter values were typical of populations analyzed earlier by Turchin (1990).

## ■ Detection of Causes of Population Change

### KEY FACTOR ANALYSIS

Many population ecologists are more interested in determining the causes of changes in density than in the causes of stability. For example, ecologists who work with animals that exhibit outbreak dynamics may want to uncover the cause of these outbreaks. Key factor analysis was developed to identify such causes, or at least the life stage on which they act (Morris 1959; Varley and Gradwell 1960). Like the density dependence techniques discussed earlier, key

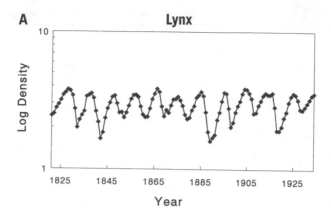

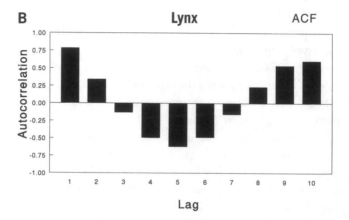

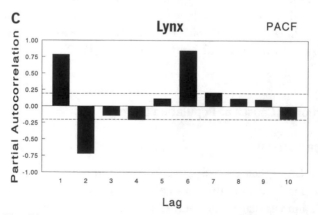

**Figure 6.4** Use of time series analysis to detect delayed density dependence. (A) Time series of the Canada lynx (data from Elton and Nicholson 1942) and the corresponding (B) ACF and (C) PACF for the lynx population (redrawn from Royama 1992). (D) Time series of the moth *Dendrolimus pini* and corresponding (E) ACF and (F) PACF (redrawn from Turchin 1990; data from Varley 1949).

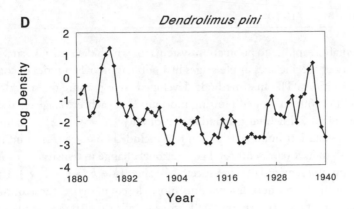

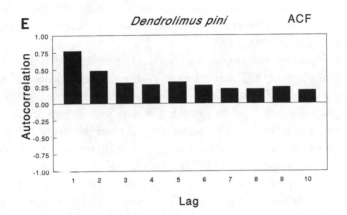

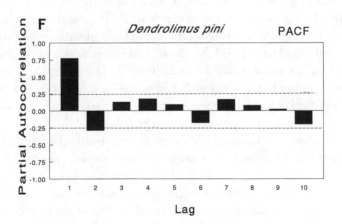

factor analysis applies to population systems for which data exist on survival or mortality of specific ages or life stages in a population over a series of consecutive generations. The first methods developed were graphic ones. Varley and Gradwell (1960) advocated plotting mortalities in each life stage expressed as $k$-values along with total generational mortality ($K = X_t - X_{t+2}$; figure 6.5) against generation number (time). They included changes in fecundity as a $k$-value so total $K$ represents total generational change in density ($K = -R$) and is the sum of $n$ sequential, stage-specific $k$-values ($K = k_1 + k_s + \ldots k_n$). The key factor was the one whose fluctuations most closely matched those of total $K$. For example, figure 6.5 shows key factor analysis of a 10-year study of partridge populations in England (Blank et al. 1967). Mortality (or loss of fecundity) in this population was attributed to eight sequential causes. Of these, mortality of chicks ($k_4$) was the one whose fluctuations clearly matched of that of total generational change ($K$).

For some species such graphic analyses might not yield a definitive answer. Podoler and Rogers (1975) advocated calculating regression lines of each $k$-value against total $K$. The key factor was the one with the largest positive slope. They applied this technique to a number of data sets, including the English partridge data in figure 6.5. Chick mortality ($k_4$) had the steepest slope, thus confirming the conclusion of the earlier graphic analyses (figure 6.5; Blank et al. 1967). Podoler and Rogers (1975) recognized that one could not test for the significance of the slope in the usual way because the axes in the regression were not independent. Manly (1977) offered a more definitive analytic approach based on partitioning the variance of $R$ into its additive components and constructing a variance–covariance matrix of all the $k$-values or causes of mortality. The key factor was the mortality or life stage with the largest variance component and was not always the same as that obtained with the earlier graphic methods (Manly 1977). Manly applied his technique to the partridge data given earlier and concluded that $k_7$ (losses in late winter due to natural causes) was the key factor. Whereas $k_4$ accounted for much of the variation in early season mortality, most of it was compensatory to the earlier mortality and thus not the main cause of population change (Manly 1977).

Key factor analysis was originally designed for univoltine insects that reproduced during a short-lived adult stage and survival data were confined to preadult stages (Morris 1959; Varley and Gradwell 1960). Problems arise when this technique is applied to organisms, including most vertebrates, for which reproduction extends over a substantial fraction of the typical life. Survival during the oldest age classes weighs equally with younger age classes in key factor analyses, yet older age classes may contribute little or even nothing to the reproduction of the next generation. Brown et al. (1993) offered a technique they called structured demographic accounting that was designed to solve this problem and others arising from key factor analysis. This method was similar to that of Manly (1977) in that the total variation in $R$ was decomposed into the variances and covariance of the component recruitment (births) and survivals during particular age classes. However, this process was done separately for each of several age classes so that variation in births as well as deaths due to each class could be properly assessed. A modification of this approach has been proposed by Silby and Smith (1998), who advocated calculating the impact of each $k$-value on population growth rate ($r$; $e^r = R$) rather than generational survival. Population growth rate is measured by standard life table methods, which account for age-specific fecundity and survival.

The limitations of key factor analysis have been documented by various authors. Kuno (1971) showed that sample error and compensatory density-dependent mortality can lead to spurious conclusions in key factor analyses. Manly (1977) acknowledges these as important limitations. Royama (1996) identified several further problems. One of these was that the factors causing population change may fluctuate on very different time scales and those responsible most of the variation in $R$ in a set of data may not be the factors responsible for the onset or decline of high-density conditions for species prone to outbreaks. Royama illustrated this with an example from spruce budworm, a major defoliator of forest trees in Canada. The rise and eventual decline of a population over a 10-year period were caused by a steady decline in larval survival, whereas recruitment (egg laying) fluctuated more from year to year and was identified by the techniques described earlier as the key factor. This example illustrated that key factor analyses may give misleading results or may not be easy to interpret. A single analytic process such as key factor analysis may not suffice to unravel the causes of density change.

### ■ Experimental Manipulation

The techniques listed in this chapter have been typically applied to naturally occurring, unmanipulated populations. Many ecologists have turned to experimental manipulations to provide proofs of the effects of various factors on population density or survival. For example, it is possible to establish experimental populations that differ in density and then to measure the mortality

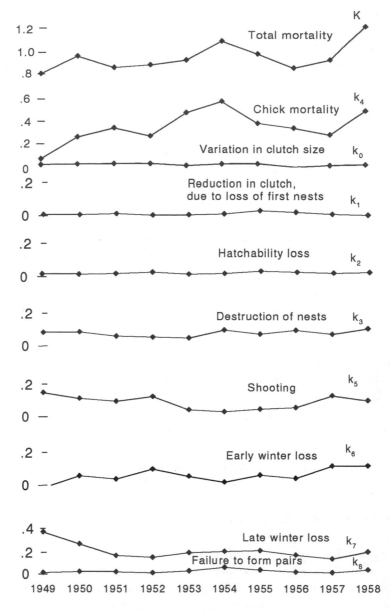

**A**

Partridge

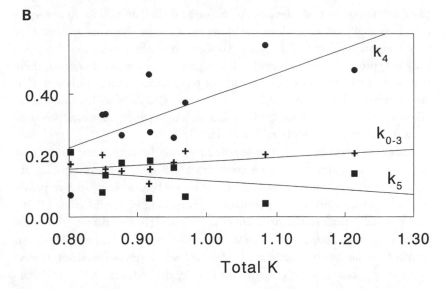

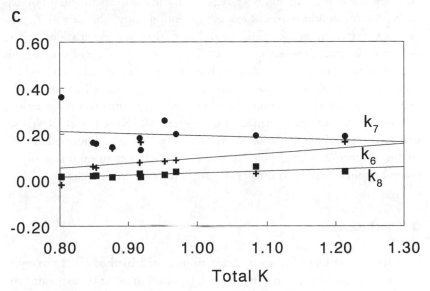

**Figure 6.5.** Key factor analysis of a population of the partridge *Perdix perdix* L. in England reported by Blank et al. (1967). In (A), generational change ($K = -R$) and individual sources of mortality are expressed as $k$-values: ($k = -\log(\text{survival})$) for each of eight sequential causes is plotted for each year of the study. In (B) and (C) $k$-values are regressed against generational change (total $K$). Redrawn from Podoler and Rogers (1975).

cause by various natural enemies. As indicated earlier, mortalities may vary in density spatially between plots, but these need not be the same as those that vary temporally in the same plots. Application of the techniques described early is appropriate only where the data represent time series. However, there is no reason why the time series could not be obtained from populations that are experimentally manipulated. A well-known example of such a study is the kilometer-scale exclosure study of the factors causing fluctuations in snowshoe hares in Yukon Territories (Krebs et al. 1995).

However, inferring the dynamic consequences of experimental results may not always be straightforward. A common problem is that experiments are almost inevitably done on a small scale and it may be inaccurate to extrapolate to the larger scale of natural populations. For example, Gould et al. (1990) manipulated gypsy moth density on 1-ha plots and showed that tachinid parasitoids decimated populations in a strongly density-dependent way. However, studies from natural populations revealed far lower levels of parasitism caused by these species and scant evidence for density dependence. Why the difference? Evidently the parasitoids aggregated to the 1-ha experimental populations from surrounding areas of low gypsy moth density. This density-dependent aggregation response would be nullified in natural populations, where densities rise simultaneously over much larger areas. The spatial scale in this study was thus crucial. Manipulation on a 1-ha spatial scale, although large relative to most ecological experiments, was not large enough to mimic the dynamics of natural populations. In other words, experimental manipulations may introduce a variety of artifacts that may be difficult to detect. Nevertheless, experimental manipulation almost always yields more information than studies of unmanipulated populations, particularly because unmanipulated control populations would usually be part of the experimental design.

## ■ Conclusions

This review has identified many limitations in the methods that ecologists have used to study the dynamics of populations. For all of these reasons the dynamics of even well-studied systems that have occupied the talents of the best minds in ecology remain unresolved and hotly debated. The best advice we can give to those who are embarking on such studies is to maintain a healthy skepticism of all the techniques and to take a multipronged approach. Wherever possible, studies of experimentally manipulated populations should be coupled with unmanipulated ones.

## Literature Cited

Andrewartha, H. G. and L. C. Birch. 1954. *The distribution and abundance of animals.* Chicago: University of Chicago Press.

Bellows, T. S. Jr., R. G. Van Driesche, and J. S. Elkinton. 1992. Life-table construction and analysis in the evaluation of natural enemies. *Annual Review of Entomology* 37: 587–614.

Berryman, A. and P. Turchin. 1997. Detection of delayed density dependence: comment. *Ecology* 78: 318–320.

Blank, T. H., T. R. E. Southwood, and D. J. Cross. 1967. The ecology of the partridge. I. Outline of population processes with particular reference to chick mortality and nest density. *Journal of Animal Ecology* 36: 549–556.

Box, G. E. P. and G. M. Jenkins. 1976. *Time series analysis: Forecasting and control.* Oakland, Calif.: Holden Day.

Brown, D., N. D. E. Alexander, R. W. Marrs, and S. Albon. 1993. Structured accounting of the variance of population change. *Journal of Animal Ecology* 62: 490–502.

Burnett, T. 1958. A model of host–parasite interaction. *Proceedings of the 10th International Congress, Entomology* 2: 679–686.

Dempster, J. P. 1983. The natural control of populations of butterflies and moths. *Biological Reviews of the Cambridge Philosophical Society* 58: 461–481.

den Boer, P. J. 1968. Spreading of risk and stabilization of animal numbers. *Acta Biotheoretica* 18: 165–194.

Dennis, B. and M. L. Taper. 1994. Density dependence in time series observations of natural populations: Estimation and testing. *Ecological Monographs* 64: 205–224.

Eberhardt, L. L. 1970. Correlation, regression, and density-dependence. *Ecology* 51: 306–310.

Elkinton, J. S., J. P. Buonaccorsi, T. S. Bellows, and R. G. van Driesche. 1992. Marginal attack rate, k-values and density dependence in the analysis of contemporaneous mortality factors. *Researches on Population Ecology* 34: 29–44.

Elkinton, J. S., W. M. Healy, J. P. Buonaccorsi, G. H. Boettner, A. M. Hazzard, H. R. Smith, and A. M. Liebhold. 1996. Interactions among gypsy moths, white-footed mice and acorns. *Ecology* 77: 2332–2342.

Elton, C. S. and M. Nicholson 1942. The ten year cycle in numbers of lynx in Canada. *Journal of Animal Ecology* 11: 215–244.

Gould, J. R., J. S. Elkinton, and W. E. Wallner. 1990. Density-dependent suppression of experimentally created gypsy moth, *Lymantria dispar* (Lepidoptera: Lymantriidae), populations by natural enemies. *Journal of Animal Ecology* 59: 213–234.

Hanski, I. 1989. Metapopulation dynamics: Does it help to have more of the same? *Trends in Ecology and Evolution* 4: 113–114.

Hanski, I. and I. Woiwod. 1991. Delayed density-dependence. *Nature* 350: 28.

Hassell, M. P. 1985. Insect natural enemies as regulating factors. *Journal of Animal Ecology* 54: 323–334.

Hassell, M. P. and C. B. Huffaker. 1969. The appraisal of delayed and direct density-dependence. *Canadian Entomologist* 101: 353–361.

Hassell, M. P., J. H. Lawton, and R. M. May. 1976. Patterns of dynamical behaviour in single-species populations. *Journal of Animal Ecology* 45: 471–486.

Hassell, M. P., T. R. E. Southwood, and P. M. Reader. 1987. The dynamics of the viburnum whitefly (*Aleurotrachelus jelinekii*): A case study of population regulation. *Journal of Animal Ecology* 56: 283–300.

Holyoak, M. 1993. New insights into testing for density dependence. *Oecologia* 93: 435–444.

Holyoak, M. 1994a. Identifying delayed density dependence in time-series data. *Oikos* 70: 296–304.

Holyoak, M. 1994b. Appropriate time scales for identifying lags in density dependent processes. *Journal of Animal Ecology* 63: 479–483.

Howard, L. O. and W. F. Fiske. 1911. *The importation into the United States of the parasites of the gipsy-moth and the brown-tail moth*. U.S. Department of Agriculture, Bureau of Entomology, bulletin no. 91.

Krebs, C. J. 1991 The experimental paradigm and long term population studies. *Ibis* 133(suppl 1): 3–8.

Krebs, C. J., S. Boutin, R. Boonstra, A. R. E. Sinclair, J. N. M. Smith, M. R. T. Dale, K. Martin, and R. Turkington. 1995. Impact of food and predation on the snowshoe hare cycle. *Science* 269: 1112–1115.

Kuno, E. 1971. Sampling error as a misleading artifact in "key factor analysis." *Researches on Population Ecology* 13: 28–45.

Lack, D. 1954. *The natural regulation of animal numbers*. New York: Oxford University Press.

Levins, R. 1969. Some demographic and genetic consequences of environmental heterogeneity for biological control. *Bulletin of the Entomological Society of America* 15: 237–240.

Manly, B. F. J. 1977. The determination of key factors from life table data. *Oecologia* 31: 111–117.

May, R. M. 1974. Biological populations with non-overlapping generations: Stable points, stable cycles, and chaos. *Science* 186: 645–647.

May, R. M. 1976. Simple mathematical models with very complicated dynamics. *Nature* 261: 459–467.

Morris, R. F. 1959. Single-factor analysis in population dynamics. *Ecology* 40: 580–588.

Nicholson, A. J. 1933. The balance of animal populations. *Journal of Animal Ecology* 2: 132–178.

Nicholson, A. J. 1957. The self-adjustment of populations to change. *Cold Springs Harbor Symposia on Quantitative Biology* 22: 153–172.

Podoler, H. and D. Rogers. 1975. A new method for the identification of key factors from life-table data. *Journal of Animal Ecology* 44: 85–114.

Pollard, E., K. H. Lakhani, and P. Rothery. 1987. The detection of density-dependence from a series of annual censuses. *Ecology* 68: 2046–2055.

Reddingius, J. 1971. Gambling for existence: A discussion of some theoretical problems in animal population ecology. *Acta Biotheoretica* 20: 1–208.

Reddingius, J. 1990. Models for testing. A secondary note. *Oecologia* 85: 50–52.

Royama, T. 1977. Population persistence and density dependence. *Ecological Monographs* 47: 1–35.

Royama, T. 1992. *Analytical population dynamics.* New York: Chapman & Hall.

Royama, T. 1996. A fundamental problem in key factor analysis. *Ecology* 77: 87–93.

St. Amant, J. L. S. 1970. The detection of regulation in animal populations. *Ecology* 51: 823.

Schaffer, W. M. and M. Kot. 1985. Nearly one dimensional dynamics in an epidemic. *Journal of Theoretical Biology* 112: 403–427.

Shenk, T. M., G. C. White, and K. P. Burnham. 1998. Sampling-variance effects on detecting density dependence from temporal trends in natural populations. *Ecological Monographs* 68: 445–463.

Silby, R. M. and R. H. Smith 1998. Identifying key factors using ( contribution analysis. *Journal of Animal Ecology* 67: 17–24.

Smith, F. E. 1961. Density-dependence in the Australian thrips. *Ecology* 42: 403–407.

Solow, A. R. 1990. Testing for density dependence. *Oecologia* 83: 47–49.

Solow, A. R. and J. H. Steele. 1990. On sample size, statistical power, and the detection of density dependence. *Journal of Animal Ecology* 59: 1073–1076.

Stiling, P. D. 1987. The frequency of density dependence in insect host–parasitoid systems. *Ecology* 68: 844–856.

Turchin, P. 1990. Rarity of density dependence or population regulation with lags? *Nature* 344: 660–663.

Turchin, P. 1995. Population regulation: Old arguments and a new synthesis .In N. Cappuccino and P. W. Price, eds., *Population dynamics: New approaches & synthesis,* 19–41. San Diego: Academic Press.

Turchin, P. and A. D. Taylor. 1992. Complex dynamics in ecological time series. *Ecology* 73: 289–305.

van Driesche, R. G., T. S. Bellows, J. S. Elkinton, J. Gould, and D. N. Ferro. 1991. The meaning of percentage parasitism revisited: Solutions to the problem of accurately estimating total losses from parasitism in a host. *Environmental Entomology* 20: 1–7.

Varley, G. C. 1949. Population changes in German forest pests. *Journal of Animal Ecology* 18: 117–122.

Varley, G. C. and G. R. Gradwell. 1960. Key factors in population studies. *Journal of Animal Ecology* 29: 399–401.

Varley, G. C. and G. R. Gradwell. 1968. Population models for the winter moth. In T. R. E. Southwood, ed., *Symposia of the Royal Entomological Society of London no. 4: Insect abundance,* 132–142. Oxford, U.K.: Blackwell Scientific.

Varley, G. C., G. R. Gradwell, and M. P. Hassell. 1973. *Insect population ecology.* Oxford, U.K.: Blackwell Scientific.

Vickery, W. L. 1991. An evaluation of bias in k-factor analysis. *Oecologia* 85: 413–418.

Watt, K. E. F. 1964. Density dependence in population fluctuations. *Canadian Entomologist* 96: 1147–1148.

Williams, D. W., R. W. Fuester, W. W. Balaam, R. J. Chianese, and R. C. Reardon. 1992.

Incidence and ecological relationships of parasitism in larval populations of *Lymantria dispar. Biological Control* 2: 35–43.

Williams, D. W. and A. M. Liebhold. 1995. Detection of delayed density dependence: Effects of autocorrelation in an exogenous factor. *Ecology* 76: 1005–1008.

Williams, D. W. and A. M. Liebhold. 1997. Detection of delayed density dependence: Reply. *Ecology* 78: 320–322.

Wolda, H. 1995. The demise of the population regulation controversy? *Researches on Population Ecology* 37: 91–93.

## Chapter 7

# Monitoring Populations

JAMES P. GIBBS

Assessing changes in local populations is the key to understanding the temporal dynamics of animal populations, evaluating management effectiveness for harvested or endangered species, documenting compliance with regulatory requirements, and detecting incipient change. For these reasons, population monitoring plays a critical role in animal ecology and wildlife conservation. Changes in abundance are the typical focus, although changes in reproductive or survival rates that are the characteristics of individuals, or other population parameters, also are monitored. Consequently, many researchers and managers devote considerable effort and resources to population monitoring. In doing so, they generally assume that systematic surveys in different years will detect the same proportion of a population in every year and changes in the survey numbers will reflect changes in population size.

Unfortunately, these assumptions are often violated. In particular, the following two questions are pertinent to any animal ecologist involved in population monitoring. First, is the index of population abundance used valid? That is, does variation in, for example, track densities of mammals, amphibian captures in sweep nets, or counts of singing birds reliably reflect changes in local populations of these organisms? Second, does the design of a monitoring program permit a reasonable statistical probability of detecting trends that might occur in the population index? In other words, are estimates of population indices obtained across a representative sampling of habitats and with sufficient intensity over time to capture the trends that might occur in the population being monitored? Failure to address these questions often results in costly monitoring programs that lack sufficient power to detect population trends (Gibbs et al. 1998).

The purpose of this chapter is to assess key assumptions made by animal ecologists attempting to identify population change and to make practical suggestions for improving the practice of population monitoring. This is done within a framework of statistical power analysis, which incorporates the explicit tradeoffs animal ecologists make when attempting to obtain statistically reliable information on population trends in a cost-effective manner (Peterman and Bradford 1987). The chapter covers five topics. First, the use and misuse of population indices are reviewed. Second, sampling issues related to the initial selection of sites for monitoring are discussed. Third, a numerical method is described for assessing the balance between monitoring effort and power to detect trends. Fourth, a review of the most critical influence on power to detect trends in local populations, the temporal variability inherent in populations, is presented, based on an analysis of over 500 published, long-term counts of local populations. Fifth, the numerical method and variability estimates are integrated to generate practical recommendations to animal ecologists for improving the practice of monitoring local populations.

## ■ Index–Abundance Relationships

### TYPES OF INDICES

Making accurate estimates of absolute population size is difficult. Animals often are difficult to capture or observe, they are harmed in the process, or the associated costs and effort of making absolute counts or censuses are prohibitive. Therefore, animal ecologists often rely on indices of population size and monitor these indices over time as a proxy for monitoring changes in actual population size. Indices may be derived from sampling a small fraction of a population using a standardized methodology, with index values expressed as individuals counted per sampling unit (e.g., fish electroshocked per kilometer of shoreline, tadpoles caught per net sweep, salamanders captured per pitfall trap, birds intercepted per mist net, or carcasses per kilometer of road). These examples involve direct counts of individuals. When individuals of a species under study are difficult to capture or observe, another class of indices makes use of indirect evidence to infer animal presence. Auditory cues are often used as indirect indices (e.g., singing birds per standard listening interval, overall sound volume produced by insect aggregations, howling frequency by packs of wild canids, or calling intensity in frog choruses). Other indirect indices are based only on evidence of animal activity (e.g., droppings per unit area, tracks per unit transect length or per bait station, or quantity of food stored per den).

## INDEX–ABUNDANCE FUNCTIONS

An index to population size (or abundance) is simply any "measurable correlative of density" (Caughley 1977) and is therefore presumably related in some manner to actual abundance. Most animal ecologists assume that the index and actual abundance are related via a positive, linear relationship with slope constant across habitats and over time. In some situations, these relationships hold true (figure 7.1a, b, c). However, the relationship often takes other forms in which changes in the index may not adequately reflect changes in the actual population (figure 7.1d, e, f).

A nonlinear (asymptotic) relationship may be common in situations where the index effectively becomes saturated at high population densities. Such may be the case for anurans monitored using an index of calling intensity (Mossman et al. 1994). The index is sensitive to changes at low densities of calling male frogs in breeding choruses because calls of individuals can be discriminated by frog counters. At higher densities, however, calls of individual frogs overlap to an extent that size variation of choruses cannot be discriminated by observers. In other words, the index increases linearly and positively with abundance to a threshold population density, and then becomes asymptotic.

Another example of a nonlinear index–abundance relationship concerns use of presence/absence as a response such that the proportion of plots occupied by a given species is the index of abundance. At low population densities, changes in population size can be reflected in changes in degree of plot occupancy. Once all plots are occupied, however, further population increases are not reflected by the index because the index becomes saturated at 100 percent occupancy. A final example involves bait stations for mammals (Conroy 1996), which may be frequented by subdominant animals more at low population densities than at high densities because of behavioral inhibition. The main implication of this type of nonlinear index–abundance relationship is that it prevents detection of population change (in any direction) above the saturation point of the index.

A threshold relationship also may occur in index–abundance relationships if the index effectively bottoms out at low population densities. For example, if sample plots are too small, listening intervals too short, or sample numbers too few, observers may simply fail to register individuals even though they are present at low densities (Taylor and Gerrodette 1993). Consequently, detection of population change below the threshold of the index is precluded. This situation probably occurs in surveys for many rare, endangered, or uncommon species (Zielinski and Stauffer 1996). The threshold and saturation phenomena can combine in some situations. For example, because calling behavior

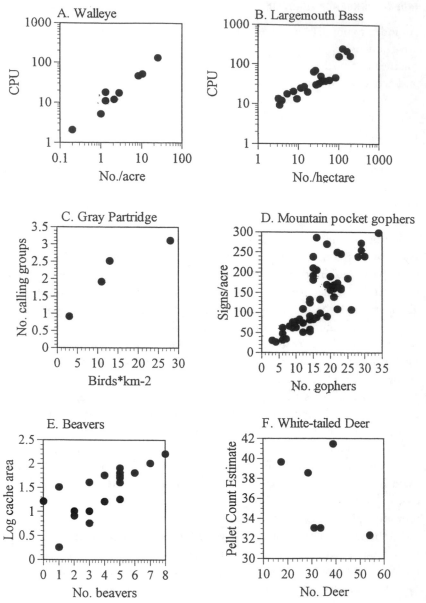

**Figure 7.1** Relationship between population indices (vertical axis) and actual animal abundance (horizontal axis). (A) From Serns (1982), (B) from Hall (1986), (C) from Rotella and Ratti (1986), (D) from Reid et al. (1966), (E) from Easter-Pilcher (1990), (F) from Ryel (1959).

may be stimulated by group size in frogs, individuals may not call (or may do so infrequently) when choruses are small and may be overlooked by frog counters, but increasing numbers of calling frogs above a certain threshold may also be indistinguishable to frog counters.

Occasionally indices used have no relationship to abundance (figure 7.1f), although sometimes an apparent lack of an index–abundance relationship may simply be a result of sampling error or too few samples taken to verify the relationship (Fuller 1992; White 1992). Nevertheless, the possibility that a seemingly reasonable, readily measured index has no relationship to the actual population must always be considered by animal ecologists using an unverified index, and preferably be examined as a null hypothesis during a pilot study.

## VARIABILITY OF INDEX–ABUNDANCE FUNCTIONS

Independent of the specific form of the index–abundance relationship, most researchers assume it to be constant among habitats and over time. However, in perhaps the most comprehensive validation study of an indirect index, a study by Reid et al. (1966) on mountain pocket gophers (*Thomomys talpoides*), the index used (numbers of mounds and earth plugs) consistently displayed a positive, linear relationship to actual gopher numbers, whereas the intercept and slope varied substantially between habitats (figure 7.2a, b). Other situations, such as electroshocking freshwater fishes, apparently yield comparable index–abundance relationships between habitats despite large differences in densities between habitats (figure 7.2c, d). In contrast, index–abundance relationships in different habitats can be reversed (figure 7.2e, f) although these examples may be compromised by sampling error. Finally, the slope, intercept, and precision of the relationship may vary among years within the same habitats (figure 7.3a, b, c).

Inferences about population change drawn from indices are also often hampered by sampling error. Whatever the form of the index–abundance relationship between habitats and over time, the precision of the relationship can be quite low (figure 7.1d, e). This is particularly true for indirect indices, in which variation is strongly influenced by environmental factors such as weather and time of day, as well as by observers (Gibbs and Melvin 1993). Such index variation can substantially reduce the power of statistical tests examining changes in index values between sites or over time (Steidl et al. 1997).

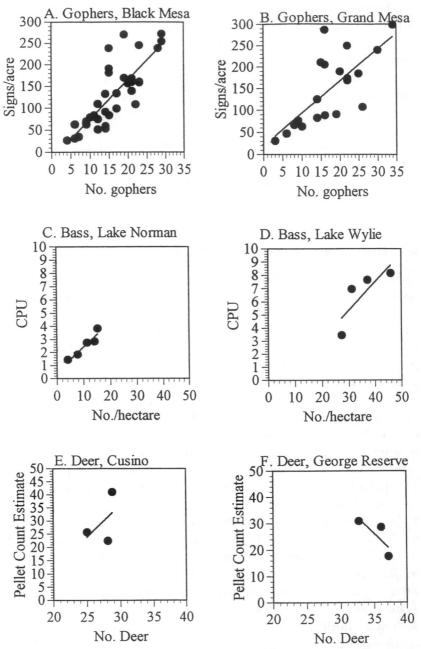

**Figure 7.2** Variation between habitats in index–abundance relationships. (A) and (B) From Reid et al. (1966), (C) and (D) from McInerny and Degan (1993), (E) and (F) from Eberhardt and Van Etten (1956).

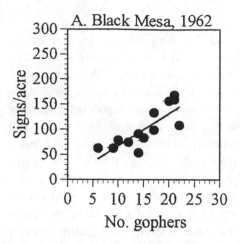

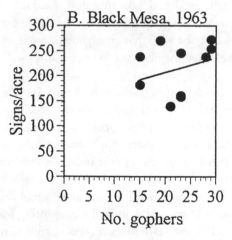

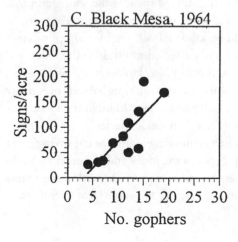

**Figure 7.3** Variation in the index–abundance relationship over time at the same site. From Reid et al. (1966).

## IMPROVING INDEX SURVEYS

The few studies attempting to validate indices suggest that population indices and absolute abundances are rarely related via a simple positive, linear relationship with slope constant across habitats and over time. Thus animal ecologists would do well to proceed cautiously when designing and implementing index surveys. In particular, index validation should be considered a necessary precursor to implementing index surveys. Some guidance on the relationship of the index to abundance may be found in the literature, but index validation studies are rare. Lacking such information, conducting a pilot study using the index in areas where abundance is known or can be estimated is useful. Such a validation study would need to be replicated across multiple sites that exhibit variation in population size or density, or over time at a site where abundance varies over time. Making multiple estimates of the index:abundance ratio at each site and time period is also useful so that the contribution of sampling error to the overall noise in the index–abundance relationship among sites can be estimated. Validation studies also may be advisable throughout a monitoring program's life span because the index may need to be periodically calibrated or updated (Conroy 1996).

Ecologists should also be aware that developing indices that have a 1:1 relationship with abundance will most reliably reflect changes in abundance. If the slope describing the index–abundance relationship is low, then large changes in abundance are reflected in small changes in the index. Such small changes in the index are more likely to be obscured by variation in the index–abundance relationship than if the slope of the index–abundance relationship were higher.

Methods of reducing index variability and increasing the precision of the index–abundance relationship include adjusting the index by accounting for auxiliary variables such as weather and observers. In practice, these factors may be overlooked if many years of data are gathered because the short-term bias they introduce typically is converted simply to error in long-term data sets. In an ideal situation, each index would be validated, adjusted for sampling error by accounting for external variables, and corrected to linearize the index and make it comparable across habitats and over years. However, this is rarely an option for regional-scale surveys conducted across multiple habitats over many years by many people and involving multiple species, although it may be possible for local monitoring programs focused on single species.

The following advice may be useful to animal ecologists for improving index surveys. First, the basic relationship between the index and abundance should be ascertained to determine whether the index might yield misleading results and therefore should not be implemented. Second, any results from trend analysis of index data should be considered in light of potential limitations imposed by the index–abundance relationship. For example, saturated indices could be the cause of a failure to detect population changes. Most importantly, animal ecologists must be cautious about concluding that a lack of trend in a time series of index data indicates population stability. Often an index may be unable to capture population change because of a flawed index–abundance relationship or simply excessive noise caused by sampling error in the index.

# ■ Spatial Aspects of Measuring Changes in Indices

Many animal ecologists are concerned with monitoring multiple local populations with the intent of extrapolating changes observed in those populations to larger, regional populations. In such a case, the sample of areas monitored must be representative of areas in a region that are not sampled if observed trends are to be extrapolated to regional populations. Selection of sites for monitoring is therefore a key consideration for animal ecologists concerned with identifying change in regional populations.

Balancing sampling needs and logistical constraints in the design of regional monitoring programs can be problematic, however. For sampling areas to be representative, random selection of sites for surveying is advised, but a purely random scheme for site selection is often unworkable in practice. For example, sites near roadsides and those on public lands are generally easier to access by survey personnel than are randomly selected sites. Also, monitoring sites that occur in clusters minimize unproductive time traveling among survey sites. Time is generally at a premium in monitoring efforts not only because of the costs of supporting survey personnel but also because the survey window each day or season for many animals is brief.

A simple random sample of sites may also produce unacceptably low encounter rates for the organisms being monitored (too many zero counts to be useful). This could be overcome by stratifying sampling according to habitat types frequented by the species being monitored. However, information on habitat distributions in a region from which a stratified random sampling scheme might be developed often is not available to researchers. Furthermore, prior knowledge of habitat associations of most species that can be used as a basis for stratification often is not available. Finally, ecologists often monitor multiple species for which a single optimal sampling strategy may simply not be identifiable.

These difficulties in implementing random sampling schemes imply that nonrandom site selection schemes may be the most practical way to organize sampling for monitoring programs. However, animal ecologists would do well to be aware of the serious and lasting potential consequences of nonrandom site selection. Researchers initiating a survey program are often drawn to sites with abundant populations, where counts are initiated under the rationale that visiting low-density or unoccupied sites will be unproductive. If the populations or habitats under study cycle, however, then initial counts may be made at cycle peaks. As time progresses, populations at the sites selected will then tend, on average, to decline. The resulting pattern of decline observed in counts is an artifact of site selection procedures and does not reflect any real population trend. This sampling artifact can lead researchers to make erroneous conclusions about regional population trends. This problem has compromised a regional monitoring program for amphibians (Mossman et al. 1994) and regional game bird surveys (Foote et al. 1958).

These examples highlight why site selection can be an important pitfall in designing monitoring programs. Unfortunately, few simple recommendations can be made for guiding the process. A detailed knowledge of habitat associations of the species under study, as well as the distribution of those habitats in a region, can provide useful guidance to animal ecologists in selecting a sampling design that is logistically feasible to monitor. Stratifying (or blocking) sampling effort based on major habitat features such as land cover type will almost always yield gains in precision of population estimates each sampling interval (see Thompson 1992). Specifically, researchers would do well to identify species–habitat associations and generate regional habitat maps before initiating surveys so that the explicit tradeoffs between alternative sampling schemes, logistical costs, and sampling bias can be evaluated. One workable solution to this problem involves two steps. First, populations at selected sites that are presumably representative of particular habitat strata in a region are rigorously monitored. Second, an independent program is established that explicitly monitors changes in the distribution and abundance of habitats in the region. Trends in habitats can then be linked to trends in populations at specific sites to extrapolate regional population trends.

## ■ Monitoring Indices Over Time

Once animal ecologists attempting to monitor populations have addressed issues of index validity and sampling schemes for selecting survey sites, another set of issues related to the intensity of monitoring over time must be considered. These issues include how many plots to monitor, how often to survey plots in any given year, the interval and duration of surveys over time, the magnitude of sampling variation that occurs in abundance indices, and the magnitude of trend variation in local populations in relation to overall trends in regional populations (Gerrodette 1987). Other less obvious but often equally important factors to be considered include $\alpha$ levels and desired effect sizes (trend strengths) set by researchers (Hayes and Steidl 1997; Thomas 1997). Specifically, researchers need to specify the probabilities at which they are willing to make statistical errors in trend detection, that is, the probability of wrongly rejecting the null hypothesis of no trend (at a probability = $\alpha$, that is, the level of significance) and of wrongly accepting the null hypothesis of no trend (at a probability = $\beta$). Furthermore, the statistical method chosen to examine trends in a count series also can influence the likelihood of detecting them (Hatfield et al. 1996). Understanding how these factors interact with the inherent sampling variation of abundance indices can provide insights into the design of statistically powerful yet labor-efficient monitoring programs (Peterman and Bradford 1987; Gerrodette 1987; Taylor and Gerrodette 1993; Steidl et al. 1997).

Statistical power underlies these issues and provides a useful conceptual framework for biologists designing studies that seek to identify population change. The key problem identifying population change is that sources of noise in sample counts obscure the signal associated with ongoing population trends. Trends represent the sustained patterns in count data (the signal) that occur independently of cycles, seasonal variations, irregular fluctuations that are sources of sampling error (the noise) in counts. Statistical power simply represents the probability that a biologist using a particular population index in conjunction with a specific monitoring protocol will detect an actual trend in sample counts, despite the noise in the count data. In a statistical context, power is the probability that the null hypothesis of no trend will be rejected when it is, in fact, false, and is calculated as $1 - \beta$.

Although statistical power is central to every monitoring effort, it is rarely assessed (Gibbs et al. 1998). Consequences of ignoring power include collecting insufficient data to reliably detect actual population trends. Occasionally, collection of more data than is needed occurs. Unfortunately, until recently few tools have been available to animal ecologists that permit assessment of statistical power for trends (Gibbs and Melvin 1997; Thomas 1997).

## POWER ESTIMATION FOR MONITORING PROGRAMS

The large numbers of factors that interact to determine the statistical power of a monitoring program make power estimation a complex undertaking. Analytical approaches are forced by the large number of variables involved to oversimplify the problem (Gerrodette 1987). Because of the complexities involved in generating power estimates for monitoring programs, the problem may be most tractable with simulation methods. Accordingly, a conceptually straight-forward Monte Carlo approach based on linear regression analysis has been devised (table 7.1; Gibbs and Melvin 1997). With this approach a researcher defines the basic structure of a monitoring program and provides a variance estimate for the population index used. Simulations are then run in which many sets of sample counts are generated based on the structure of the monitoring program with trends of varying strength underlying them. The frequency with which trends are detected in the counts, despite the sampling error imposed by the population index and the structure of the monitoring program, reflects the power of the monitoring design to detect trends. The simulation program is particularly useful for evaluating the tradeoffs between sampling effort, logistical constraints, and power to detect trends. The simulation software ("monitor.exe") has been adapted for general use on DOS-based microcomputers, and is available from the author or via the Internet at http://www.im.nbs.gov/powcase/powcase.html.

## VARIABILITY OF INDICES OF ANIMAL ABUNDANCE

A key influence on power to detect a given population trend is the variability of the population index used. Power to detect trends is inversely related to the magnitude of index variability and monitoring programs must be designed around the component of index variability that cannot be controlled (Gerrodette 1987). In other words, sufficient numbers of plots must be monitored frequently enough to capture trends despite the inherent variability of the population index. Without pilot studies, however, researchers often have no estimate of population variability. Lacking estimates of this critical parameter impairs the ability of animal ecologists to design statistically powerful monitoring programs.

A ready source of data on the variability of population indices can be found in published time series of population counts. Hundreds of long-term population studies for a variety of taxa have been published in the last century, albeit mostly for temperate-zone organisms. Because most of these population series were generated using population indices, not population censuses, presumably variation in these count series reflects both environmental variation in the populations and sampling error associated with the counting methodology. As long as the time series are of sufficient and comparable duration, significant trends have been removed from them, and sufficient numbers of studies have been made, approximations of index variability can be estimated. Further-

**Table 7.1    Monte Carlo Simulation Procedure Used to Estimate the Power of Population-Monitoring Programs to Detect Trends**

| Step | Procedure |
|---|---|
| 1. | Basic structure of the monitoring program is defined (i.e., number of plots surveyed, survey frequency, and a series of survey years). |
| 2. | Deterministic linear trends are projected from the initial abundance index on each plot over the series of survey years. |
| 3. | Sample counts are generated at each survey occasion across all plots and for each trend. Sample counts are random deviates drawn from a normal distribution (truncated at 0) with mean equal to the deterministic projection on a particular monitoring occasion and with a variance approximated by the standard deviation in initial abundance (constant variances over time). |
| 4. | The slope of a least-squares regression of sample abundances versus survey occasion is determined for each plot and each trend. |
| 5. | The mean and variance for slope estimates are calculated across plots for each trend. |
| 6. | Whether the mean slope estimate is statistically different from zero for each trend is determined. |
| 7. | Steps 1 through 6 are repeated many times, whereupon the proportion of repetitions in which the mean slope estimate was different from zero is determined. The resulting proportion represents the power estimate, which ranges from 0 (low power) to 1 (high power) and indicates how often the survey program correctly detected an ongoing trend. |

more, these estimates can be integrated with power analyses to provide general guidance on sampling protocols that animal ecologists can use to design robust monitoring programs for local populations.

To this end, count series of local animal and plant populations that extended more than 5 years were obtained by examining 25 major ecology journals published from 1940 to the present (nonwoody plants are also presented here because animal ecologists often must monitor plant populations in the course of their animal studies). Variability of each count series thus obtained was estimated by dividing the standard deviation of the counts by the mean count to determine the coefficient of variation (CV). To remove trends in the counts (which might have inflated variance estimates), the standard deviation was determined from the standardized residuals of a linear regression of counts against time. Furthermore, because the variability of a time series is related in part to its length (Warner et al. 1995), a 5-year moving CV (similar in concept to a moving average) was calculated for each count series. (However, most studies of birds, moths, and butterflies failed to present raw counts that could be detrended and standardized, so the means and error terms as presented in these studies were used. The index variabilities for these groups are therefore potentially biased high in relation to those estimates for other taxa). CVs were subsequently averaged within groups of taxonomically and ecologically related species.

A total of 512 time series for local animal and plant populations were analyzed (appendix 7.1), which provided estimates to calculate average index variabilities for each of 24 separate taxonomic and ecological groups (table 7.2). Few groups had low variability indices (CV below 25 percent), including large mammals, grasses and sedges, and herbs. A larger number had intermediate variability indices (CV 25–50 percent), including turtles, terrestrial salamanders, large birds, lizards, salmonid fishes, and caddis flies. Most groups had indices with CVs between 50–100 percent, including snakes, dragonflies, small-bodied birds, beetles, small mammals, spiders, medium-sized mammals, nonsalmonid fishes, pond-breeding salamanders, moths, frogs and toads, and bats. Finally, only butterflies and drosophilid flies had average indices with CVs above 100 percent. Although a pilot study is clearly preferable, lacking one of their own animal ecologists can refer to the specific studies (appendix 7.1) or to the summary (table 7.2) for information useful for designing monitoring programs for a particular species.

It is important to note that index variabilities (table 7.2) reflect temporal variation inherent in populations as well as sampling error associated with the counting methods. For example, direct count methods were used most often for those groups with the lowest index variability, including large mammals, all plants, terrestrial salamanders, and large-bodied birds. An exception was butterflies, which typically were counted with time-constrained visual searches. Nets and traps were used to capture individuals in most remaining groups. Trapping methods that sampled only a segment of a population (e.g., frogs, toads, and pond-breeding salamanders on breeding migrations) or that relied on attractants (e.g., most small- and medium-sized mammals at bait stations, moths and caddis flies at light traps, and drosophilid flies at fruit baits) were associated with high index variabilities. Similarly, most studies of small-bodied birds were based on counts of singing individuals and also displayed high variability. Both method-associated sampling error and inherent population variability clearly make important contributions to overall index variability, and the recommendations that follow assume that researchers will use the same standardized counting methods used by the researchers who generated the count series analyzed here (appendix 7.1).

**Table 7.2   Variability Estimates for Local Populations**

| Group | N | CV |
|---|---|---|
| Mammals, large | 17 | 0.142 |
| Grasses and sedges | 16 | 0.209 |
| Herbs, Compositae | 9 | 0.213 |
| Herbs, non-Compositae | 32 | 0.225 |
| Turtles | 7 | 0.333 |
| Terrestrial salamanders | 8 | 0.354 |
| Large-bodied birds | 25 | 0.363 |
| Lizards | 11 | 0.420 |
| Fishes, salmonids | 42 | 0.473 |
| Caddis flies | 15 | 0.497 |
| Snakes | 9 | 0.541 |
| Dragonflies | 8 | 0.566 |
| Small-bodied birds | 73 | 0.569 |
| Beetles | 20 | 0.580 |
| Small mammals | 14 | 0.597 |
| Spiders | 10 | 0.643 |
| Medium-sized mammals | 22 | 0.647 |
| Fishes, nonsalmonids | 30 | 0.709 |
| Pond-breeding salamanders | 10 | 0.859 |
| Moths | 63 | 0.903 |
| Frogs and toads | 21 | 0.932 |
| Bats | 24 | 0.932 |
| Butterflies | 13 | 1.106 |
| Flies, drosophilids | 13 | 1.314 |

CV = coefficient of variation, N = number of detrended count series of at least 5 years' duration obtained from the literature. Values are average coefficients of variation (standard deviation/mean) for standardized 5-year count series. Data sources are listed in appendix 7.1.

## SAMPLING REQUIREMENTS FOR ROBUST MONITORING PROGRAMS

Estimates of index variabilities (table 7.2) were incorporated into a power analysis (table 7.1) to generate sampling recommendations for animal ecologists for designing effective programs for monitoring local populations. The power analysis assumed the following logistical constraints. Resources available for a local or regional monitoring program would permit surveys of up to 500 plots or subpopulations on one to five occasions annually over a monitoring period of 10 years. Average plot counts for all groups were assumed to equal 10, with count variances comparable to the average value calculated for each group based on the literature survey (table 7.2). Trends in the population index were assumed to be linear, $\alpha$ and $\beta$ were set at 0.05, and tests of significance were two-sided. Within this framework, sampling requirements to detect overall changes in population indices of 10 percent, 25 percent, and 50 percent for each group were estimated.

This analysis (table 7.3) indicated that infrequent monitoring (for example, once or twice per year) on a small number of sites or plots (10 or less) would reliably detect strong population trends (that is, a 50 percent change over 10 years) in most groups. Even for highly variable groups frequent monitoring (three to five times per year) of a small number of plots (30 or less) would permit detection of a trend of this magnitude. However, more intensive monitoring is needed to detect weaker trends of 25 percent and 10 percent, but nevertheless is still at a logistically feasible level (100 or fewer plots) for animal ecologists to undertake for most groups. The sampling requirements become more modest if significance levels are relaxed. For example, setting $\alpha = \beta = 0.10$ reduced the sampling requirements in table 7.3 by, on average, 20 percent The main utility of these results (table 7.3) is to provide a reference for animal ecologists to consult when planning monitoring activities or assessing the effectiveness of existing programs. Note that stringent $\alpha$ and $\beta$ levels (0.05) were used to generate these results. Less stringent levels may well be more appropriate in a monitoring context (Gibbs et al. 1998). Sampling recommendations using other combinations of $\alpha$ and $\beta$ are provided over the Internet at http://www.im.nbs.gov/powcase/powcase.html.

A caveat is that these recommendations are based on the assumption that trends in populations are fixed and linear. This is appropriate in certain situations, such as declining endangered species or increasing introduced species, whose populations often follow deterministic trends. However, most populations monitored follow an irregular trajectory. Furthermore, trends in a particular local population probably represent a random sample of a spatially variable, regional population trend. The simulation software described (table 7.1) can accommodate random trend variation among plots or sites if estimates of its magnitude are available.

## SETTING OBJECTIVES FOR A MONITORING PROGRAM

It is important to emphasize that conclusions drawn from these analyses are contingent on the initial statement of a monitoring program's objectives. Power estimates are influenced by many factors controlled by researchers, such as duration and interval of monitoring, count means and variances, and number of sites and counts made per season. Several other, somewhat arbitrary factors also exert an important influence on power estimates. These include trend strength (effect size), significance level (type I error rate), and the number of tails to use in statistical tests. It is therefore critical that animal ecologists establish explicit and well-reasoned monitoring objectives before the initiation of any monitoring program (Steidl et al. 1997; Thomas 1997). These goals should address what magnitude of change in the population index is sought for detection, what probability of false detections will be tolerated (a type I error = $\alpha$), and what frequency of true declines can go undetected (a type II error = $\beta$, with power = $1 - \beta$). An initial statement of objectives is important because subsequent efforts to judge the success or failure of a monitoring program are made in terms of those objectives.

## ■ Conclusions

Identifying change in local populations is fraught with difficulties. Dubious population indices, bias in selection of survey sites, and weak design of monitoring programs can undermine trend detection. The practice of assessing population change in animal ecology could therefore be improved substantially. First, one should not blindly assume that any readily measured population index can serve as a valid proxy for estimating actual abundance. As an alternative, performing simple pilot studies to ascertain the basic relationship between the index used and actual abundance will give animal ecologists much insight. Such pilot studies can indicate whether the index used might yield misleading results, how it might be modified, and how it could potentially compromise trend detection. Second, animal ecologists must be aware of the potential pitfalls of nonrandom schemes for selecting sites for monitoring. A major challenge is to devise sampling methods that permit unbiased and statistically powerful surveys to be made in a logistically feasible manner. Finally, conducting power analyses during the pilot phase of a monitoring program is critical because it permits an assessment of a program's potential for meeting its stated goals while the opportunity for altering the program's structure is still available. The simulation method outlined and the summary of taxon-specific index variabilities can provide animal ecologists just such an option.

Successful monitoring of populations is based on making the best choices among sampling designs that yield precise estimates of a population index, statistical power considerations (trend strength, sample size, index variability, $\alpha$,

**Table 7.3  Sampling Intensities Needed to Detect Overall Population Changes of 50%, 25%, and 10% over 10 Years of Annual Monitoring of Animal Populations**

| Group | 50% | | | | | 25% | | | | | 10% | | | | |
|---|---|---|---|---|---|---|---|---|---|---|---|---|---|---|---|
| | 5 | 4 | 3 | 2 | 1 | 5 | 4 | 3 | 2 | 1 | 5 | 4 | 3 | 2 | 1 |
| Mammals, large | 10 | 10 | 10 | 10 | 10 | 10 | 10 | 10 | 10 | 10 | 10 | 10 | 10 | 20 | 40 |
| Grasses and sedges | 10 | 10 | 10 | 10 | 10 | 10 | 10 | 10 | 10 | 20 | 20 | 20 | 30 | 40 | 70 |
| Herbs, Compositae | 10 | 10 | 10 | 10 | 10 | 10 | 10 | 10 | 10 | 20 | 20 | 20 | 30 | 40 | 80 |
| Herbs, non-Compositae | 10 | 10 | 10 | 10 | 10 | 10 | 10 | 10 | 10 | 20 | 20 | 30 | 30 | 50 | 80 |
| Turtles | 10 | 10 | 10 | 10 | 10 | 10 | 10 | 20 | 20 | 40 | 40 | 50 | 70 | 90 | 170 |
| Terrestrial salamanders | 10 | 10 | 10 | 10 | 10 | 10 | 10 | 20 | 20 | 40 | 50 | 50 | 70 | 110 | 190 |
| Large-bodied birds | 10 | 10 | 10 | 10 | 10 | 10 | 20 | 20 | 20 | 40 | 50 | 60 | 70 | 110 | 210 |
| Lizards | 10 | 10 | 10 | 10 | 20 | 20 | 20 | 20 | 30 | 50 | 60 | 70 | 100 | 150 | 280 |
| Fishes, salmonids | 10 | 10 | 10 | 10 | 20 | 20 | 20 | 30 | 40 | 70 | 80 | 90 | 120 | 180 | 370 |
| Caddis flies | 10 | 10 | 10 | 20 | 20 | 20 | 20 | 30 | 40 | 70 | 80 | 100 | 120 | 200 | 380 |
| Snakes | 10 | 10 | 10 | 20 | 30 | 20 | 30 | 30 | 50 | 80 | 90 | 120 | 150 | 220 | 460 |
| Dragonflies | 10 | 10 | 10 | 20 | 30 | 20 | 30 | 30 | 50 | 80 | 100 | 130 | 170 | 230 | 470 |

| | | | | | | | | | | | | | | |
|---|---|---|---|---|---|---|---|---|---|---|---|---|---|---|
| Small-bodied birds | 10 | 10 | 20 | 30 | 20 | 30 | 30 | 50 | 80 | 110 | 130 | 180 | 240 | 490 |
| Beetles | 10 | 10 | 20 | 30 | 20 | 30 | 30 | 50 | 90 | 110 | 150 | 180 | 250 | >500 |
| Small mammals | 10 | 10 | 20 | 30 | 30 | 40 | 40 | 60 | 90 | 120 | 150 | 210 | 300 | >500 |
| Spiders | 10 | 10 | 20 | 30 | 30 | 30 | 40 | 60 | 120 | 120 | 170 | 210 | 320 | >500 |
| Medium-sized mammals | 10 | 20 | 20 | 40 | 30 | 40 | 40 | 70 | 120 | 130 | 160 | 230 | 330 | >500 |
| Fishes, nonsalmonids | 10 | 20 | 20 | 40 | 40 | 40 | 50 | 70 | 130 | 150 | 190 | 290 | 390 | >500 |
| Pond-breeding salamanders | 20 | 20 | 30 | 60 | 40 | 50 | 60 | 120 | 190 | 240 | 320 | 400 | >500 | >500 |
| Moths | 20 | 20 | 30 | 60 | 40 | 60 | 80 | 120 | 210 | 280 | 320 | 440 | >500 | >500 |
| Frogs and toads | 20 | 20 | 30 | 70 | 50 | 60 | 80 | 120 | 230 | 280 | 360 | 460 | >500 | >500 |
| Bats | 20 | 30 | 40 | 70 | 50 | 70 | 80 | 120 | 230 | 280 | 360 | 480 | >500 | >500 |
| Butterflies | 20 | 30 | 50 | 90 | 70 | 90 | 120 | 170 | 350 | 400 | 500 | >500 | >500 | >500 |
| Flies, drosophilids | 30 | 40 | 50 | 120 | 90 | 130 | 160 | 260 | 440 | 500 | >500 | >500 | >500 | >500 |

Values are the number of plots or subpopulations that must be monitored to detect the change at $p = 0.05$ with a likelihood (power) of $>0.95$, given 5, 4, 3, 2, or 1 annual counts or surveys per year of each plot or subpopulation. All estimates were made with the simulation program described in table 7.1 (with 250 replications) in conjunction with estimates of population index variation described in table 7.2.

and β), and the statistical method used to analyze a count series. The primary consequence of failing to make the best choices and thereby improve methods for identifying population change in animal ecology will be a chronic failure to detect population change. Unfortunately, these errors will often be misinterpreted as reflecting population stability, lack of treatment effect, or ineffectiveness of management. Neither the science of animal ecology nor the wild resources under our surveillance should be expected to bear the consequences of these errors.

### Acknowledgments

I am grateful to L. Boitani and T. K. Fuller for the invitation to make a presentation at the conference in Erice, Sicily, in December 1996. That opportunity provided me with the impetus to assemble the information and ideas about population monitoring that are presented in this chapter. S. M. Melvin and S. Droege have also provided important encouragement and guidance to me on monitoring issues. The chapter was improved by comments from M. R. Fuller, R. J. Steidl, and an anonymous reviewer.

**Appendix 7.1    Variability Estimates for Local Populations of Plants and Animals**

| Publication | Organism | Length of Time Series (years) | CV of Counts |
|---|---|---|---|
| *Plants* | | | |
| GRASSES AND SEDGES | | | |
| Dodd et al. (1995) | *Agrostis capillaris* | 60 | 0.61 |
| Symonides (1979) | *Agrostis vulgaris* | 8 | 0.06 |
| Dodd et al. (1995) | *Anthoxanthum odoratum* | 60 | 0.25 |
| Fitch and Bentley (1949) | *Bromus mollis* | 7 | 0.14 |
| Symonides (1979) | *Bromus mollis* | 8 | 0.21 |
| Fitch and Bentley (1949) | *Bromus rigidus* | 7 | 0.61 |
| Fitch and Bentley (1949) | *Bromus rubens* | 7 | 0.60 |
| Symonides (1979) | *Carex caryophyllea* | 8 | 0.13 |
| Symonides (1979) | *Carex ericetorum* | 8 | 0.07 |
| Symonides (1979) | *Carex hirta* | 8 | 0.08 |
| Symonides (1979) | *Corynephorus canescens* | 8 | 0.02 |
| Symonides (1979) | *Digitaria sanguinalis* | 8 | 0.05 |
| Symonides (1979) | *Festuca duriuscula* | 8 | 0.01 |
| Fitch and Bentley (1949) | *Festuca megalura* | 7 | 0.39 |
| Symonides (1979) | *Festuca psammophila* | 8 | 0.04 |
| Symonides (1979) | *Koeleria glauca* | 8 | 0.07 |
| HERBS, COMPOSITAE | | | |
| Symonides (1979) | *Achillea millefolium* | 8 | 0.08 |
| Symonides (1979) | *Artemisia campestris* | 8 | 0.05 |
| Symonides (1979) | *Centaurea rhenana* | 8 | 0.10 |
| Symonides (1979) | *Chondrilla juncea* | 8 | 0.12 |
| Fitch and Bentley (1949) | *Hemizonia virgata* | 7 | 0.94 |
| Symonides (1979) | *Hieracium pilosella* | 8 | 0.05 |
| Symonides (1979) | *Scorzonera humilis* | 8 | 0.15 |
| Dodd et al. (1995) | *Veronica chamaedrys* | 60 | 0.40 |
| Symonides (1979) | *Veronica spicata* | 8 | 0.03 |
| HERBS, NON-COMPOSITAE | | | |
| Wells (1981) | *Aceras anthropophorum* | 13 | 0.49 |
| Symonides (1979) | *Alium verum* | 8 | 0.02 |
| Symonides (1979) | *Arabis arenosa* | 8 | 0.08 |
| Symonides (1979) | *Arenaria serpyllifolia* | 8 | 0.11 |
| Symonides (1979) | *Armeria elongata* | 8 | 0.09 |
| Symonides (1979) | *Cerastium semidecandrum* | 8 | 0.01 |
| Dodd et al. (1995) | *Chamerion angustifolium* | 60 | 1.15 |

| Publication | Organism | Length of Time Series (years) | CV of Counts |
|---|---|---|---|
| Dodd et al. (1995) | *Conopodium majus* | 60 | 0.21 |
| Symonides (1979) | *Dianthus carthusianorum* | 8 | 0.04 |
| Symonides (1979) | *Dianthus deltoides* | 8 | 0.00 |
| Fitch and Bentley (1949) | *Erodium botrys* | 7 | 0.58 |
| Symonides (1979) | *Euphorbia cyparissias* | 8 | 0.09 |
| Symonides (1979) | *Hernaria glabra* | 8 | 0.07 |
| Symonides (1979) | *Hypericum perforatum* | 8 | 0.07 |
| Symonides (1979) | *Jasione montana* | 8 | 0.04 |
| Symonides (1979) | *Knautia arvensis* | 8 | 0.05 |
| Fitch and Bentley (1949) | *Lotus americanus* | 7 | 0.50 |
| Fitch and Bentley (1949) | *Lupinus bicolor* | 7 | 0.79 |
| Hutchings (1987) | *Ophrys sphegodes* | 10 | 0.24 |
| Symonides (1979) | *Peucedanum oreoselinum* | 8 | 0.07 |
| Svensson et al. (1993) | *Pinguicula alpina* | 8 | 0.09 |
| Svensson et al. (1993) | *Pinguicula villosa* | 8 | 0.31 |
| Svensson et al. (1993) | *Pinguicula vulgaris* | 8 | 0.15 |
| Fitch and Bentley (1949) | *Plagiobothrys nothofulvus* | 7 | 0.91 |
| Symonides (1979) | *Potentilla arenaria* | 8 | 0.03 |
| Symonides (1979) | *Potentilla argentea* | 8 | 0.04 |
| Symonides (1979) | *Rumex acetosella* | 8 | 0.01 |
| Symonides (1979) | *Thymus serphyllum* | 8 | 0.03 |
| Dodd et al. (1995) | *Tragopogon pratensis* | 60 | 0.37 |
| Symonides (1979) | *Trifolium arvense* | 8 | 0.02 |
| Fitch and Bentley (1949) | *Trifolium microcephalum* | 7 | 0.43 |
| Dodd et al. (1995) | *Trifolium pratense* | 60 | 0.11 |

*Insects*

DRAGONFLIES

| | | | |
|---|---|---|---|
| Macan (1977) | *Aeshna juncea* | 18 | 0.56 |
| Macan (1977) | *Enallagma cyathigerum* | 18 | 0.60 |
| Moore (1991) | *Ischnura elegans* | 27 | 0.46 |
| Macan (1977) | *Lestes sponsa* | 14 | 0.46 |
| Moore (1991) | *Libellula quadrimaculata* | 27 | 0.39 |
| Moore (1991) | *Libellula sponsa* | 27 | 0.64 |
| Macan (1977) | *Pyrrhosma nymphula* | 18 | 1.09 |
| Moore (1991) | *Sympetrum striolatum* | 27 | 0.33 |

CADDIS FLIES

| | | | |
|---|---|---|---|
| Critchton (1971) | *Anabolia nervosa* | 5 | 0.45 |
| Critchton (1971) | *Halesus digitatus* | 5 | 1.23 |
| Critchton (1971) | *Limnephilus affinis* | 5 | 0.72 |

| Publication | Organism | Length of Time Series (years) | CV of Counts |
|---|---|---|---|
| Critchton (1971) | *Limnephilus auricula* | 5 | 0.49 |
| Critchton (1971) | *Limnephilus centralis* | 5 | 0.45 |
| Critchton (1971) | *Limnephilus lunatus* | 5 | 0.35 |
| Critchton (1971) | *Limnephilus lunatus* | 5 | 0.24 |
| Critchton (1971) | *Limnephilus lunatus* | 5 | 0.31 |
| Critchton (1971) | *Limnephilus marmoratus* | 5 | 0.33 |
| Macan (1977) | *Limnephilus marmoratus* | 17 | 0.69 |
| Critchton (1971) | *Limnephilus sparsus* | 5 | 0.33 |
| Critchton (1971) | *Limnephilus vittatus* | 5 | 0.24 |
| Critchton (1971) | *Stenophylax permistus* | 5 | 0.33 |
| Critchton (1971) | *Stenophylax vibex* | 5 | 0.37 |
| Macan (1977) | *Triaenodes bicolor* | 18 | 0.92 |

BEETLES

| Publication | Organism | Length of Time Series (years) | CV of Counts |
|---|---|---|---|
| Clark (1994) | *Thymallus arcticus* | 6 | 0.19 |
| Jones (1976) | *Acupalpus meridianus* | 5 | 0.09 |
| Jones (1976) | *Agonum dorsale* | 5 | 0.03 |
| Jones (1976) | *Bembidion lampros* | 5 | 0.24 |
| Den Boer (1971) | *Calanthus melanocephalus* | 7 | 1.23 |
| Hill and Kinsley (1994) | *Cincindela dorsalis* | 7 | 0.15 |
| Hill and Kinsley (1993) | *Cincindela puritana* | 7 | 0.25 |
| Jones (1976) | *Clivina fossor* | 5 | 1.00 |
| Macan (1977) | *Dermestes assimilis* | 18 | 1.20 |
| Macan (1977) | *Deronectes duodecimpustulatus* | 18 | 0.65 |
| Macan (1977) | *Halpinus confinus* | 18 | 1.03 |
| Macan (1977) | *Halpinus confinus* | 16 | 1.48 |
| Macan (1977) | *Halpinus fulvus* | 18 | 0.72 |
| Jones (1976) | *Harpalus rufripes* | 5 | 0.75 |
| Jones (1976) | *Nebria brevicollis* | 5 | 0.51 |
| Jones (1976) | *Notiophilus biguttatus* | 5 | 0.42 |
| Den Boer (1971) | *Pterostichus coerulescens* | 7 | 0.39 |
| Jones (1976) | *Pterostichus madidus* | 5 | 0.36 |
| Jones (1976) | *Pterostichus melanarius* | 5 | 0.42 |
| Jones (1976) | *Trechus quadristriatus* | 5 | 0.50 |

FLIES, DROSOPHILIDAE

| Publication | Organism | Length of Time Series (years) | CV of Counts |
|---|---|---|---|
| Momma (1965) | *Drosophila auraria* | 10 | 1.41 |
| Momma (1965) | *Drosophila bifasciata* | 10 | 0.98 |
| Momma (1965) | *Drosophila brachynephros* | 10 | 0.76 |
| Momma (1965) | *Drosophila coracina* | 10 | 0.77 |
| Momma (1965) | *Drosophila histrio* | 10 | 0.69 |
| Momma (1965) | *Drosophila histriodes* | 10 | 0.78 |

| Publication | Organism | Length of Time Series (years) | CV of Counts |
|---|---|---|---|
| Momma (1965) | *Drosophila immigrans* | 10 | 2.40 |
| Momma (1965) | *Drosophila lacertosa* | 10 | 1.35 |
| Momma (1965) | *Drosophila lutea* | 10 | 2.25 |
| Momma (1965) | *Drosophila nigromaculata* | 10 | 1.57 |
| Momma (1965) | *Drosophila sordidula* | 10 | 1.30 |
| Momma (1965) | *Drosophila suzukii* | 10 | 2.09 |
| Momma (1965) | *Drosophila testacea* | 10 | 0.73 |

MOTHS

| | | | |
|---|---|---|---|
| Spitzer and Leps (1988) | *Acronicta rumicus* | 15 | 1.06 |
| Spitzer and Leps (1988) | *Agrochola circellaris* | 15 | 1.18 |
| Spitzer and Leps (1988) | *Agrochola litura* | 15 | 0.80 |
| Spitzer and Leps (1988) | *Agrochola lota* | 15 | 0.92 |
| Spitzer and Leps (1988) | *Agrostis exclamationis* | 15 | 0.88 |
| Spitzer and Leps (1988) | *Agrostis ipsilon* | 15 | 1.69 |
| Spitzer and Leps (1988) | *Amphipoea fucosa* | 15 | 0.86 |
| Spitzer and Leps (1988) | *Anaplectoides prasina* | 15 | 0.92 |
| Spitzer and Leps (1988) | *Apamea crenata* | 15 | 0.76 |
| Spitzer and Leps (1988) | *Apamea monoglypha* | 15 | 1.00 |
| Spitzer and Leps (1988) | *Apamea ophiograma* | 15 | 0.57 |
| Spitzer and Leps (1988) | *Autographa gamma* | 15 | 0.69 |
| Spitzer and Leps (1988) | *Autographa pulchrina* | 15 | 0.92 |
| Spitzer and Leps (1988) | *Axylia putris* | 15 | 0.74 |
| Spitzer and Leps (1988) | *Caradrina morpheus* | 15 | 0.81 |
| Spitzer and Leps (1988) | *Celaena leucostigma* | 15 | 0.62 |
| Spitzer and Leps (1988) | *Cerapteryx graminis* | 15 | 0.73 |
| Spitzer and Leps (1988) | *Cerastis rubricosa* | 15 | 0.57 |
| Spitzer and Leps (1988) | *Chlogophera meticulosa* | 15 | 1.37 |
| Spitzer and Leps (1988) | *Cosmia trapzina* | 15 | 1.95 |
| Spitzer and Leps (1988) | *Diachrysia chrysitis* | 15 | 0.62 |
| Spitzer and Leps (1988) | *Diarsia brunnea* | 15 | 1.09 |
| Spitzer and Leps (1988) | *Diarsia ruoi* | 15 | 1.44 |
| Spitzer and Leps (1988) | *Dicestra trifolii* | 15 | 1.25 |
| Spitzer and Leps (1988) | *Eupsilia transversa* | 15 | 1.23 |
| Spitzer and Leps (1988) | *Eustrotia uncula* | 15 | 0.68 |
| Spitzer and Leps (1988) | *Graphiophora augur* | 15 | 0.74 |
| Spitzer and Leps (1988) | *Hoplodrina alsines* | 15 | 0.69 |
| Spitzer and Leps (1988) | *Hoplodrina blanda* | 15 | 0.99 |
| Spitzer and Leps (1988) | *Hydraecia micacea* | 15 | 0.50 |
| Spitzer and Leps (1988) | *Mamestra brassicae* | 15 | 1.78 |
| Spitzer and Leps (1988) | *Mamestra oleracea* | 15 | 0.90 |
| Spitzer and Leps (1988) | *Mamestra pisi* | 15 | 0.64 |

| Publication | Organism | Length of Time Series (years) | CV of Counts |
|---|---|---|---|
| Spitzer and Leps (1988) | *Mamestra suasa* | 15 | 1.33 |
| Spitzer and Leps (1988) | *Mamestra thalassina* | 15 | 0.69 |
| Spitzer and Leps (1988) | *Mythimna impura* | 15 | 0.57 |
| Spitzer and Leps (1988) | *Mythimna pallens* | 15 | 0.84 |
| Spitzer and Leps (1988) | *Mythimna pudorina* | 15 | 0.45 |
| Spitzer and Leps (1988) | *Mythimna turca* | 15 | 0.54 |
| Spitzer and Leps (1988) | *Noctua pronuba* | 15 | 0.73 |
| Spitzer and Leps (1988) | *Ochropleura plecta* | 15 | 1.00 |
| Spitzer and Leps (1988) | *Oligia latruncula* | 15 | 1.20 |
| Spitzer and Leps (1988) | *Oligia strigilis* | 15 | 0.54 |
| Spitzer and Leps (1988) | *Opigena polygona* | 15 | 1.34 |
| Spitzer and Leps (1988) | *Orthosia cruda* | 15 | 1.91 |
| Spitzer and Leps (1988) | *Orthosia gothica* | 15 | 0.43 |
| Spitzer and Leps (1988) | *Orthosia gracilis* | 15 | 0.94 |
| Spitzer and Leps (1988) | *Orthosia incerta* | 15 | 0.69 |
| Spitzer and Leps (1988) | *Photodes fluxa* | 15 | 0.59 |
| Spitzer and Leps (1988) | *Photodes minima* | 15 | 0.41 |
| Spitzer and Leps (1988) | *Photodes pygmina* | 15 | 0.70 |
| Spitzer and Leps (1988) | *Phragmitiphila nexa* | 15 | 0.33 |
| Spitzer and Leps (1988) | *Plusia putnami* | 15 | 0.80 |
| Spitzer and Leps (1988) | *Rasina feruginea* | 15 | 0.81 |
| Spitzer and Leps (1988) | *Rhizadra lutosa* | 15 | 0.84 |
| Spitzer and Leps (1988) | *Rivula sericealis* | 15 | 0.93 |
| Spitzer and Leps (1988) | *Tholera decimalis* | 15 | 0.79 |
| Spitzer and Leps (1988) | *Xanthia icteritia* | 15 | 1.11 |
| Spitzer and Leps (1988) | *Xestia baja* | 15 | 0.84 |
| Spitzer and Leps (1988) | *Xestia nigrum* | 15 | 1.20 |
| Spitzer and Leps (1988) | *Xestia ditrapezium* | 15 | 0.82 |
| Spitzer and Leps (1988) | *Xestia triangulum* | 15 | 0.93 |
| Spitzer and Leps (1988) | *Xestia xanthographa* | 15 | 1.00 |
| | | | |
| BUTTERFLIES | | | |
| | | | |
| Sutcliffe et al. (1996) | *Anthocharis cardamines* | 10 | 1.20 |
| Sutcliffe et al. (1996) | *Aphantopus hyperantus* | 10 | 1.12 |
| Sutcliffe et al. (1996) | *Coenonympha pamphilus* | 10 | 1.18 |
| Ehrlich and Murphy (1987) | *Euphydryas editha* | 25 | 0.81 |
| Sutcliffe et al. (1996) | *Gonepteryx rhamni* | 10 | 1.10 |
| Sutcliffe et al. (1996) | *Inachis io* | 10 | 1.10 |
| Sutcliffe et al. (1996) | *Maniola jurtina* | 10 | 0.90 |
| Sutcliffe et al. (1996) | *Pararge aegeria* | 10 | 1.17 |
| Sutcliffe et al. (1996) | *Pieris brassicae* | 10 | 1.20 |
| Sutcliffe et al. (1996) | *Pieris napi* | 10 | 1.10 |

| Publication | Organism | Length of Time Series (years) | CV of Counts |
|---|---|---|---|
| Sutcliffe et al. (1996) | *Pieris rapae* | 10 | 1.00 |
| Sutcliffe et al. (1996) | *Polyommatus icarus* | 10 | 1.27 |
| Sutcliffe et al. (1996) | *Pyronia tithonus* | 10 | 1.23 |

*Arachnids*

SPIDERS

| Publication | Organism | Length of Time Series (years) | CV of Counts |
|---|---|---|---|
| Schoener and Spiller (1992) | *Argiope argentata* | >5 | 0.86 |
| Schoener and Spiller (1992) | *Eustala cazieri* | >5 | 0.72 |
| Schoener and Spiller (1992) | *Gasteracantha cancriformis* | >5 | 0.76 |
| Schoener and Spiller (1992) | *Metepeira datona* | >5 | 0.70 |
| Renault and Miller (1972) | *Ceraticelus atriceps* | 9 | 0.81 |
| Renault and Miller (1972) | *Dictynid phylax* | 9 | 0.65 |
| Renault and Miller (1972) | *Grammonota angusta* | 9 | 0.37 |
| Renault and Miller (1972) | *Pityohyphantes costatus* | 9 | 0.46 |
| Renault and Miller (1972) | *Philodromus placidus* | 9 | 0.50 |
| Renault and Miller (1972) | *Theridion montanum* | 9 | 0.60 |

*Fishes*

FISHES, NONSALMONIDS

| Publication | Organism | Length of Time Series (years) | CV of Counts |
|---|---|---|---|
| Rainwater and Houser (1982) | *Aplodinotus grunniens* | 15 | 0.48 |
| Rainwater and Houser (1982) | *Cyprinus carpio* | 18 | 0.97 |
| Rainwater and Houser (1982) | *Dorosoma cepedianum* | 18 | 0.60 |
| Rainwater and Houser (1982) | *Dorosoma petense* | 13 | 1.16 |
| Willis et al. (1984) | *Esox lucius* | 6 | 0.23 |
| Kipling (1983 | *Esox lucius* | 22 | 0.17 |
| Rainwater and Houser (1982) | *Ictalurus punctatus* | 18 | 0.41 |
| Rainwater and Houser (1982) | *Labidesthes sicculus* | 14 | 1.73 |
| Rainwater and Houser (1982) | *Lepomis cyanellus* | 18 | 0.76 |
| Rainwater and Houser (1982) | *Lepomis gulosus* | 16 | 0.80 |
| Rainwater and Houser (1982) | *Lepomis macrochirus* | 18 | 0.88 |
| Rainwater and Houser (1982) | *Lepomis megalotis* | 18 | 0.78 |
| Rainwater and Houser (1982) | *Micropterus punctulatus* | 18 | 0.49 |
| Rainwater and Houser (1982) | *Micropterus salmoides* | 18 | 0.59 |
| Rainwater and Houser (1982) | *Morone chrysops* | 18 | 0.81 |
| Moore et al. (1991) | *Morone saxatilis* | 11 | 0.34 |
| Rainwater and Houser (1982) | *Moxostoma duquesnei* | 17 | 0.49 |
| Rainwater and Houser (1982) | *Moxostoma erythrurum* | 18 | 0.48 |
| Kallemeyn (1987) | *Perca flavescens* | 5 | 0.62 |
| Kallemeyn (1987) | *Perca flavescens* | 5 | 0.73 |
| Rainwater and Houser (1982) | *Percina caprodes* | 18 | 0.64 |
| Rainwater and Houser (1982) | *Pomoxis annularis* | 16 | 1.53 |
| Rainwater and Houser (1982) | *Pomoxis nigromaculatus* | 18 | 1.45 |

| Publication | Organism | Length of Time Series (years) | CV of Counts |
|---|---|---|---|
| Rainwater and Houser (1982) | *Pylodictus olivaris* | 16 | 0.65 |
| Lyons and Welke (1996) | *Stizostedion canadense* | 8 | 0.82 |
| Lyons and Welke (1996) | *Stizostedion canadense* | 7 | 0.74 |
| Lyons and Welke (1996) | *Stizostedion vitreum* | 7 | 0.45 |
| Kallemeyn (1987) | *Stizostedion vitreum* | 5 | 0.11 |
| Kallemeyn (1987) | *Stizostedion vitreum* | 5 | 0.13 |
| Lyons and Welke (1996) | *Stizostedion vitreum* | 8 | 1.23 |

FISHES, SALMONIDS

| Publication | Organism | Length of Time Series (years) | CV of Counts |
|---|---|---|---|
| House (1995) | *Oncorhynchus clarkii* | 11 | 0.29 |
| Platts and Nelson (1988) | *Salmo clarkii* | 8 | 0.31 |
| Platts and Nelson (1988) | *Salmo gairdneri* | 5 | 0.41 |
| Platts and Nelson (1988) | *Salmo gairdneri* | 9 | 0.66 |
| Platts and Nelson (1988) | *Salmo gairdneri* | 5 | 1.24 |
| Platts and Nelson (1988) | *Salmo gairdneri* | 6 | 0.91 |
| Platts and Nelson (1988) | *Salmo trutta* | 5 | 0.26 |
| Rieman and McIntyre (1996) | *Salvelinus confluentus* | 12 | 0.71 |
| Rieman and McIntyre (1996) | *Salvelinus confluentus* | 13 | 0.58 |
| Rieman and McIntyre (1996) | *Salvelinus confluentus* | 12 | 0.41 |
| Rieman and McIntyre (1996) | *Salvelinus confluentus* | 12 | 0.23 |
| Rieman and McIntyre (1996) | *Salvelinus confluentus* | 12 | 0.25 |
| Platts and Nelson (1988) | *Salvelinus confluentus* | 5 | 0.69 |
| Platts and Nelson (1988) | *Salvelinus confluentus* | 11 | 0.14 |
| Rieman and McIntyre (1996) | *Salvelinus confluentus* | 13 | 0.51 |
| Rieman and McIntyre (1996) | *Salvelinus confluentus* | 13 | 0.39 |
| Rieman and McIntyre (1996) | *Salvelinus confluentus* | 11 | 0.97 |
| Rieman and McIntyre (1996) | *Salvelinus confluentus* | 15 | 0.58 |
| Rieman and McIntyre (1996) | *Salvelinus confluentus* | 14 | 0.61 |
| Rieman and McIntyre (1996) | *Salvelinus confluentus* | 13 | 0.54 |
| Rieman and McIntyre (1996) | *Salvelinus confluentus* | 16 | 0.61 |
| Rieman and McIntyre (1996) | *Salvelinus confluentus* | 14 | 0.42 |
| Rieman and McIntyre (1996) | *Salvelinus confluentus* | 16 | 0.75 |
| Rieman and McIntyre (1996) | *Salvelinus confluentus* | 12 | 1.08 |
| Rieman and McIntyre (1996) | *Salvelinus confluentus* | 14 | 0.61 |
| Rieman and McIntyre (1996) | *Salvelinus confluentus* | 16 | 0.70 |
| Rieman and McIntyre (1996) | *Salvelinus confluentus* | 16 | 0.61 |
| Platts and Nelson (1988) | *Salvelinus fontinalis* | 9 | 0.16 |
| Platts and Nelson (1988) | *Salvelinus fontinalis* | 5 | 0.41 |
| Platts and Nelson (1988) | *Salvelinus fontinalis* | 7 | 0.33 |
| Platts and Nelson (1988) | *Salvelinus fontinalis* | 7 | 0.22 |
| Hansen et al. (1995) | *Salvelinus namaycush* | 15 | 0.33 |
| Hansen et al. (1995) | *Salvelinus namaycush* | 15 | 0.28 |
| Hansen et al. (1995) | *Salvelinus namaycush* | 15 | 0.23 |

| Publication | Organism | Length of Time Series (years) | CV of Counts |
|---|---|---|---|
| Hansen et al. (1995) | *Salvelinus namaycush* | 15 | 0.19 |
| Hansen et al. (1995) | *Salvelinus namaycush* | 14 | 0.35 |
| Hansen et al. (1995) | *Salvelinus namaycush* | 15 | 0.50 |
| Hansen et al. (1995) | *Salvelinus namaycush* | 15 | 0.20 |
| Hansen et al. (1995) | *Salvelinus namaycush* | 15 | 0.23 |
| Hansen et al. (1995) | *Salvelinus namaycush* | 15 | 0.21 |
| Hansen et al. (1995) | *Salvelinus namaycush* | 15 | 0.43 |
| Hansen et al. (1995) | *Salvelinus namaycush* | 15 | 0.33 |

*Amphibians*

FROGS AND TOADS

| | | | |
|---|---|---|---|
| Dodd (1992) | *Acris gryllus* | 6 | 0.05 |
| Gittins (1983) | *Bufo bufo* | 5 | 0.55 |
| Beebee et al. (1996) | *Bufo calamita* | 9 | 0.5 |
| Dodd (1992) | *Bufo quercicus* | 6 | 0.82 |
| Dodd (1992) | *Bufo terrestris* | 6 | 1.03 |
| Semlitsch et al. (1996) | *Bufo terrestris* | 16 | 0.87 |
| Dodd (1992) | *Eleutherodactylus planirostris* | 6 | 0.37 |
| Dodd (1992) | *Gastrophryne carolinensis* | 5 | 0.49 |
| Semlitsch et al. (1996) | *Gastrophryne carolinensis* | 16 | 0.48 |
| Dodd (1992) | *Hyla femoralis* | 6 | 0.85 |
| Semlitsch et al. (1996) | *Pseudacris crucifer* | 16 | 1.04 |
| Dodd (1992) | *Pseudacris ocularis* | 6 | 0.84 |
| Semlitsch et al. (1996) | *Pseudacris ornata* | 16 | 1.27 |
| Semlitsch et al. (1996) | *Pseudacris nigrita* | 16 | 1.72 |
| Weitzel and Panik (1993 | *Pseudacris regilla* | 15 | 0.53 |
| Semlitsch et al. (1996) | *Rana clamitans* | 16 | 0.87 |
| Berven (1990) | *Rana sylvatica* | 6 | 0.65 |
| Semlitsch et al. (1996) | *Rana utricularia* | 16 | 1.25 |
| Dodd (1992) | *Scaphiopus holbrooki* | 6 | 1.56 |
| Semlitsch et al. (1996) | *Scaphiopus holbrooki* | 16 | 2.78 |

SALAMANDERS, POND-BREEDERS

| | | | |
|---|---|---|---|
| Husting (1965) | *Ambystoma maculatum* | 5 | 0.45 |
| Semlitsch et al. (1996) | *Ambystoma opacum* | 16 | 0.62 |
| Semlitsch et al. (1996) | *Ambystoma talpoideum* | 16 | 0.94 |
| Semlitsch et al. (1996) | *Ambystoma tigrinum* | 16 | 0.95 |
| Semlitsch et al. (1996) | *Eurycea quadridigitata* | 16 | 2.31 |
| Dodd (1992) | *Eurycea quadridigitata* | 6 | 0.58 |
| Dodd (1992) | *Notophthalmus perstriatus* | 6 | 0.63 |
| Semlitsch et al. (1996) | *Notophthalmus viridescens* | 16 | 1.07 |

| Publication | Organism | Length of Time Series (years) | CV of Counts |
|---|---|---|---|
| Macan (1977) | *Triturus helviticus* | 16 | 0.52 |
| Macan (1977) | *Triturus helviticus* | 18 | 0.52 |

TERRESTRIAL SALAMANDERS

| | | | |
|---|---|---|---|
| Hairston and Wiley (1993) | *Desmognathus monticola* | 15 | 0.46 |
| Hairston and Wiley (1993) | *Desmognathus ochrophaeus* | 15 | 0.26 |
| Hairston and Wiley (1993) | *Desmognathus oeneus* | 15 | 0.34 |
| Hairston and Wiley (1993) | *Desmognathus quadramaculatus* | 15 | 0.49 |
| Jaeger (1980) | *Plethodon cinereus* | 13 | 0.17 |
| Hairston and Wiley (1993) | *Plethodon glutinosus* | 15 | 0.41 |
| Hairston (1983) | *Plethodon jordani* | 8 | 0.51 |
| Hairston and Wiley (1993) | *Plethodon jordani* | 15 | 0.19 |

*Reptiles*

LIZARDS

| | | | |
|---|---|---|---|
| Andrews and Rand (1982) | *Anolis limifrons* | 10 | 0.73 |
| Dodd (1992) | *Cnemidopherus sexlineatus* | 6 | 0.21 |
| Turner (1977) | *Cnemidopherus tigris* | 8 | 0.24 |
| Dodd (1992) | *Eumeces egregius* | 6 | 0.15 |
| Dodd (1992) | *Ophisaurus ventralis* | 6 | 0.90 |
| Tinkle (1993) | *Sceloporus graciosus* | 11 | 0.25 |
| Turner (1977) | *Sceloporus olivaceous* | 5 | 0.26 |
| Dodd (1992) | *Sceloporus undulatus* | 6 | 0.35 |
| Dodd (1992) | *Scincella lateralis* | 6 | 0.93 |
| Turner (1977) | *Uta stansburiana* | 8 | 0.27 |
| Turner (1977) | *Xanthusia vigilis* | 6 | 0.33 |

SNAKES

| | | | |
|---|---|---|---|
| Fitch (1960) | *Agkistrodon contortrix* | 6 | 0.55 |
| Dodd (1992) | *Cemophora coccinea* | 6 | 0.73 |
| Dodd (1992) | *Coluber constrictor* | 6 | 0.34 |
| Fitch (1963) | *Coluber constrictor* | 7 | 0.61 |
| Dodd (1992) | *Diadophis punctatus* | 6 | 0.27 |
| Dodd (1992) | *Seminatrix pygea* | 6 | 0.58 |
| Dodd (1992) | *Tantilla relicta* | 6 | 0.94 |
| Dodd (1992) | *Thamnophis sirtalis* | 6 | 0.36 |
| Fitch (1965) | *Thamnophis sirtalis* | 6 | 0.49 |

TURTLES

| | | | |
|---|---|---|---|
| Congdon and Gibbons (1996) | *Chelydra serpentina* | 20 | 0.26 |
| Congdon and Gibbons (1996) | *Chrysemys pictapicta* | 20 | 0.07 |

| Publication | Organism | Length of Time Series (years) | CV of Counts |
|---|---|---|---|
| Garber and Burger (1995) | *Clemmys insculpta* | 8 | 0.16 |
| Congdon and Gibbons (1996) | *Emydoidea blandingi* | 20 | 0.21 |
| Dodd (1992) | *Kinosternon subrubrum* | 6 | 0.59 |
| Doroff and Keith (1990) | *Terrapene ornata* | 9 | 0.27 |
| Frazer et al. (1989) | *Trachemys scripta* | 12 | 0.77 |

*Birds*

SMALL BIRDS

| | | | |
|---|---|---|---|
| Pimm et al. (1988) | *Acanthis cannabina* | 5 | 0.44 |
| Hogstad (1993) | *Aegithalos caudatus* | 12 | 2.33 |
| Pimm et al. (1988) | *Aluada arvensis* | 5 | 0.41 |
| Pimm et al. (1988) | *Anthus pratensis* | 5 | 0.41 |
| Hogstad (1993) | *Anthus trivialis* | 12 | 0.29 |
| Hogstad (1993) | *Carduelis chloris* | 12 | 1.23 |
| Hogstad (1993) | *Carduelis spinus* | 12 | 0.29 |
| Holmes and Sherry (1988) | *Catharus fucescens* | 17 | 0.47 |
| Holmes and Sherry (1988) | *Catharus guttatus* | 17 | 0.68 |
| Holmes and Sherry (1988) | *Catharus ustulatus* | 17 | 0.59 |
| Hogstad (1993) | *Certhia familiaris* | 12 | 0.97 |
| Hogstad (1993) | *Corvus corone* | 12 | 0.73 |
| Holmes and Sherry (1988) | *Dendroica caerulescens* | 17 | 0.24 |
| Holmes and Sherry (1988) | *Dendroica fusca* | 17 | 0.63 |
| Holmes and Sherry (1988) | *Dendroica virens* | 17 | 0.26 |
| Hogstad (1993) | *Emberiza citrinella* | 12 | 0.74 |
| Pimm et al. (1988) | *Emberiza citrinella* | 5 | 0.38 |
| Pimm et al. (1988) | *Emberiza schoenichus* | 5 | 0.67 |
| Holmes and Sherry (1988) | *Empidonax minimus* | 17 | 0.89 |
| Hogstad (1993) | *Erithacus rubecula* | 12 | 0.14 |
| Pimm et al. (1988) | *Erithacus rubecula* | 5 | 0.26 |
| Hogstad (1993) | *Ficedula hypoleuca* | 12 | 0.37 |
| Lack (1969) | *Ficedula hypoleuca* | 16 | 0.15 |
| Hogstad (1993) | *Fringilla coelebs* | 12 | 0.11 |
| Pimm et al. (1988) | *Fringilla coelebs* | 5 | 0.42 |
| Hogstad (1993) | *Fringilla montrifringilla* | 12 | 0.66 |
| Hogstad (1993) | *Garrulus glandarius* | 12 | 0.49 |
| Holmes and Sherry (1988) | *Hylocichla mustelina* | 17 | 0.61 |
| Roth and Johnson (1993) | *Hylocichla mustelina* | 16 | 0.13 |
| Holmes and Sherry (1988) | *Junco hyemalis* | 17 | 1.15 |
| Pimm et al. (1988) | *Motacilla alba* | 5 | 0.28 |
| Hogstad (1993) | *Moticilla alba* | 12 | 1.04 |

| Publication | Organism | Length of Time Series (years) | CV of Counts |
|---|---|---|---|
| Hogstad (1993) | *Musicapa striata* | 12 | 2.48 |
| Pimm et al. (1988) | *Oenanthe oenanthe* | 5 | 0.33 |
| Lack (1969) | *Parus ater* | 8 | 0.31 |
| Hogstad (1993) | *Parus ater* | 12 | 0.63 |
| Lack (1969) | *Parus caeruleus* | 18 | 0.30 |
| Hogstad (1993) | *Parus cristatus* | 12 | 0.14 |
| Pimm et al. (1988) | *Parus major* | 5 | 0.35 |
| Hogstad (1993) | *Parus major* | 12 | 0.49 |
| Lack (1969) | *Parus major* | 18 | 0.48 |
| Hogstad (1993) | *Parus montanus* | 12 | 0.16 |
| Pimm et al. (1988) | *Passer domesticus* | 5 | 0.27 |
| Holmes and Sherry (1988) | *Pheucticus ludovicianus* | 17 | 0.33 |
| Hogstad (1993) | *Phoenecurus phoenicurus* | 12 | 0.27 |
| Hogstad (1993) | *Phylloscopus collybita* | 12 | 0.15 |
| Hogstad (1993) | *Phylloscopus trochilus* | 12 | 0.41 |
| Hogstad (1993) | *Pica pica* | 12 | 1.49 |
| Holmes and Sherry (1988) | *Piranga olivacea* | 17 | 0.46 |
| Hogstad (1993) | *Prunella modularis* | 12 | 0.12 |
| Pimm et al. (1988) | *Prunella modularis* | 5 | 0.41 |
| Hogstad (1993) | *Pyrrhula pyrrhula* | 12 | 0.17 |
| Hogstad (1993) | *Regulus regulus* | 12 | 0.39 |
| Hogstad (1993) | *Saxicola rubetra* | 12 | 1.04 |
| Pimm et al. (1988) | *Saxicola torquata* | 5 | 0.49 |
| Holmes and Sherry (1988) | *Seiurus aurocapillus* | 17 | 0.31 |
| Holmes and Sherry (1988) | *Setophaga ruticilla* | 17 | 0.31 |
| Holmes and Sherry (1988) | *Sitta carolinensis* | 17 | 0.72 |
| Pimm et al. (1988) | *Sturnus vulgaris* | 5 | 0.41 |
| Hogstad (1993) | *Sylvia atricapilla* | 12 | 1.32 |
| Hogstad (1993) | *Sylvia borin* | 12 | 1.81 |
| Pimm et al. (1988) | *Sylvia communis* | 5 | 0.48 |
| Hogstad (1993) | *Sylvia curruca* | 12 | 1.23 |
| Hogstad (1993) | *Troglodytes troglodytes* | 12 | 0.26 |
| Holmes and Sherry (1988) | *Troglodytes troglodytes* | 17 | 1.45 |
| Pimm et al. (1988) | *Troglodytes troglodytes* | 5 | 0.62 |
| Hogstad (1993) | *Turdus iliacus* | 12 | 0.16 |
| Hogstad (1993) | *Turdus merula* | 12 | 0.26 |
| Pimm et al. (1988) | *Turdus merula* | 5 | 0.44 |
| Hogstad (1993) | *Turdus philomelos* | 12 | 0.21 |
| Hogstad (1993) | *Turdus pilaris* | 12 | 0.98 |
| Holmes and Sherry (1988) | *Vireo olivaceus* | 17 | 0.20 |
| Petrinovich and Patterson (1982) | *Zonotrichia leucophrys* | 6 | 0.26 |

| Publication | Organism | Length of Time Series (years) | CV of Counts |
|---|---|---|---|
| LARGE BIRDS | | | |
| Pimm et al. (1988) | *Asio flammeus* | 5 | 0.70 |
| Pimm et al. (1988) | *Athene noctua* | 5 | 0.27 |
| Pimm et al. (1988) | *Buteo buteo* | 5 | 0.23 |
| Pimm et al. (1988) | *Charadrius hiaticula* | 5 | 0.43 |
| Lack (1969) | *Ciconia ciconia* | 35 | 0.13 |
| Pimm et al. (1988) | *Columa livia* | 5 | 0.18 |
| Lack (1969) | *Columba palumbus* | 5 | 0.04 |
| Pimm et al. (1988) | *Corvus corax* | 5 | 0.30 |
| Pimm et al. (1988) | *Corvus corrone* | 5 | 0.24 |
| Pimm et al. (1988) | *Corvus monedula* | 5 | 0.28 |
| Pimm et al. (1988) | *Crex crex* | 5 | 0.70 |
| Pimm et al. (1988) | *Falco perigrinus* | 5 | 0.33 |
| Pimm et al. (1988) | *Falco sparverius* | 5 | 0.17 |
| Pimm et al. (1988) | *Fulica atra* | 5 | 0.00 |
| Pimm et al. (1988) | *Gallinago gallinago* | 5 | 0.26 |
| Pimm et al. (1988) | *Gallinula chloropus* | 5 | 0.44 |
| Lack (1969) | *Lagopus scoticus* | 5 | 0.44 |
| Pimm et al. (1988) | *Perdix perdix* | 5 | 0.68 |
| Pimm et al. (1988) | *Phasianus colchicus* | 5 | 0.35 |
| Pimm et al. (1988) | *Pica pica* | 5 | 0.46 |
| Holmes and Sherry (1988) | *Picoides pubescens* | 17 | 0.58 |
| Holmes and Sherry (1988) | *Picoides villosus* | 17 | 0.40 |
| Holmes and Sherry (1988) | *Sphyrapicus varius* | 17 | 0.42 |
| Pimm et al. (1988) | *Tringa totanus* | 5 | 0.46 |
| Pimm et al. (1988) | *Vanellus vanellus* | 5 | 0.58 |
| *Mammals* | | | |
| BATS | | | |
| Kalko et al. (1996) | *Artibeus jamaicensis* | 8 | 0.20 |
| Kalko et al. (1996) | *Artibeus lituratus* | 8 | 0.39 |
| Kalko et al. (1996) | *Artibeus phaeotis* | 8 | 0.89 |
| Kalko et al. (1996) | *Artibeus watsoni* | 8 | 0.30 |
| Kalko et al. (1996) | *Carollia castanea* | 8 | 0.34 |
| Kalko et al. (1996) | *Carollia perspicillata* | 8 | 0.39 |
| Kalko et al. (1996) | *Chiroderma villosum* | 8 | 1.33 |
| Kalko et al. (1996) | *Glossophaga soricina* | 8 | 1.20 |
| Kalko et al. (1996) | *Micronycteris brachyotis* | 8 | 0.20 |
| Kalko et al. (1996) | *Micronycteris hirsuta* | 8 | 0.45 |
| Kalko et al. (1996) | *Micronycteris megalotis* | 8 | 0.61 |
| Kalko et al. (1996) | *Micronycteris nicefori* | 8 | 0.76 |

| Publication | Organism | Length of Time Series (years) | CV of Counts |
|---|---|---|---|
| Kalko et al. (1996) | *Mimon cranulatum* | 8 | 0.65 |
| Kalko et al. (1996) | *Phyllostomus discolor* | 8 | 0.86 |
| Kalko et al. (1996) | *Phyllostomus hastatus* | 8 | 0.75 |
| Kalko et al. (1996) | *Platyrrhinus helleri* | 8 | 1.00 |
| Kalko et al. (1996) | *Pteronotus parnellii* | 8 | 0.32 |
| Kalko et al. (1996) | *Tonatia bidens* | 8 | 0.24 |
| Kalko et al. (1996) | *Tonatia silvicola* | 8 | 0.21 |
| Kalko et al. (1996) | *Trachops cirrhosus* | 8 | 0.41 |
| Kalko et al. (1996) | *Uroderma bilobatum* | 8 | 1.13 |
| Kalko et al. (1996) | *Vampyressa nymphaea* | 8 | 0.78 |
| Kalko et al. (1996) | *Vampyressa pusilla* | 8 | 0.67 |
| Kalko et al. (1996) | *Vampyrodes caraccioli* | 8 | 0.41 |
| | | | |
| SMALL MAMMALS | | | |
| | | | |
| Grant et al. (1985) | *Baiomys taylori* | 6 | 0.64 |
| Getz (1989) | *Blarina brevicauda* | 7 | 0.53 |
| Swihart and Slade (1990) | *Blarina brevicauda* | 15 | 0.33 |
| Swihart and Slade (1990) | *Microtus ochrogaster* | 15 | 0.89 |
| Swihart and Slade (1990) | *Mus musculus* | 15 | 0.74 |
| Grant et al. (1985) | *Peromyscus leucopus* | 6 | 0.33 |
| Swihart and Slade (1990) | *Peromyscus leucopus* | 15 | 0.63 |
| Swihart and Slade (1990) | *Peromyscus maniculatus* | 15 | 0.88 |
| Grant et al. (1985) | *Reithrodontomys fulvescens* | 6 | 0.31 |
| Swihart and Slade (1990) | *Reithrodontomys megalotis* | 15 | 0.48 |
| Odum (1955) | *Sigmodon hispidus* | 11 | 0.41 |
| Grant et al. (1985) | *Sigmodon hispidus* | 6 | 0.41 |
| Swihart and Slade (1990) | *Sigmodon hispidus* | 15 | 1.09 |
| Swihart and Slade (1990) | *Synaptomys cooperi* | 15 | 0.69 |
| | | | |
| MEDIUM-SIZED MAMMALS | | | |
| | | | |
| Angerbjorn et al. (1995) | *Alopex lagopus* | 19 | 1.05 |
| Kaikusalo and Angerbjorn (1995) | *Alopex lagopus* | 30 | 0.89 |
| Messier (1991) | *Canis lupus* | 10 | 0.25 |
| Messier (1991) | *Canis lupus* | 26 | 0.21 |
| Busher (1987) | *Castor canadensis* | 40 | 0.32 |
| Keith and Cary (1991) | *Erethizon dorsatum* | 10 | 0.54 |
| Keith and Cary (1991) | *Glaucomys sabrinus* | 10 | 0.47 |
| Hewson (1971) | *Lepus timidus* | 13 | 0.46 |
| Pulliainen (1981) | *Martes martes* | 13 | 0.75 |
| Keith and Cary (1991) | *Mephitis mephitis* | 10 | 0.44 |
| Fuller and Kuehn (1985) | *Mephitis mephitis* | 8 | 1.24 |

| Publication | Organism | Length of Time Series (years) | CV of Counts |
|---|---|---|---|
| Keith and Cary (1991) | *Mustela erminea* | 10 | 0.75 |
| Erlinge (1983) | *Mustela erminea* | 6 | 0.41 |
| Pulliainen (1981) | *Mustela erminea* | 13 | 0.90 |
| Keith and Cary (1991) | *Mustela freneta* | 10 | 1.28 |
| Keith and Cary (1991) | *Mustela vison* | 10 | 1.38 |
| Southwick et al. (1986) | *Ochotona princeps* | 8 | 0.22 |
| Danell (1978) | *Ondatra zibethicus* | 7 | 0.38 |
| Gurnell (1996) | *Sciurus carolinensis* | 9 | 0.45 |
| Keith and Cary (1991) | *Spermophilus franklinii* | 10 | 0.36 |
| Keith and Cary (1991) | *Tamiasciurus hudsonicus* | 10 | 0.76 |
| Pulliainen (1981) | *Vulpes vulpes* | 13 | 0.73 |
| LARGE MAMMALS | | | |
| Nicholls et al. (1996) | *Aepyceros melampus* | 10 | 0.03 |
| Messier (1991) | *Alces alces* | 27 | 0.13 |
| Nicholls et al. (1996) | *Ceratotherium simum* | 10 | 0.03 |
| Nicholls et al. (1996) | *Conochaetes taurinus* | 10 | 0.07 |
| Nicholls et al. (1996) | *Damaliscus lunatus* | 10 | 0.11 |
| Nicholls et al. (1996) | *Equus burchelli* | 10 | 0.17 |
| Nicholls et al. (1996) | *Giraffa camelopardalis* | 10 | 0.06 |
| Nicholls et al. (1996) | *Hippotragus equinus* | 10 | 0.18 |
| Nicholls et al. (1996) | *Hippotragus niger* | 10 | 0.08 |
| Nicholls et al. (1996) | *Kobus ellipsiprynus* | 10 | 0.07 |
| Bayliss (1985) | *Macropus fuliginosus* | 5 | 0.02 |
| Bayliss (1985) | *Macropus rufus* | 5 | 0.27 |
| Messier (1991) | *Odocoileus virginianus* | 11 | 0.20 |
| Nicholls et al. (1996) | *Phacochoerus aethiopicus* | 10 | 0.14 |
| Nicholls et al. (1996) | *Taurotragus oryx* | 10 | 0.62 |
| Nicholls et al. (1996) | *Tragelaphus strepsiceros* | 10 | 0.18 |
| Wielgus and Bunnell (1994) | *Ursus arctos* | 5 | 0.05 |

CV = coefficient of variation.

## Literature Cited

Andrews, R. M. and S. Rand. 1982. Seasonal breeding and long-term population fluctuations in the lizard *Anolis limifrons*. In E. Leigh, A. S. Rand, and D. M. Windsor, eds., *The ecology of a tropical forest, seasonal rhythms and long-term changes*, 405–412. Washington, D.C.: Smithsonian Institution Press.

Angerbjorn, A., M. Tannerfeldt, A. Bjarvall, M. Ericson, J. From, and E. Noren. 1995. Dynamics of the arctic fox population in Sweden. *Annales Zoologici Fennici* 32: 55–68.

Bayliss, P. 1985. The population dynamics of red and western grey kangaroos in arid New South Wales, Australia. I. Population trends and rainfall. *Journal of Animal Ecology* 54: 111–125.

Beebee, T. J. C., J. S. Denton, and J. Buckley. 1996. Factors affecting population densities of adult natterjack toads *Bufo calamita* in Britain. *Journal of Applied Ecology* 33: 263–268.

Berven, K. A. 1990. Factors affecting population fluctuations in larval and adult stages of the wood frog (*Rana sylvatica*). *Ecology* 71: 1599–1608.

Busher, P. E. 1987. Population parameters and family composition of beaver in California. *Journal of Mammalogy* 68: 860–864.

Caughley, G. 1977. *Analysis of vertebrate populations*. New York: Wiley.

Clark, R. A. 1994. Population dynamics and potential utility per recruit of arctic grayling in Fielding Lake, Alaska. *North American Journal of Fisheries Management* 14: 500–515.

Congdon, J. D. and J. W. Gibbons. 1996. Structure and dynamics of a turtle community over two decades. In M. L. Cody and J. A. Smallwood, eds., *Long-term studies of vertebrate communities*, 137–160. San Diego: Academic Press.

Conroy, M. J. 1996. Abundance indices. In D. E. Wilson, F. R. Cole, J. D. Nicholls, R. Rudran, and M. S. Foster, eds., *Measuring and monitoring biological diversity, standard methods for mammals*, 179–191. Washington, D.C.: Smithsonian Institution Press.

Critchton, M. I. 1971. A study of caddis flies (*Trichoptera*) of the family Limnephilidae, based on the Rothamsted Insect Survey, 1964–68. *Zoology* 163: 533–563.

Danell, K. 1978. Population dynamics of the muskrat in a shallow Swedish Lake. *Journal of Animal Ecology* 47: 697–709.

Den Boer, P. J. 1971. Stabilization of animal numbers and the heterogeneity of the environment: The problem of the persistence of sparse populations. In P. J. den Boer and G. R. Gradwell, eds., *Dynamics of populations*, 77–97. Wageningen, Netherlands: Centre for Agricultural Publishing and Documentation.

Dodd, C. K. 1992. Biological diversity of a temporary pond herpetofauna in north Florida sandhills. *Biodiversity and Conservation* 1: 125–142.

Dodd, M., J. Silvertown, K. McConway, J. Potts, and M. Crawley. 1995. Community stability: A 60-year record of trends and outbreaks in the occurrence of species in the Park Grass Experiment. *Journal of Ecology* 83: 277–285.

Doroff, A. M. and A. B. Keith. 1990. Demography and ecology of an ornate box turtle *Terrapene ornata* population in south-central Wisconsin, USA. *Copeia* 1990: 387–389.

Easter-Pilcher, A. 1990. Cache size as an index to beaver colony size in northwestern Montana. *Wildlife Society Bulletin* 18: 110–113.

Eberhardt, L. and R. C. Van Etten. 1956. Evaluation of the pellet group count as a deer census method. *Journal of Wildlife Management* 20: 70–74.

Ehrlich, P. R. and D. D. Murphy. 1987. Conservation lessons from long-term studies of checkerspot butterflies. *Conservation Biology* 1: 122–131.

Erlinge, S. 1983. Demography and dynamics of a stoat *Mustela erminea* population in a diverse community of vertebrates. *Journal of Animal Ecology* 52: 705–726.

Fitch, H. S. 1960. Autecology of the copperhead. Miscellaneous publication of the University of Kansas Museum of Natural History 13: 85–288.

Fitch, H. S. 1963. Natural history of the racer *Coluber constrictor*. Miscellaneous publication of the University of Kansas Museum of Natural History 15: 351–468.

Fitch, H. S. 1965. An ecological study of the garter snake, *Thamnophis sirtalis*. Miscellaneous publication of the University of Kansas Museum of Natural History 15: 493–564.

Fitch, H. S. and J. R. Bentley. 1949. Use of California annual-plant forage by range rodents. *Ecology* 30: 306–321.

Foote, L. E., H. S. Peters, and A. L. Finkner. 1958. Design test for mourning dove call-count sampling in seven southeastern states. *Journal of Wildlife Management* 22: 402–408.

Frazer, N. B., J. W. Gibbons, and J. L. Greene. 1989. Life tables of a slider turtle population. In J. W. Gibbons, ed., *Life history and ecology of the slider turtle,* 183–195. Washington, D.C.: Smithsonian Institution Press.

Fuller, T. K. 1992. Do pellet counts index white-tailed deer numbers and population change?: A reply. *Journal of Wildlife Management* 56: 613.

Fuller, T. K. and D. W. Kuehn. 1985. Population characteristics of striped skunks in north-central Minnesota. *Journal of Mammalogy* 66: 813–815.

Garber, S. D. and J. Burger. 1995. A 20-yr study demonstrates the relationship between turtle decline and human disturbance. *Ecological Applications* 5: 1151–1162.

Gerrodette, T. 1987. A power analysis for detecting trends. *Ecology* 68: 1364–1372.

Getz, L. L. 1989. A 14-year study of *Blarina brevicauda* populations in east-central Illinois. *Journal of Mammalogy* 70: 58–66.

Gibbs, J. P., S. Droege, and P. A. Eagle. 1998. Monitoring populations of plants and animals. *BioScience* 48: 935–940.

Gibbs, J. P. and S. M. Melvin. 1993. Call–response surveys for monitoring breeding waterbirds. *Journal of Wildlife Management* 57: 27–34.

Gibbs, J. P. and S. M. Melvin. 1997. Power to detect trends in waterbird abundance with call–response surveys. *Journal of Wildlife Management* 61: 1262–1267.

Gittins, S. P. 1983. Population dynamics of the common toad (*Bufo bufo*) at a lake in mid-Wales. *Journal of Animal Ecology* 52: 981–988.

Grant, W. E., P. E. Carothers, and L. A. Gidley. 1985. Small mammal community structure in the postoak savanna of east-central Texas. *Journal of Mammalogy* 66: 589–594.

Gurnell, J. 1996. The effects of food availability and winter weather on the dynamics of a grey squirrel population in southern England. *Journal of Applied Ecology* 33: 325–338.

Hairston, N. G. 1983. Growth, survival and reproduction of *Plethodon jordani*: Trade-offs between selective pressures. *Copeia* 1983: 1024–1035.

Hairston, N. G. and R. H. Wiley. 1993. No decline in salamander (Amphibia: Caudata) populations: A twenty-year study in the southern Appalachians. *Brimleyana* 18: 59–64.

Hall, T. J. 1986. Electrofishing catch per hour as an indicator of largemouth bass density in Ohio impoundments. *North American Journal of Fisheries Management* 6: 397–400.

Hansen, M. J., R. G. Schorfhaar, J. W. Peck, J. H. Selgeby, and W. W. Taylor. 1995. Abundance indices for determining the status of lake trout restoration in Michigan waters of Lake Superior. *North American Journal of Fisheries Management* 15: 830–837.

Hatfield, J. S., W. R. Gould IV, B. A. Hoover, M. R. Fuller, and E. L. Lundquist. 1996. Detecting trends in raptor counts: Power and type I error rates of various statistical tests. *Wildlife Society Bulletin* 24: 505–515.

Hayes, J. P. and R. J. Steidl. 1997. Statistical power analysis and amphibian population trends. *Conservation Biology* 11: 273–275.

Hewson, R. 1971. A population study of mountain hares (*Lepus timidus*) in north-east Scotland from 1956–1969. *Journal of Ecology* 40: 395–417.

Hill, J. M. and C. B. Kinsley. 1993. *Puritan tiger beetle (Cincindela puritana G. Horn) recovery plan*. Hadley, Mass.: U.S. Fish and Wildlife Service.

Hill, J. M. and C. B. Kinsley. 1994. *Northeastern beach tiger beetle (Cincindela dorsalis dorsalis Say) recovery plan*. Hadley, Mass.: U.S. Fish and Wildlife Service.

Hogstad, O. 1993. Structure and dynamics of a passerine bird community in a spruce-dominated boreal forest. A 12-year study. *Annales Zoologici Fennici* 30: 43–54.

Holmes, R. T. and T. W. Sherry. 1988. Assessing population trends of New Hampshire forest birds: Local vs. regional patterns. *Auk* 105: 756–768.

House, R. 1995. Temporal variation in abundance of an isolated population of cutthroat trout in western Oregon, 1981–1991. *North American Journal of Fisheries Management* 15: 33–41.

Husting, E. L. 1965. Survival and breeding structure in a population of *Ambystoma maculatum*. *Copeia* 1965: 352–359.

Hutchings, M. J. 1987. The population biology of the early spider orchid, *Ophrys sphegodes* Mill. I. A demographic study from 1975–1984. *Journal of Ecology* 75: 711–727.

Jaeger, R. G. 1980. Density-dependent and density-independent causes of extinction of a salamander population. *Evolution* 34: 617–621.

Jones, M. G. 1976. The carabid and staphylinid fauna of winter wheat and fallow on a clay with flints soil. *Journal of Applied Ecology* 13: 775–792.

Kaikusalo, A. and A. Angerbjorn. 1995. The arctic fox population in Finnish Lapland during 30 years, 1964–93. *Annales Zoologici Fennici* 32: 69–77.

Kalko, E. K. V., C. O. Handley, and D. Handley. 1996. Organization, diversity, and long-term dynamics of a Neotropical bat community. In M. L. Cody and J. A. Smallwood, eds., *Long-term studies of vertebrate communities,* 503–554. San Diego: Academic Press.

Kallemeyn, L. W. 1987. Correlations of regulated lake levels and climatic factors with abundance of young-of-the-year walleye and yellow perch in four lakes in Voyageurs National Park. *North American Journal of Fisheries Management* 7: 513–521.

Keith, L. B. and J. H. Cary. 1991. Mustelid, squirrel, and porcupine trends during a snow-shoe hare cycle. *Journal of Mammalogy* 72: 373–378.

Kipling, C. 1983. Changes in the population of pike (*Esox lucius*) in Windermere from 1944 to 1981. *Journal of Animal Ecology* 52: 989–999.

Lack, D. 1969. *Population studies of birds*. Oxford, U.K.: Clarendon.

Lyons, J. and K. Welke. 1996. Abundance and growth of young-of-year walleye (*Stizostedion vitreum*) and sauger (*S. canadense*) in Pool 10, Upper Mississippi River, and at Prairie du Sac Dam, Lower Wisconsin River, 1987–1994. *Journal of Freshwater Ecology* 11: 39–47.

Macan, T. T. 1977. A twenty-year study of the fauna in the vegetation of a moorland fish-pond. *Archiv für Hydrobiologie* 81: 1–24.

McInerny, M. C. and D. J. Degan. 1993. Electrofishing catch rates as an index of large-mouth bass population density in two large reservoirs. *North American Journal of Fisheries Management* 13: 223–228.

Messier, F. 1991. The significance of limiting and regulating factors on the demography of moose and white-tailed deer. *Journal of Animal Ecology* 60: 377–393.

Momma, E. 1965. The dynamic aspects of Drosophila populations in semi-natural areas. *Japanese Journal of Genetics* 40: 275–295.

Moore, C. M., R. J. Neves, and J. J. Ney. 1991. Survival and abundance of stocked striped bass in Smith Mountain Lake, Virginia. *North American Journal of Fisheries Management* 11: 393–399.

Moore, N. W. 1991. The development of dragonfly communities and the consequences of territorial behaviour: A 27 year study on small ponds at Woodwalton Fen, Cambridgeshire, United Kingdom. *Odonatologica* 20: 203–231.

Mossman, M. J., L. M. Hartman, J. Huff, and R. Hay. 1994. Auditory frog and toad surveys: Wisconsin population trends 1984–1993 and suggestions for regional monitoring. *Proceedings of the North Dakota Academy of Sciences* 48: 14.

Nicholls, A. O., P. C. Viljoen, M. H. Knight, A. S. van Jarrsveld. 1996. Evaluating population persistence of censused and unmanaged herbivore populations from the Kruger National Park, South Africa. *Biological Conservation* 76: 57–67.

Odum, E. P. 1955. An eleven year history of a *Sigmodon* population. *Journal of Mammalogy* 36: 368–378.

Peterman, R. M., and M. J. Bradford. 1987. Statistical power of trends in fish abundance. *Canadian Journal of Fisheries and Aquatic Sciences* 44: 1879–1889.

Petrinovich, L. and T. L. Patterson. 1982. The white-crowned sparrow: stability, recruitment, and population structure in the *Nuttall* subspecies (1975–1980). *Auk* 99: 1–14.

Pimm, S. L., H. L. Jones, and J. M. Diamond. 1988. On the risk of extinction. *American Naturalist* 132: 757–785.

Platts, W. S. and R. L. Nelson. 1988. Fluctuations in trout populations and their implications for land-use evaluation. *North American Journal of Fisheries Management* 8: 333–345.

Pulliainen, E. 1981. A transect survey of small land carnivore and red fox populations on a subarctic fell in Finnish Forest Lapland over 13 winters. *Annales Zoologici Fennici* 18: 270–278.

Rainwater, W. C. and A. Houser. 1982. Species composition and biomass of fish in selected coves in Beaver Lake, Arkansas, during the first 18 years of impoundment (1963–1980). *North American Journal of Fisheries Management* 4: 316–325.

Reid, V. H., R. M. Hansen, and R. L. Ward. 1966. Counting mounds and earth plugs to census mountain pocket gophers. *Journal of Wildlife Management* 30: 327–334.

Rieman, B. E. and J. D. McIntyre. 1996. Spatial and temporal variability in bull trout redd counts. *North American Journal of Fisheries Management* 16: 132–141.

Renault, T. R. and C. A. Miller. 1972. Spiders in a fir-spruce biotype: Abundance, diversity, and influence on spruce budworm densities. *Canadian Journal of Zoology* 50: 1039–1047.

Rotella, J. J. and J. T. Ratti. 1986. Test of a critical density index assumption: A case study with gray partridge. *Journal of Wildlife Management* 50: 532–539.

Roth, R. R. and R. K. Johnson. 1993. Long-term dynamics of a Wood Thrush population breeding in a forest fragment. *Auk* 110: 37–48.

Ryel, L. A. 1959. *Deer pellet group surveys on an area of known herd size.* Lansing: Michigan Department of Conservation, Game Division Report 2252.

Schoener, T. W. and D. A. Spiller. 1992. Is extinction rate related to temporal variability in population size? An empirical answer for orb spiders. *American Naturalist* 139: 1176–1207.

Semlitsch, R. D., D. E. Scott, J. H. K. Pechmann, and J. W. Gibbons. 1996. Structure and dynamics of an amphibian community. In M. L. Cody and J. A. Smallwood, eds., *Long-term studies of vertebrate communities,* 217–250. San Diego: Academic Press.

Serns, S. L. 1982. Relationship of walleye fingerling density and electrofishing catch per effort in northern Wisconsin lakes. *North American Journal of Fisheries Management* 2: 38–44.

Southwick, C. H., S. C. Golian, M. R. Whitworth, J. C. Halfpenny, and R. Brown. 1986. Population density and fluctuations of pikas (*Ochotona princeps*) in Colorado. *Journal of Mammalogy* 67: 149–153.

Spitzer, K. and J. Leps. 1988. Determinants of temporal variation in moth abundance. *Oikos* 53: 31–36.

Steidl, R. J., J. P. Hayes, and E. Schauber. 1997. Statistical power analysis in wildlife research. *Journal of Wildlife Management* 61: 270–279.

Sutcliffe, O. L., C. D. Thomas, and D. Moss. 1996. Spatial synchrony and asynchrony in butterfly population dynamics. *Journal of Animal Ecology* 65: 85–95.

Svensson, B. M., B. A. Carlsson, P. S. Karlsson, and K. O. Nordell. 1993. Comparative long-term demography of three species of *Pinguicula. Journal of Ecology* 81: 635–645.

Swihart, R. K. and N. A. Slade. 1990. Long-term dynamics of an early successional small mammal community. *American Midland Naturalist* 123: 372–382.

Symonides, E. 1979. The structure and population dynamics of psammophytes on inland dunes III. Populations of compact psammophyte communities. *Ekologia Polska* 27: 235–257.

Taylor, B. L. and T. Gerrodette. 1993. The uses of statistical power in conservation biology: The vaquita and northern spotted owl. *Conservation Biology* 7: 489–500.

Thomas, L. 1997. Retrospective power analysis. *Conservation Biology* 11: 276–280.

Thompson, S. K. 1992. *Sampling.* New York: Wiley.

Tinkle, D. W. 1993. Life history and demographic variation in the lizard *Sceloperus graciousus*: A long-term study. *Ecology* 74: 2413–2429.

Turner, F. B. 1977. The dynamics of populations of squamates, crocodilians and rhynchocephalians. In C. Gans, ed., *Biology of the reptilia,* 157–264. London: Academic Press.

Warner, S. C., K. E. Limberg, A. H. Arino, M. Dodd, J. Dushoff, K. I. Stergiou, and J. Potts. 1995. Time series compared across the land-sea gradient. In T. M. Powell and J. H. Steele, eds., *Ecological time series,* 242–273. New York: Chapman & Hall.

Weitzel, N. H. and H. R. Panik. 1993. Long-term fluctuations of an isolated population of the Pacific chorus frog (*Pseudacris regilla*) in northwestern Nevada. *Great Basin Naturalist* 53: 379–384.

Wells, T. C. E. 1981. Population ecology of terrestrial orchids. In H. Synge, ed., *The biological aspects of rare plant conservation,* 281–295. Chichester, U.K.: Wiley.

White, G. C. 1992. Do pellet counts index white-tailed deer numbers and population change?: A comment. *Journal of Wildlife Management* 56: 611–612.

Wielgus, R. B. and F. L. Bunnell. 1994. Dynamics of a small, hunted brown bear *Ursus arctos* population in southwestern Alberta, Canada. *Biological Conservation* 67: 161–166.

Willis, D. W., J. F. Smeltzer, and S. A. Flickinger. 1984. Characteristics of a crappie population in an unfished small impoundment containing northern pike. *North American Journal of Fisheries Management* 4: 385–389.

Zielinski, W. J. and H. B. Stauffer. 1996. Monitoring *Martes* populations in California: Survey design and power analysis. *Ecological Applications* 6: 1254–1267.

## Chapter 8

# Modeling Predator–Prey Dynamics

MARK S. BOYCE

Our gathering in Sicily from which contributions to this volume developed coincided with the continuing celebration of 400 years of modern science since Galileo Galilei (1564–1642). Although Galileo is most often remembered for his work in astronomy and physics, I suggest that his most fundamental contributions were to the roots of rational approaches to conducting science. An advocate of mathematical rationalism, Galileo made a case against the Aristotelian logicoverbal approach to science (Galilei 1638) and in 1623 insisted that the "Book of Nature is written in the language of mathematics" (McMullin 1988). Backed by a rigorous mathematical basis for logic and hypothesis building, Galileo founded the modern experimental method. The method of Galileo was the combination of calculation with experiment, transforming the concrete into the abstract and assiduously comparing results (Settle 1988).

Studies of predator–prey dynamics will benefit if we follow Galileo's rigorous approach. We start with logical mathematical models for predator–prey interactions. This logical framework then should provide the stimulus by which we design experiments and collect field data. Science is the iteration between observation and theory development that gradually, even ponderously, enhances our understanding of nature. Like Galileo, I insist that the book of predator–prey dynamics is written in mathematical form.

In wildlife ecology, the interface between theory and empiricism is poorly developed. For predator–prey systems, choosing appropriate model structure is key to anticipating dynamics and system responses to management. Predator–prey interactions can possess remarkably complex dynamics, including various routes to chaos (Schaffer 1988). This presents several problems for the empiricist, including the difficulty of estimating all of the parameters in a complex model and distinguishing stochastic variation from deterministic dynamics.

Wildlife biologists in particular seem to suffer from what I call the techniques syndrome: They are preoccupied with resolving how to compile reliable field data, often at the expense of understanding what one might do with the data once obtained. This became particularly apparent to me during my tenure as editor-in-chief of the *Journal of Wildlife Management,* where I was surprised to discover that fully 40 percent of the manuscripts submitted to the journal in 1995–1996 were on techniques rather than wildlife management. Such a preoccupation with techniques has been symptomatic of wildlife curricula in the United States. For example, the capstone course in my undergraduate training at Iowa State University in 1972 was a course in wildlife techniques; principles were presumed to have emerged from lower-level courses in animal and plant ecology.

In this context, one might find it curious that a chapter on predator–prey modeling would appear in a book on techniques. Modeling is indeed viewed by some as a technique. I prefer to consider modeling as a way of thinking and structuring ideas rather than a technique. We sometimes use modeling as a technique; for example, we might use predator–prey modeling to predict the nature of population fluctuations and to forecast future population sizes. In this vein, predator–prey modeling can be used as a technique for assisting managers with decision making. Modeling also can be used to test our assumptions about predator–prey interactions and to guide the collection of data. Modeling provides the impetus for what Galileo called the "cimento" (experiment). To my mind, most fundamentally, predator–prey modeling is used to improve our understanding of system dynamics emerging from trophic-level interactions.

### ■ Modeling Approaches for Predator–Prey Systems

Approaches and objectives for modeling predator–prey interactions can vary a great deal. I classify predator–prey models into three classes: noninteractive models in which one or the other of a predator–prey interaction is assumed to be constant, true predator–prey models in which two trophic levels interact, and statistical models for characterizing the dynamics of populations that may be driven by a predator–prey interaction. Predator–prey interactions are similar to plant–herbivore interactions, and indeed, the same models have been used to characterize plant–herbivore interactions (Caughley 1976) as have been used to characterize predator–prey interactions (Edelstein-Keshet 1988).

In this review I touch only briefly on more complex models involving multiple species, but of course, seldom is a two-species interaction sufficient to capture the complexity of biological interactions that occur in ecosystems.

## NONINTERACTIVE MODELS

Predator–prey models are by definition based on a predator having a negative effect on a prey population while the predator benefits from consuming the prey. Yet to simplify the system, many ecologists choose to ignore the interaction by assessing only the dynamics of a single species. This can take at least four forms: single-species models of predators or prey, demographic trajectories of prey anticipating the consequences of predator-imposed mortality, attempts to assess whether predator-imposed mortality on prey is compensatory or additive, and habitat capability models. Each of these approaches circumvents the issue of predator–prey interactions; consequently, noninteractive models are less likely to capture the dynamic behaviors of a predator–prey system. However, these approaches pervade the wildlife ecology literature and deserve to be placed into context.

### Single-species models

We can model the effect of a predator population on a prey population with a single equation for the prey. For example, consider a population of prey governed by the differential equation

$$dV/dt = r \times V(1 - V/K) - P \times F(V) \tag{8.1}$$

where $V \equiv V(t)$ is the victim or prey population size at time $t$, $r$ is the potential per capita growth rate for the prey, $K$ is the prey carrying capacity (i.e., where $dV/dt = 0$ in the absence of predators), $P$ is the number of predators, and the function $F(\cdot)$ is the functional response characterizing the number of prey killed per predator (figure 8.1). This simple single-species model is useful because it can be used to illustrate the consequences of variation in the functional response and how multiple equilibria can emerge when $F(\cdot)$ is logistic in shape (see Yodzis 1989:16–17). But we must assume that the number of predators is constant and there is no opportunity to anticipate the dynamics of the predator population without another equation for $dP/dt$.

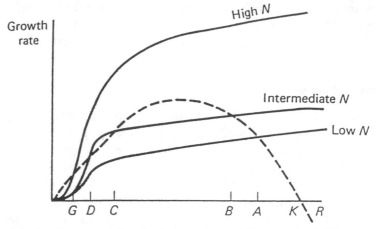

**Figure 8.1** Graphic representation of a single-species model (see equation 8.1) for prey abundance (R) given low, intermediate, and high predator abundance (N). Dashed line is the growth rate for prey and the solid lines are the rate of killing of prey by predators. This predicts prey population density as a consequence of predators. Equilibrium population size for prey occurs where the two curves intersect. Low equilibriums are predicted at G and D. C is an unstable critical point, and A, B, and K are stable equilibria at high populations of prey. From Yodzis (1989:17).

A related approach is to estimate the potential rate of increase for the prey and to assume that predator numbers could be increased to a level that they could consume up to this rate through predation. For example, Fryxell (1988) concluded that moose (*Alces alces*) in Newfoundland could sustain a maximum human predation of 25 percent. Likewise, the amount of wolf (*Canis lupus*) predation on blackbuck (*Antelope cervicapra*) in India was calculated to balance potential population growth rate of the prey (Jhala 1993).

Alternatively, we might anticipate the dynamics for a predator population while ignoring the dynamics of the prey. A typical approach would be to assume equilibrial dynamics for the predator, presumably depending on a continuously renewing resource of prey (e.g., the logistic and related models). The same criticism that Caughley (1976) articulated for single-species models of herbivores might be leveled against this approach for a predator population. In particular, trophic-level interactions create dynamic patterns that can be trivialized or destroyed by collapsing the system to a single-species model, but not necessarily. Incorporation of a time lag in density dependence (see Lotka 1925) is a surrogate for a trophic-level interaction from which complex dynamics can emerge (cf. Takens' theorem, Broomhead and Jones 1989; Royama 1992).

Likewise, difference equations possess implicit time lags; that is, the population cannot respond between $t$ and $t + 1$, thereby creating complex dynamics of the same sort observed in more complete predator–prey models (Schaffer 1988). The actual biological interactions that create implicit or explicit time lags are disguised in such models. Consider McKelvey et al.'s (1980) model for the dynamics of the Dungeness crab (*Cancer magister*) off the California coast. An age-structured difference equation was constructed that oscillated in a fashion that mimicked fluctuations in the harvest of crabs. But because the mechanisms creating the fluctuations in harvest were implicit in the discrete-time nature of the model rather than explicit trophic-level interactions, we gained little knowledge about the biology that yielded the pattern of dynamics.

Although we easily can be critical of assumptions associated with a single-species model, in many cases this may be the best that we can do. Imagine the difficulty trying to construct a model for grizzly bear (*Ursus arctos horribilis*) populations that included all of the predator–prey and plant–herbivore interactions that form the trophic-level interactions of this omnivore. We might make the assumption that food resources are renewable and diverse and then proceed to use a density-dependent model for the bears, essentially ignoring the vast diversity of food resources on which individual bears depend. Variability in the resources can be covered up by making the resources stochastic variables, for example, enforcing a stochastic carrying capacity, $K(t)$, as in the time-dependent logistic

$$dN/dt = rN[1 - N/K(t)] \qquad (8.2)$$

An alternative perspective is to accept the deterministic dynamics as representing a trophic-level interaction that we might not understand, but that might well be modeled using time-delay models. There are direct links between the complex dynamics of multispecies continuous-time systems and those of discrete-time difference equations. For example, one can reconstruct a difference equation from a Poincaré section of a strange attractor (Schaffer 1988). In this way one can envisage models of biological populations that exploit the complex dynamics from single-species models as appropriate ways to capture higher-dimensional complexity in ecosystems.

*Demographic trajectories*

Another single-population approach to predator–prey modeling includes attempts to model the demographic consequences of a predator. For example,

Vales and Peek (1995) modeled elk (*Cervus elaphus*) and mule deer (*Odocoileus hemionus*) populations on the Rocky Mountain East Front of Montana, attempting to anticipate the consequences of wolf predation. So for a given number of wolves and an estimate of the number of elk eaten per wolf, Vales and Peek estimated the effect of wolf predation and hunter kill on population growth rate for the elk and deer. This is akin to a sensitivity analysis for elk population growth in which the effect of predation mortality is figured, holding all else constant. But such a modeling approach cannot possibly anticipate the rich dynamic behaviors known to emerge from predator–prey interactions simply because the model structure precludes interaction between populations. Mack and Singer (1993) generated a similarly restricted model using the software POPII for conducting demographic projections for ungulate populations. POPII projections are structurally identical to the Leslie matrix projection approach followed by Vales and Peek (1993).

## Compensatory versus additive mortality

Field studies of predation (and hunter harvest) on bobwhite (*Colinus virginianus*), cottontail rabbits (*Sylvilagus floridanus*), muskrats (*Ondatra zibethicus*), wood pigeons (*Columba palumbus*), and waterfowl have shown that fall and overwinter mortality can be compensated by a reduction in other sources of "natural" mortality yielding constant spring breeding densities for prey irrespective of predation mortality (Errington 1946, 1967; Murton et al. 1974; Anderson and Burnham 1976). The principle of compensatory mortality has led some biologists to question whether wolf recovery in Yellowstone National Park will actually have any measurable effect on elk population size (Singer et al. 1997).

On the surface compensatory mortality appears to be at odds with the predictions of classic predator–prey or harvest models because increased predation or harvest mortality should always reduce equilibrium population size. This apparent contradiction is simply a consequence of not modeling the details of within-year seasonality and the timing of mortality. Compensatory mortality emerges, of course, as a consequence of density dependence whereby reduced prey numbers results in heightened survival among the individuals that escaped predation or harvest. But these seasonal details are all ignored in the classic predator–prey models in continuous time with no explicit seasonality. Likewise, if the models are difference equations, the within-year details of the seasonality usually are not incorporated into the models.

Seasonal models are certainly possible. In continuous time we can make relevant parameters to be periodic functions of time. For example, we can rewrite equation (8.1) with time-varying $r$ or $K$:

$$dV/dt = rV[1 - V/K(t)] - P \times F(V) \tag{8.3}$$

where $K(t)$ varies periodically, say according the seasonal forcing function:

$$K(t) = \overline{K} + K_a \times \cos(2\pi t/\tau) \tag{8.4}$$

with $\overline{K}$ equal to the mean $K(t)$, $K_a$ the amplitude variation in $K(t)$, and $\tau$ the period length in units of time, $t$ (Boyce and Daley 1980). If density dependence is strong enough in such a seasonal regimen, we can observe spring breeding densities that do not change with seasonal predation or harvest. Necessarily, however, the integral of population size over the entire year must decline to evoke the density-dependent response, even though spring breeding densities need not be reduced.

## Habitat capability models

In a study of blackbuck and wolves in Velavadar National Park, Gujarat, India, Jhala (in press) modeled the relationship between habitat and abundance for each species. The primary habitat variable was the areal extent of a tenacious exotic shrub, *Prosopis juliflora,* which provided denning and cover habitats for wolves, as well as nutritious seed pods eaten by blackbuck during periods of food shortage. Jhala (in press) established a desired ratio of wolves to blackbucks in advance and then modeled the amount of *Prosopis* habitat that would achieve the desired ratio of wolves to blackbuck. The model afforded no opportunity for a dynamic interaction between the wolves and the blackbuck, despite the fact that wolves are major predators on blackbuck. Instead, the number of blackbuck per wolf to maintain a stable blackbuck population was computed using Keith's (1983) model:

$$N = [k/(\lambda - 1)] \times W \tag{8.5}$$

where $N$ is the number of blackbuck, $k$ is the number blackbuck killed per wolf per year, $\lambda$ is the finite growth factor for the blackbuck population estimated using life table analysis, and $W$ is the number of wolves in the park. The condition of the population at the time that the demographic data were estimated will be crucial to determining $\lambda$, so the vital rates estimated during 1988–1990 will establish how many wolves the population of blackbuck can sustain.

Although attempting to model the differential habitat requirements for blackbuck and wolves in an area is a novel approach, the interaction between predator and prey is not sufficiently known to offer an ecological basis for setting the desired ratio of predators to prey. Nor do we have sufficient data on the predator–prey interaction to know that establishing certain amounts of preferred habitats for each species would yield the target numbers of each species when they are allowed to interact dynamically. An implicit assumption with Jhala's (in press) model is that both the predator and the prey have equilibrium dynamics set by the amount of habitat.

The Jhala (in press) paper illustrates the dangers of using Keith's (1983) model, which assumes no functional response. This application of Keith's model assumes that wolf predation is the only source of mortality, it is not compensatory, and wolf numbers can increase to a level at which the entire prey production is removed by the predator. I believe that these assumptions are usually violated.

Habitat capability models are usually focused on just one species (e.g., habitat suitability indices). Methods for extrapolating distribution and abundance have improved with the use of geographic information systems (Mladenoff et al. 1997) and resource selection functions (Manly et al. 1993).

## TRUE PREDATOR–PREY MODELS

### Lotka–Volterra models

The structure of modern predator–prey models in ecology was outlined by Italian mathematician Vito Volterra (1926), who held the Chair of Mathematical Physics in Rome (Kingsland 1985). Volterra's interest in predator–prey interactions was piqued by Umberto D'Ancona, a marine biologist who was engaged to marry Volterra's daughter, Luisa. D'Ancona suggested to Volterra that there might be a mathematical explanation for the fact that several species of predaceous fish increased markedly during World War I, when fishing by humans almost ceased. Volterra suggested the use of two simultaneous differential equations to model the dynamics for interacting populations of predator and prey. The model had potential for cyclic fluctuations in predator and prey that were driven entirely by the interaction between the two species. The model is

$$dV/dt = bV - aVP \qquad (8.6)$$

$$dP/dt = cVP - dP \qquad (8.7)$$

where $b$ is the potential growth rate for the prey in the absence of predation, $a$ is the attack rate, $c$ is the rate of amelioration of predator population decline afforded by eating prey, and $d$ is the per capita death rate for the predator in the absence of prey. The right-hand portion of the prey equation (equation 8.6) models the rate at which prey are removed from the population by predation. The product of $a \times V$ is often called the functional response. Note that in the first portion of the predator equation we see a similar function of $V \times P$ that models how the rate of predator decline is ameliorated by the conversion of prey into predator population growth. This portion of the model, $c \times V \times P$, is what we usually call the numerical response.

Although Volterra developed his model independently from basic principles, an American, Alfred J. Lotka (1925), had already suggested the same mathematical structure for two-species interactions and presented a full mathematical treatment. Lotka was quick to advise Volterra of his priority (Kingsland 1985). Consequently, most ecologists call the two-species system of differential equations the Lotka–Volterra models. Nevertheless, Volterra developed the analysis of predator–prey interactions in more detail, offered more examples, and published in several languages, doing much to bring attention to the approach.

Despite the valuable insight that this simple model affords, the Lotka–Volterra model has been mercilessly attacked for its unrealistic assumptions and dynamics (Thompson 1937). The dynamics include neutrally stable oscillations with period length, $T \approx 2\pi/\sqrt{bd}$, for which the amplitude of oscillations depends on initial conditions (Lotka 1925). Assumptions include a linear functional response that essentially says that the number of prey killed per predator will increase with increasing prey abundance without bound. Yet at some level we must expect that the per capita rate at which prey are killed would level off because of satiation or time limitations (Holling 1966). Another assumption is that neither the predator nor prey has density-dependent limitations other than that afforded by the abundance of the other species. Furthermore, we have a number of assumptions that are symptomatic of most simple predator–prey models (i.e., they have no age or sex structure) and the model is deterministic, whereas fundamentally all ecological systems are inherently stochastic (Maynard Smith 1974).

Rather than dwelling further on the Lotka–Volterra model, I believe that we can dismiss it as an early effort that gave useful insight. Not only do the neutrally stable oscillations appear peculiar and inconsistent with ecological intuition, but the model is structurally unstable, meaning that small variations in the model destroy the neutrally stable oscillations, leading to convergence to equilibrium, divergence to extinction, or even stable limit cycles (Edelstein-Keshet 1988).

Volterra was aware of certain limitations to his predator–prey model and later proposed a form in which prey were limited by density dependence:

$$dV/dt = V[b - (b/K)V - a \times P] \tag{8.8}$$

$$dP/dt = P(c \times V - d) \tag{8.9}$$

Now in the absence of predators the prey population converges asymptotically on a carrying capacity, $K$. But the model still suffers from the assumption of prey being eaten proportionally to the product of the two population sizes; similarly, the numerical response remains linear. However, instead of neutrally stable cycles, the populations now oscillate while converging on an equilibrium number of predator and prey (Volterra 1931).

*Kolmogorov's equations*

More useful than the Lotka–Volterra model is the more general analysis by Kolmogorov (1936), who studied predator–prey models of the general form

$$dV/dt = V \times f(V, P) \tag{8.10}$$

$$dP/dt = P \times g(V, P) \tag{8.11}$$

where we assume that the functions $f$ and $g$ have several properties that are generally consistent with the ecology of predator–prey interactions. These include

$$\partial f / \partial V < 0 \text{ (for large } V)$$

$$\partial f / \partial P < 0$$

$$\partial g / \partial V > 0$$

and

$$\partial g / \partial P < 0 \tag{8.12}$$

Biologically Kolmogorov's assumptions seem reasonable. For example, we assume that an increase in the predator population results in a decrease in the per capita growth rate for the prey. Conversely, we assume that increases in prey enhance the per capita growth rate for the predator. Kolmogorov requires that there be some predator density that will check the growth of the prey population and that some minimal number of prey are necessary for the predator population to increase. In contrast with the original Lotka–Volterra model (equations 8.6 and 8.7), which invokes exponential population growth except as modified by the species' interactions, Kolmogorov requires density dependence, at least for the prey population. Density dependence for the predator can be explicit, as might be caused by territoriality or simply by a limitation in the availability of prey.

When coefficients are such that the critical point ($dV/dt = dP/dt = 0$) is unstable, the interaction between predator and prey can lead to stable limit cycles. Biologically, stable limit cycles seem more reasonable than neutrally stable cycles because perturbations to the system dampen out and when unperturbed the system returns to the same perpetual oscillation between the two species (figure 8.2, top). Rather than dependence on initial conditions, systems with stable limit cycles converge on the same dynamics irrespective of the starting population sizes.

The exact form of the Kolmogorov equations is quite flexible. For example, prey density dependence can be of quadratic form, $f(V) = r(1 - V/K)$, as in Pielou (1969) and Caughley (1976); $f(V) = r[(K/V)^{-\theta} - 1]$ ($1 \geq \theta > 0$), used by Rosenzweig (1971); or $f(V) = r(K/V - 1)$, as suggested by Schoener (1973).

The rate at which prey are taken by predators is known as the functional response, depending on the behavior of both the predator and the prey. A remarkable variety of functions has been proposed to characterize the functional response, with Gutierrez (1996) listing 14 equations that focus largely on killing rates as functions of density of prey. Included among these models

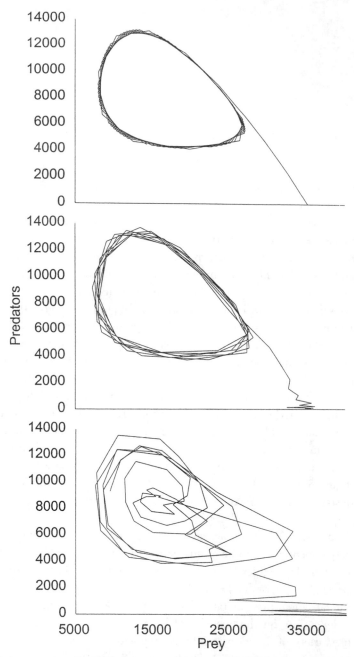

**Figure 8.2** Stable-limit cycle from a two-species predator–prey model with density dependence for prey and Ivlev functional and numerical response. *Top:* Deterministic simulation. *Middle:* Stochastic variation in carrying capacity for prey of $\sigma = 500$. *Bottom:* Stochastic variation in $K$ where $\sigma = 5{,}000$.

are the familiar type I, II, III, and IV functional responses (figure 8.3) proposed by Holling (1966). Among arthropods most functional responses fit a type II or III (Hassell 1978).

Although some have claimed that mammals often have type III functional responses, apparently due to learning (Holling 1966; Maynard Smith 1974), Messier (1994) and Dale et al. (1994) present evidence that wolves preying on moose and caribou *(Rangifer tarandus)* better fit a type II response.

But the rate of prey capture is much more complex than just a dependence on prey density (i.e., a dependence on the physical environment, vulnerability of prey, condition of the predator, prey group size, and a number of other variables). Indeed, much of the theory of optimal foraging (Stephens and Krebs 1986; Fryxell and Lundberg 1994) deals with understanding adaptations to factors that influence the rate of prey capture, and much of this theory is relevant to the development of sound models for functional response. Students of herbivory (Spalinger et al. 1988) appear to have a more mechanistic and enlightened perspective on the structure of the functional response than those studying predation.

The numerical response is usually modeled as a simple multiple of the functional response, so the numerical response assumes the same shape as the functional response. Indeed, there is an empirical basis for this relationship (Emlen 1984) that is especially well documented among invertebrates. But vertebrate examples also exist. For example, Maker (1970) found a logistic-shaped plot (type III) of the density of pomarine jaeger *(Stercorarius pomarinus)* nests as a function of the density of brown lemmings *(Lemmus trimucronatus)* in Alaska. Messier (1994) found what appeared to be a type II curve for both the functional response and numerical responses of wolves preying on moose (figure 8.4).

Numerical response is defined in different ways. As noted earlier a numerical response can be defined to predict the response in population growth rate for the predator afforded by the killing of prey. Alternatively, a numerical response may be defined to be the number of predators at equilibrium at a given prey population size (Holling 1959; Messier 1994). The latter definition is convenient because when this quantity is multiplied by the functional response it yields the total number of prey consumed for a given prey abundance. Dividing this quantity by prey population size yields the predation rate (figure 8.4).

As an example of the Kolmogorov equations, we will consider specifically the pair of equations that Caughley (1976) offered as a plant–herbivore model:

$$dV/dt = rV(1 - V/K) - Va(\exp(-c_1 V)) \qquad (8.13)$$

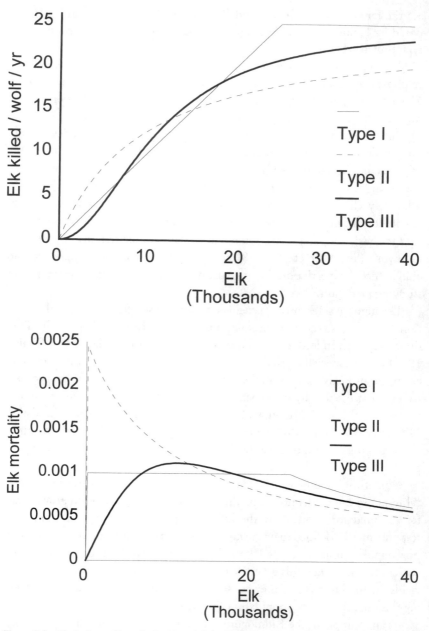

**Figure 8.3** Illustrations of hypothetical type I, II, and III functional responses for wolves preying on elk (top). *Bottom:* The proportional mortality attributable to each of the functional response types is plotted as a function of prey density (see Boyce and Anderson 1999), assuming no changes in wolf numbers with increasing elk density.

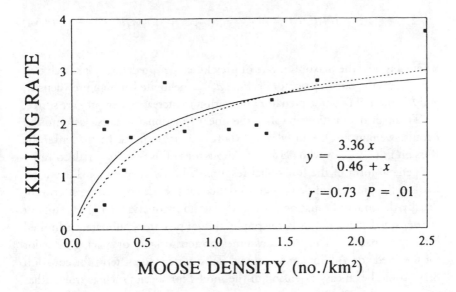

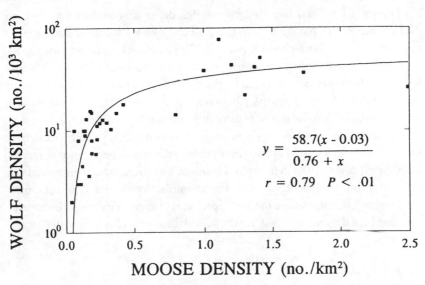

**Figure 8.4** Functional (top) and numerical (bottom) responses for wolves preying on moose. Messier (1994) fit Michaelis–Menton equations (type II) to these data from 27 wolf/moose studies. Sparsity of data make distinction between a type II and a type III (logistic) response impossible to assess. The product of the functional and numerical responses yields the total moose killed, which when divided by moose density gives the predation rate. From Messier (1995).

$$dP/dt = P \times h[1 - \exp(-c_2{}^* V)] - d_1 P \qquad (8.14)$$

where $a$ is now the maximum rate of prey killed per predator, $c_1$ is search efficiency, $c_2$ is the rate of predator decline, $d_1$ is ameliorated at high prey density, and $h$ is the ability of the predator population to increase when prey are scarce.

This pair of equations resolves the linear functional response assumption because we now assume an Ivlev (1961) functional response (named after the Russian fish ecologist who performed thousands of fish-feeding trials to verify the general form of the functional response). Likewise, the model explicitly resolves the problem of density dependence for the prey by adding a term for density-dependent limitation for the $V$ equation (equation 8.13). Because of the density dependence in $V$, the population of predators ultimately is limited by prey availability. This model assumes no territoriality or spacing behavior for the predator. Adding an additional density-dependent term for equation 8.14 would be an easy extension of the model for species such as wolves that are territorial.

This model can have interesting dynamics, depending on the values for each of the seven model parameters. In the simplest case we see rapid convergence to equilibrium for both predator and prey. But as model parameters are tuned, we can witness overshoots and convergent oscillations to equilibrium (Caughley 1976). Tuning parameters even further leads to the emergence of stable limit cycles resulting from an interplay between the destabilizing effect of satiation and the stabilizing influence of density dependence (figure 8.2, top).

According to the Poincaré–Bendixson theorem, the most complex behavior possible from a system of two simultaneous differential equations is a stable limit cycle (Edelstein-Keshet 1988). However, complications to the model can result in more complex dynamics. For example, Inoue and Kamifukumoto (1984) showed that seasonal forcing of prey carrying capacity results in remarkably complex dynamics, including the toroidal route to chaos (Schaffer 1988).

*Graphic models*

Graphic approaches have proven to be powerful ways to anticipate the outcome of predator–prey interactions. A simple approach was shown in figure 8.1, where the growth rate and predation rate are plotted simultaneously. This approach was used effectively by Messier (1994) to characterize population regulation in moose–wolf systems. Alternatively, Rosenzweig and MacArthur (1963) and Noy-Mier (1975) illustrate the use of static plots for predator and prey, allowing prediction of the dynamics (Edelstein-Keshet 1988). These graphic models are useful ways to anticipate the range of dynamics given only rough approximations for the system parameters.

## Ratio-dependent models

An energized debate has waged recently over the use of ratio-dependent models for predator–prey systems (Matson and Berryman 1992). A ratio-dependent model assumes that the functional response is determined by the ratio of predators to prey. On the surface this seems reasonable because an increasing prey:predator ratio implies that each predator will have available more potential prey. In practice, the ratio-dependent models have some strange properties and dynamic behaviors that should be avoided (Abrams 1994). For example, the functional response for a wolf–moose system is confounded by taking ratios, and Messier (1994) recommends against using the predator:prey ratios (see also Oksanen et al. 1990; Theberge 1990).

## Multispecies systems

Adding another species to the system provides raw material for chaos on a strange attractor (Gilpin 1979). A three-species system of differential equations representing, for example, a three–trophic level system can be collapsed to a single-species difference equation by taking a Poincaré section and plotting population sizes for any one of the three species after single rotations of the model (Schaffer 1985). This is a very important observation that justifies studying population models even when data may not exist for all the biologically important species.

### STOCHASTIC MODELS

Any of these models can be made stochastic by defining parameters or variables to be random variables. Computer simulation makes evaluation of the consequences of stochasticity fairly easily. But generalizing about the consequences of randomness is not easy. Because of the pathological structure of the original Lotka–Volterra model, stochastic versions of the model invariably result in the extinction of one or the other species (Renshaw 1991). But this result is not general for predator–prey models.

May (1976) suggested that the addition of stochastic variation in population models generally has the consequence of destabilizing the dynamics. Indeed, I suspect that this is often the pattern, but this is not true generally because certain population models actually can become more stable with the addition of noise (Markus et al. 1987). The notion is similar to the observation that whitening can actually enhance a signal (Bezrukov and Vodyanoy 1997).

The consequence of randomness in a predator–prey system depends on the magnitude of noise, the autocorrelation structure and distribution of the stochastic process, and nonlinearities in the model. To illustrate the consequences of stochastic variation in a predator–prey model I have plotted the outcome of adding normally distributed white noise to populations in a stable limit cycle (figure 8.2). Using Caughley's (1976) two-species model I simulated trajectories with variable $K(t)$ (prey population). With $\sigma = 500$ the population follows the stable limit cycle of the deterministic model, but with $\sigma = 5,000$ the underlying stable limit cycle is difficult to see (figure 8.2, bottom). Still, the population persists.

In the predator–prey model I developed for wolf recovery in the Greater Yellowstone Ecosystem (Boyce 1992b), an interesting pattern emerged with the addition of stochastic variation. In the deterministic model, we found convergence to equilibrium. With the addition of noise, however, autocorrelated oscillations emerged in the number of wolves, with the population apparently fluctuating on an attractor (Boyce 1992b). Tuning parameters in the model permits detection of such an unrealized attractor, but this attractor is visited only when stochastic variance causes the system to take excursions away from the simple equilibrium behavior predicted for the mean parameter values.

## AUTOREGRESSIVE MODELS

Our ability to collapse the essential dynamics of a multiple-equation set of differential equations into a difference equation has important ramifications. Even though one might not have data for all components of a complex ecological system, Poincaré's results suggest that we can capture the essential dynamics in a much simpler model in discrete time (Schaffer 1985). This idea is fundamental to the use of autoregressive models for characterizing the dynamic features of the system. Royama (1992) provides a useful introduction to this statistical mechanics approach to modeling that can easily embrace predator–prey dynamics.

The general form of the model is

$$\ln[N(t + 1)/N(t)] = a_0 + a_1 \ln[N(t)] + a_2 \ln[N(t - 1)] \qquad (8.15)$$

or, equivalently,

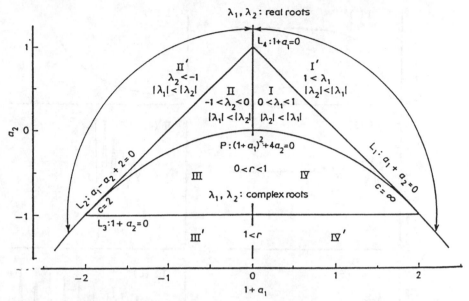

**Figure 8.5** Stability map for a second-order autoregressive process as in equations 8.15 and 8.16. Different regions in parameter space numbered with Roman numerals correspond to dynamics illustrated in figure 8.6. From Royama (1992).

$$\ln[N(t + 1)] = a_0 + (1 + a_1)\ln[N(t)] + a_2\ln[N(t - 1)] \quad (8.16)$$

Royama (1992) has mapped the regions of parameter space with different stability properties (figures 8.5 and 8.6).

Bjørnstad et al. (1995) used autoregressive procedures to estimate $(1 + a_1)$ and $a_2$ from a number of vole populations and then studied geographic variation in the autoregressive coefficients. This approach holds promise for revealing the ecological correlates of predator–prey dynamics. The usual interpretation is that the first autoregressive term represents density dependence and the second and higher-order terms are a consequence of trophic-level interactions. Certainly the autoregression coefficients do not yield to such simple interpretations, but work is just beginning in this area.

Akaike's Information Criterion (AIC) has been used to optimize the dimensionality and magnitude of coefficients for autoregressive models (Bjørnstad et al. 1995). Although such models are primarily descriptive rather than mechanistic, attempts to interpret the autoregressive coefficients have been encouraging.

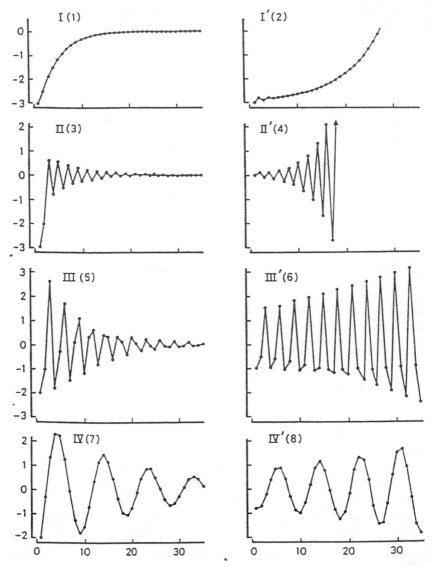

**Figure 8.6** Population dynamics emerging from second-order autoregressive models. Each plot is representative of the patterns coming from regions plotted with corresponding Roman numerals in figure 8.5. The horizontal axis is time for each of these plots. From Royama (1992).

## ■ Fitting the Model to Data

*"Data analysis through model building and selection should begin with an array of models that seem biologically reasonable."*

Burnham and Anderson (1992)

Seldom do we have sufficient data to do justice to estimating all of the parameters necessary to reconstruct the interesting dynamics of predator–prey interactions. One approach that has met with limited success is to build predator–prey models piecewise. For example, the predator–prey model in equations 8.13 and 8.14 contains a density-dependent portion for the prey in the absence of predators, a functional response, a numerical response, and a rate of decline in the absence of prey. For some populations each of these pieces could be constructed independently with field data and then combined to study the model dynamics.

The traditional approach to model building has been to use goodness-of-fit tests or between-model tests. As an alternative, Burnham and Anderson (1992) advocate the use of AIC for models that can be posed in a maximum likelihood framework or Mallow's $C_p$ for regression models. For building the model piecewise, I believe that information criteria procedures could prove useful.

### BAYESIAN STATISTICS

Bayesian statistics holds promise for accommodating complex models and uncertainty about model parameters. Recent contributions in the application of Bayesian techniques appear to have few reservations about tackling problems at the level of complexity outlined among the models described earlier in this chapter (McAllister et al. 1994). A recent issue of *Ecological Applications* featured a series of papers on Bayesian statistics (Ellison 1996). Given the current excitement over these techniques, I am confident that useful applications to predator–prey modeling will be developed in the near future.

### BEST GUESS FOLLOWED BY ADAPTIVE MANAGEMENT

Most detailed predator–prey models are too complex for the available data. Modeling by the principle of parsimony fails because any principled ecologist is unwilling to give up the structural details of the model that give it the dynamics of interest. Even though the data may not justify the complexity of the model, our understanding of ecology demands that we insist on the more complex model, despite the objections of statisticians. When the objective is to model the dynamics, models can be too simple if they cannot yield the dynamics seen in nature.

So we build models where parameter estimates are poor and sometimes outright guesses. And we fiddle with the parameter estimates until the behavior of the system matches what we have observed or believe to be true. The model is perhaps a figment of our imagination, but it is probably the most rigorous statement of hypothesis about how this ecological system functions that has ever been constructed.

I have met many wildlife ecologists who believe that such modeling exercises are fruitless, and even dangerous, speculation. But I disagree. Instead, I argue that such a modeling exercise is the fundamental building block on which one should build adaptive management (Walters 1986). The process of building the model requires compiling available data on the system and the process of developing a model structure involves outlining many of the ecological mechanisms that underlie the population dynamics. Surely the model is wrong. Indeed, all models are wrong at some level. But many models are useful for framing our data and explicitly identifying our understanding of how we see that it all fits together.

In adaptive management, this stage of hypothesis building is followed by monitoring to see how well the model predicted the future dynamics of the system. As new data become available, the model can be evaluated. If the data are sufficient, the model probably must be adjusted to accommodate the new information. If the patterns are drastically different from those predicted by the original model, alterations to the structure of the model may be necessary. But key to the process of adaptive management is that the model must be updated and revised to make a new prediction of the future.

Active adaptive management involves manipulating the system (Walters and Holling 1990). Rather than simply observing the system's dynamics, by intervening one is essentially imposing a management experiment on the system. The model can predict the system response that again is evaluated by monitoring. And the process of perpetual modeling, manipulating, monitoring, evaluating, and revising the model continues indefinitely. Given the complexity of ecological systems, we may never get the model predictions just right. Updating and revising the model probably always will be necessary as we gain improved knowledge and gradually learn how to manage the populations better.

## ■ Choosing a Good Model

I am loath to suggest that any of the models I have reviewed are wrong. Any model can be a useful construct for understanding, and no model is truly realistic. Models are always abstractions of nature. The issue is choosing a model that best meets one's needs. The author of the model is not necessarily the one who will determine the need for the model. In management applications, careful attention must be paid to identifying the audience and understanding how the model might be applied.

### HOW MUCH DETAIL?

Often the selection of an appropriate model is determined by the extent of data available. Detailed models might include details of age specificity or, in the extreme, individually based models can take advantage of high-speed computers to track the fate of each individual in a population through time. Because radiotelemetry allows us to collect extensive and detailed data on the movements, behavior, and demography of individuals, building individually based models is often feasible (DeAngelis and Gross 1992).

### *Age and sex structure?*

Vulnerability of prey is often a function of age, with young and old individuals more vulnerable to predation (Mech 1995). Similarly, prime-age predators are often more efficient at finding and killing prey, so that both the functional and numerical responses may be a function of the age of both the predator and the prey. If differences by age are not large, however, changes to dynamics may be small, so adding age structure to a model may be difficult to justify given the difficulty of estimating all of the additional parameters.

Age structure can either stabilize or destabilize the interaction between predator and prey (Beddington and Free 1976). Populations undergoing predator–prey oscillations are perpetually perturbed out of a stable age distribution, vastly complicating the demography for the populations (Tuljapurkar 1989; Nations and Boyce 1997). Age structure introduces irregularities into regular oscillations, making the dynamics more complex. Age structure may also change the period length of oscillations that result from predator–prey interactions (Oster and Takahashi 1974).

When building a model to anticipate the consequences of wolf predation on ungulates in the greater Yellowstone area (Boyce 1992b), I explored the dynamics of an age-structured model for elk and wolves. However, qualitatively the dynamics did not appear very different from those emerging from the model without age structure. My model was designed to encourage users to experiment with management alternatives, so speed of computation was an important consideration. The age-structured model was much slower, so I decided to abandon age structure to enhance the user interface. Dixon et al. (1997) questioned my conclusion that the dynamics were not altered much, partly because of the strong age-specificity of wolf predation (Mech 1995). But after reconstructing an age-structured model, Dixon et al. (1997) ultimately came to the same conclusion that the dynamic patterns were similar in the structured and unstructured models.

### Functional response

The shape of the functional response has major influence on the ultimate dynamics between predator and prey; therefore, accurately characterizing the functional response is one of the most important steps in developing a useful predator–prey model.

Functional responses generally have been viewed too simply. The focus has mostly been on characterizing the relationship between rate of prey capture and the density of prey or, in the case of ratio-dependent models, the ratio of prey to predators. In nature, however, the rate at which killing occurs is as much a function of the conditions for predation and the vulnerability of prey as it is a function of the abundance of prey. For example, wolves prey on individuals in poor condition or those rendered more vulnerable because of deep snow (Mech 1995).

### Spatial structure

Spatial heterogeneity is argued to stabilize predator–prey interactions (Huffaker 1959), but this generalization is overly simplistic. Recent attempts to model parasitoid–host interactions in a cellular automaton revealed that dispersal can result in complex spatial patterns in the abundance of parasitoids and their prey (Commins et al. 1992). Similar results have been described for integrodifference equations for predator and prey (Neubert et al. 1995). Predators can be responsible for sinks in source–sink systems (Pulliam 1988), and dispersal capability of predator and prey can have strong implications for their dynamics. Raptors with high dispersal capability tend to show strong regional fluctuations in abundance, whereas species that disperse less show weaker fluctuations (Galushin 1974). Nomadic raptors can help to synchronize small mammal populations geographically by moving to areas of high prey abundance, thereby making predator dispersal an important component in regional population fluctuations (Norrdahl and Korpimäki 1996).

Many predator–prey models are unstable. Indeed, such instability may be inherent in many biological predator–prey interactions and only through spatial structure can such species persist. Metapopulation structures may emerge as a consequence of predator–prey interactions (McCullough 1996).

Models for spatially structured populations are a complex topic that I cannot address sufficiently here. Suffice it to say that spatial structure can have profound consequences for predator–prey dynamics. Because of the complexity of spatially structured population models, these are necessarily computer simulation models. With the development of geographic information systems (GIS) into which one can build models to superimpose on maps, the ability to develop spatially structured population models is greatly facilitated. Software is available that permits a direct interface between population models and GIS data layers that characterize habitat (Boyce 1996).

## MODEL VALIDATION

To validate means to verify or substantiate. In the context of model validation, Grant (1986) proposed the following four considerations:

• *Does the model address the problem?* Often the problem is not well articulated, meaning that it is the modeler who defines a precise statement of the problem. The modeler may feel compelled to focus on questions that are mathematically tractable or that can be reliably addressed given the available data. The true natural resource problem being confronted by management may be a difficult one to answer using modeling, yet failure to confront issues directly has led to a distrust of modelers by managers (Boyce 1992a).

• *Does the model have reasonable structure and behavior?* Dynamics such as those predicted by the Lotka–Volterra model are simply too weird to be of biological interest. Often population data may exist for similar species or populations that can be used for qualitative assessment of whether the dynamics emerging from a model are reasonable.

• *Sensitivity analysis.* Sensitivity analysis involves evaluating the response of a selected response variable to change in system parameters. For example, population growth rate is often explored as a function of perturbations to vital rates.

Understanding of the sensitivity structure of the model helps one to better anticipate the driving variables that most influence the behavior of the model; thus Grant (1986) encourages sensitivity analysis as a fundamental step in a validation of the model. This view is echoed by Dixon et al. (1997:472), who suggest that "particular effort should be made to ensure that parameters with high sensitivity are accurately known and that critical parts of the model structure are appropriately specified."

Two considerations limit the utility of sensitivity analysis, however. First, some parameters may not vary much in nature; even if they have high sensitivities, they may not contribute much to the dynamics of the system. For example, adult survival is invariably a highly sensitive parameter in age-structured models for long-lived species (Meyer and Boyce 1994). Yet adult survival often varies much less than juvenile or subadult survival, so variation in adult survival may not account for much of the variation in population growth rate. A solution to this problem can be to decompose the variance in population growth rate into that attributable to variation in selected parameters. Numerically this is approximately equivalent to multiplying the variance times the sensitivity squared, adjusted for covariance (Brault and Caswell 1993).

My biggest concern about the suggested utility of sensitivity analysis is the conclusion by Dixon et al. (1997:510) that "sensitivities indicate where management actions to change vital rates have large effects on the population growth rate." In practice, I do not believe that sensitivities will provide much insight into how well the system can respond to management actions. Yet usually our objective in applied population modeling is being able to anticipate how much the system will respond to management actions. Even though population growth rate may have a high sensitivity to adult survival for long-lived species, management might be at a loss to manipulate adult survival. On the other hand, protection of breeding habitats might accomplish a dramatic increase of reproductive output. What we really need to know is how effectively management actions can achieve a response in a system parameter of interest. Simple demographic sensitivities or elasticities may not provide this insight.

• *Quantitative assessment of accuracy and precision of model's outputs and behavior.* Ultimately we want to know how well the model actually compares with the ecological system. This can be evaluated based on how well the model describes the data that were used to build the model, but ideally the model is validated when predictions are compared with future behavior of the system or when the model is applied to data from another population that was not used for building the model.

# ■ Recommendations

My motivation for writing this review is to encourage wildlife ecologists to begin focusing their research on testing ecological principles. When inconsistencies emerge between the models and field observations, the solution is not to reject modeling and modelers, but to fix the models. Improved models will emerge if field ecologists work with modelers to change models to accommodate the peculiarities of their study organisms.

An obvious area needing attention is the development of models for functional response. Although we know that functional response has important consequences for population dynamics (Sherratt and MacDougall 1995), seldom do carnivore food habit studies relate results to the predictions of alternative functional response models. This seems to be particularly rich ground for multispecies functional response models that are poorly understood in theory or practice. I am unaware of any models that go beyond linear substitutability among prey (Abrams 1987), yet I suspect that this assumption is often violated.

Likewise, improved understanding of numerical response is essential to our understanding of population dynamics. For equilibrium systems, Messier (1994) presents a convincing illustration of how the combination of numerical and functional responses allow us to anticipate the role of predators in the regulation of prey numbers. The outcome is largely a consequence of the shapes of the numerical and functional response curves. For dynamic systems, obtaining reliable estimates of the numerical response will be more complex and should receive careful thought.

Fitting complex models to data is an area needing careful attention. Even the simplest predator–prey model requires the estimation of several parameters, and estimating these parameters directly can be difficult without detailed laboratory studies. Yet Bayesian procedures hold promise for accommodating models with moderate numbers of parameters (McAllister et al. 1994).

## REMEMBER THE AUDIENCE

Ecologists usually approach models as tools for understanding ecological processes. But models are becoming essential tools in natural resource conservation and management. In this context, targeting the audience carefully is crucial to the success of a model. Dueling models have dominated courtroom debates over the management of striped bass in eastern United States (Barthouse et al. 1984). Detailed courtroom testimony in trials related to lawsuits over management of the northern spotted owl (*Strix occidentalis caurina*) was ignored in the decision by Judge Helen Frye because she did not understand the models. Models presented by scientists are often viewed as a smokescreen by decision makers who fail to understand the complexities of the mathematics or computer code (Boyce 1992a). Most of these problems can be overcome by careful attention to producing models that are accessible to managers and other people who might be able to benefit from the models.

In 1988 I was asked to develop a model for wolf recovery in the greater Yellowstone ecosystem. I made three trips to Yellowstone National Park before developing my model with the express purpose of trying to understand how the model would be put to use. No one seemed to have a clear picture of why the government had asked for a model, other than hoping that it would give them some insight into the possible consequences of wolf recovery to ungulate populations in the greater Yellowstone area. I decided that the most useful model would be one that would help to educate the managers and interested public about predator–prey dynamics. Interface between the model and the user was crucial to how well the model would be received. I programmed a model that was user friendly and simple enough that almost anyone could use it (Boyce 1993). I learned through experience that giving careful attention to presentation is much more time-consuming and tedious than the modeling itself.

It was clear to me that the Yellowstone ungulate populations fluctuate substantially among years depending on winter severity. It was also clear that I was dealing with considerable uncertainty in my estimates of functional responses because I had meager data on wolf predation on elk. So uncertainty was a major issue that would shape my model. I confronted this issue in part by building a stochastic model that never gave the same result twice.

Another major source of uncertainty was anticipating how the wolves would be managed once they were introduced. Indeed, how the wolves would be managed was a current issue of debate in the process of developing a plan for the reintroduction. I confronted this problem by allowing the user of the model to choose how he or she would manage the wolves given the chance. This helped build confidence in the model because the user was helping to parameterize the model and thereby had more confidence in the results. User inputs were kept simple and the actual consequences of user-defined management decisions were transparent only to those who wanted to read the program code.

I believe that the model was successful in accomplishing my objectives, and it was used by many managers as well as regional high school and university students. Some of the predictions gleaned from the model were incorporated directly into the wolf recovery program. For example, stochastic population projections revealed that the small number of wolves proposed for releases were likely to go extinct within a year or two, but that the success of wolf releases could be enhanced considerably by scheduling repeated releases (Boyce 1992b). Indeed, this is exactly what the government did, and apparently the program has been highly successful (Fritts et al. 1995).

## ■ Conclusion

Seventy years ago, when Lotka (1925) and Volterra (1926) first introduced predator–prey models, a debate raged about whether mathematical formulations could be trusted to provide insight into real-life ecological systems (Kingsland 1985). During the golden era of rapid theoretical development in ecology and population genetics of the 1920s and 1930s, some feared that biological facts would give way to mathematical theory (Thompson 1937). Amazingly, the same battle between modelers and empiricists still rages today. Field ecologists and managers still make disparaging remarks about models that they do not understand (Mech 1995; Schullery 1995) while theoretical ecologists continue to explore the dynamics of models for which we have yet to find clear examples in nature (Schaffer 1988).

I hold the basic premise of Galileo that the "Book of Nature is written in mathematical form." Like Galileo I believe that our understanding of nature (read predator–prey dynamics) will proceed only through a rigorous iteration of mathematical models with "cimento" (experiments). We have an extensive body of literature on predator–prey models. We also have an extensive body of literature on the behavior and ecology of predators and prey (Gittleman 1989, 1996). However, empirical studies of predators and prey seem preoccupied with techniques and are seldom placed in the context of the theory. Only when the theory and the fieldwork become integrated will we begin to develop a science on the level of sophistication that Galileo envisioned 400 years ago.

### Acknowledgments

Thanks to Todd Fuller and Luigi Boitani for inviting my participation in the conference and to Danilo Mainardi for support from the International School of Ethology. I benefited from suggestions and discussions with Gary White, Todd Fuller, Charles Krebs, Joe Elkinton, and David Garshelis. Thanks to Arild Gautestad for serving as the discussion leader following my presentation at the Ettore Majorana Center.

## Literature Cited

Abrams, P. 1987. The functional response of adaptive consumers of two resources. *Theoretical Population Biology* 32: 262–288.

Abrams, P. 1994. The fallacies of "ratio-dependent" predation. *Ecology* 75: 1842–1850.

Anderson, D. R. and K. P. Burnham. 1976. *Population ecology of the mallard. VI. The effect of exploitation on survival.* U.S. Fish and Wildlife Service Research Publication 128.

Barthouse, L. W., J. Boreman, S. W. Christensen, C. P. Goodyear, W. VanWinkle, and D. S. Vaughan. 1984. Population biology in the courtroom: The Hudson River controversy. *BioScience* 34: 14–19.

Beddington, J. R. and C. A. Free. 1976. Age-structure effects in predator-prey interactions. *Theoretical Population Biology* 9: 15–24.

Bezrukov, S. M. and I. Vodyanoy. 1997. Stochastic resonance in nondynamic systems without thresholds. *Nature* 385: 319–321.

Bjørnstad, O. N., W. Falck, and N. C. Stenseth. 1995. A geographic gradient in small rodent density fluctuations: a statistical modelling approach. *Proceedings of the Royal Society of London* B262: 127–133.

Boyce, M. S. 1992a. Simulation modelling and mathematics in wildlife management and conservation. In N. Maruyama et al., eds., *Wildlife conservation: Present trends and perspectives for the 21st century,* 116–119. Tokyo: Japan Wildlife Research Center.

Boyce, M. S. 1992b. Wolf recovery for Yellowstone National Park: A simulation model. In D. R. McCullough and R. H. Barrett, eds., *Wildlife 2001: Populations,* 123–138. London: Elsevier.

Boyce, M. S. 1993. Predicting the consequences of wolf recovery to ungulates in Yellowstone National Park. In R. Cook, ed. *Ecological issues on reintroducing wolves into Yellowstone National Park,* 234–269. U.S. National Park Service Scientific Monograph NPS/NR YELL/NRSM-93/22.

Boyce, M. S. 1996. RAMAS/GIS: Linking landscape data with population viability analysis. *Quarterly Review of Biology* 71: 167–168.

Boyce, M. S, and E. M. Anderson. (1999). Evaluating the role of carnivores in the Greater Yellowstone Ecosystem. In T. W. Clark, A. P. Curlee, S. C. Minta, and P. M. Kareiva, eds., *Carnivores in ecosystems: The Yellowstone Experience,* 265–283. New Haven, Conn.: Yale University Press.

Boyce, M. S. and D. J. Daley. 1980. Population tracking of fluctuating environments and natural selection for tracking ability. *American Naturalist* 115: 480–491.

Brault, S. and H. Caswell. 1993. Pod-specific demography of killer whales (*Orcinus orca*). *Ecology* 74: 1444–1455.

Broomhead, D. S. and R. Jones. 1989. Time-series analysis. *Proceedings of the Royal Society of London* A423: 103–121.

Burnham, K. P. and D. R. Anderson. 1992. Data-based selection of an appropriate biological model: The key to modern data analysis. In D. R. McCullough and R. H. Barrett, eds., *Wildlife 2001: Populations,* 16–30. London: Elsevier.

Caughley, G. 1976. Wildlife management and the dynamics of ungulate populations. In T. Coaker, ed., *Applied biology,* vol. 1, 183–246. New York: Academic Press.

Commins, H. N., M. P. Hassell, and R. M. May. 1992. The spatial dynamics of host-parasitoid systems. *Journal of Animal Ecology* 61: 735–748.

Dale, B. W., L. G. Adams, and R. T. Bowyer. 1994. Functional response of wolves preying on barren-ground caribou in a multiple-prey ecosystem. *Journal of Animal Ecology* 63: 644–652.

DeAngelis, D. A. and L. J. Gross. 1992. *Individually based models and approaches in ecology.* London: Chapman & Hall.

Dixon, P., N. Friday, P. Ang, S. Heppell, and M. Kshatriya. 1997. Sensitivity analysis of structured-population models for management and conservation. In S. Tuljapurkar and H. Caswell, eds., *Structured-population models in marine, terrestrial, and freshwater systems,* 471–513, New York: Chapman & Hall.

Edelstein-Keshet, L. 1988. *Mathematical models in biology.* New York: Random House.

Ellison, A. M. 1996. An introduction to Bayesian inference for ecological research and environmental decision-making. *Ecological Applications* 6: 1036–1046.

Emlen, J. M. 1984. *Population biology: The coevolution of population dynamics and behavior.* New York: Macmillan.

Errington, P. L. 1946. Predation and vertebrate populations. *Quarterly Review of Biology* 21: 144–177, 221–245.

Errington, P. L. 1967. *Of Predation and life.* Ames: Iowa State University Press.

Fritts, S. H., E. E. Bangs, J. A. Fontaine, W. G. Brewster, and J. F. Gore. 1995. Restoring wolves to the northern Rocky Mountains of the United States. In L. N. Carbyn, S. H. Fritts, and D. R. Seip, eds., *Ecology and conservation of wolves in a changing world,* 107–126. Edmonton, Alberta: Canadian Circumpolar Institute, occasional publication 35.

Fryxell, J. M. 1988. Population dynamics of Newfoundland moose using cohort analysis. *Journal of Wildlife Management* 52: 14–21.

Fryxell, J. M. and P. Lundberg. 1994. Diet choice and predator–prey dynamics. *Evolutionary Ecology* 8: 407–421.

Galilei, G. 1638. *Discorsi e dimostrazioni mathematiche intorno a due nuove scienze attenenti alla meccanica.* Leiden, the Netherlands: Louis Elzevirs.

Galushin, V. M. 1974. Synchronous fluctuations in populations of some raptors and their prey. *Ibis* 116: 127–134.

Gilpin, M. E. 1979. Spiral chaos in a predator–prey model. *American Naturalist* 107: 306–308.

Gittleman, J. L., ed. 1989. *Carnivore behavior, ecology, and evolution.* Ithaca, N.Y.: Cornell University Press.

Gittleman, J. L., ed. 1996. *Carnivore behavior, ecology, and evolution,* vol. 2. Ithaca, N.Y.: Cornell University Press.

Grant, W. E. 1986. *Systems analysis and simulation in wildlife and fisheries science.* New York: Wiley.

Gutierrez, A. P. 1996. *Applied population ecology: A supply–demand approach.* New York: Wiley.

Hassell, M. P. 1978. *The dynamics of arthropod predator–prey systems.* Princeton, N.J.: Princeton University Press.

Holling, C. S. 1959. The components of predation as revealed by a study of small-mammal predation of the European pine sawfly. *Canadian Entomologist* 91: 293–320.

Holling, C. S. 1966. The functional response of invertebrate predators to prey density. *Memoirs of the Entomological Society of Canada* 48: 1–86.

Huffaker, C. B. 1959. Experimental studies on predation: Dispersion factors and predator–prey oscillations. *Hilgardia* 27: 343–383.

Inoue, M. and H. Kamifukumoto. 1984. Scenarios leading to chaos in a forced Lotka–Volterra model. *Progress in Theoretical Physics* 71: 930–937.

Ivlev, V. S. 1961. *Experimental ecology of the feeding of fishes.* New Haven, Conn.: Yale University Press.

Jhala, Y. 1993. Predation on blackbuck by wolves in Velvadar National Park, Gujarat, India. *Conservation Biology* 7: 874–881.

Jhala, Y. In press. Optimization for the management of an endangered predator (*Canis lupus pallipes*) and prey (*Antelope cervicapra*) system. *Journal of Wildlife Research.*

Keith, L. B. 1983. Population dynamics of wolves. In L. N. Carbyn, ed., *Wolves in Canada and Alaska: Their status, biology, and management,* 66–77. Canadian Wildlife Service, Report Series 45.

Kingsland, S. E. 1985. *Modeling nature.* Chicago: University of Chicago Press.

Kolmogorov, A. N. 1936. On Volterra's theory of the struggle for existence [Sulla teoria di Volterra della lotta per l'esistenza. *Giornale dell' Instituto Italiano degli Attuari* 7: 74–80]. Reprinted in F. M. Scudo and J. R. Ziegler, eds. 1978. *The golden age of theoretical ecology: 1923–1940.* New York: Springer-Verlag.

Lotka, A. J. 1925. *Elements of physical biology.* Baltimore, Md.: Williams & Wilkins. Reprinted with corrections and bibliography in 1956 as *Elements of mathematical biology.* New York: Dover.

Mack, J. A. and F. J. Singer. 1993. Using POPII models to predict effects of wolf predation and hunter harvests on elk, mule deer, and moose on the northern range. In R. Cook, ed., *Ecological issues on reintroducing wolves into Yellowstone National Park,* 49–74, National Park Service Scientific Monograph 22.

Maker, W. J. 1970. The pomerine jaeger as a brown lemming predator in northern Alaska. *Wilson Bulletin* 82: 130–157.

Manly, B. F. J., L. L. McDonald, and D. L. Thomas. 1993. *Resource selection by animals.* London: Chapman & Hall.

Markus, M., B Hess, J. Rössler, and K. Kiwi. 1987. Populations under periodically and randomly varying growth conditions. In H. Degn, A. V. Holden, and L. F. Olsen, eds., *Chaos in biological systems,* 267–277, NATO ASI Series 138. New York: Plenum.

Matson, P. A. and A. A. Berryman, eds. 1992. Ratio-dependent predator–prey theory. *Ecology* 73: 1529–1566.

May, R. M. 1976. Models for single populations. In R. M. May, ed. *Theoretical ecology: Principles and applications,* 4–25. Oxford, U.K.: Blackwell Scientific.

Maynard Smith, J. 1974. *Models in ecology.* Cambridge, U.K.: Cambridge University Press.

McAllister, M. K., E. K. Pikitch, A. E. Punt, and R. Hilborn. 1994. A Bayesian approach to stock assessment and harvest decisions using the sampling/importance resampling algorithm. *Canadian Journal of Fisheries and Aquatic Sciences* 51: 2673–2687.

McCullough, D. R. 1996. *Metapopulations and wildlife conservation.* Washington, D.C.: Island Press.

McKelvey, R., D. Hankin, K. Yanosko, and S. Snygg. 1980. Stable cycles in multistage recruitment models: An application to the northern California Dungeness crab (*Cancer magister*) fishery. *Canadian Journal of Fisheries and Aquatic Sciences* 37: 2323–2345.

McMullin, E. 1988. Introduction. In E. McMullin, ed., *Galileo: Man of science,* 3–51. Princeton Junction, N.J.: Scholar's Bookshelf.

Mech, L. D. 1995. What do we know about wolves and what more do we need to learn? In L. N. Carbyn, S. H. Fritts, and D. R. Seip, eds., *Ecology and conservation of wolves in a changing world,* 537–545. Edmonton, Alberta: Canadian Circumpolar Institute, occasional publication 35.

Messier, F. 1994. Ungulate population models with predation: A case study with the North American moose. *Ecology* 75: 478–488.

Messier, F. 1995. On the functional and numerical responses of wolves to changing prey density. In L. N. Carbyn, S. H. Fritts, and D. R. Seip, eds., *Ecology and conservation of wolves in a changing world,* 187–198, Edmonton, Alberta: Canadian Circumpolar Institute, occasional publication 35.

Meyer, J. S. and M. S. Boyce. 1994. Life historical consequences of pesticides and other insults to vital rates. In R. J. Kendall and T. E. Lacher, eds., *Wildlife toxicology and population modeling: Integrated studies of agroecosystems,* 349–363, Washington, D.C.: Lewis.

Mladenoff, D. J., R. G. Haight, T. A. Sickley, and A. P. Wydeven. 1997. Causes and implications of species restoration in altered ecosystems. *BioScience* 47: 21–31.

Murton, R. K., N. J. Westwood, and A. J. Isaacson. 1974. A study of wood pigeon shooting: The exploitation of a natural animal population. *Journal of Applied Ecology* 11: 61–81.

Nations, C. and M. S. Boyce. 1997. Stochastic demography for conservation biologists. In S. D. Tuljapurkar and H. Caswell, eds., *Structured population models in marine, terrestrial, and freshwater systems,* 461–469. New York: Chapman & Hall.

Neubert, M. G., M. Kot, and M. A. Lewis. 1995. Dispersal and pattern formation in a discrete-time predator–prey model. *Theoretical Population Biology* 48: 7–43.

Norrdahl, K, and E. Korpimäki. 1996. Do nomadic avian predators synchronize population fluctuations of small mammals? A field experiment. *Oecologia* 107: 478–483.

Noy-Meir, I. 1975. Stability of grazing systems: An application of predator–prey graphs. *Journal of Ecology* 63: 459–481.

Oksanen, L., J. Moen, and P. A. Lundberg. 1990. The time scale problem in exploiter–victim models: Does the solution lie in ratio-dependent exploitation? *American Naturalist* 140: 938–960.

Oster, G. and Y. Takahashi. 1974. Models for age-specific interactions in a periodic environment. *Ecological Monographs* 44: 483–501.

Pielou, E. C. 1969. *An Introduction To Mathematical Ecology*. New York: Wiley-Interscience.

Pulliam, R. 1988. Sources, sinks and population regulation. *American Naturalist* 132: 652–661.

Renshaw, E. 1991. *Modelling biological populations in space and time*. Cambridge, U.K.: Cambridge University Press.

Rosenzweig, M. L. 1971. Paradox of enrichment: Destabilization of exploitation ecosystems in ecological time. *Science* 171: 385–387.

Rosenzweig, M. L. and R. H. MacArthur. 1963. Graphical representation and stability conditions of predator–prey interactions. *American Naturalist* 97: 209–223.

Royama, T. 1992. *Analytical population dynamics*. New York: Chapman & Hall.

Schaffer, W. M. 1985. Order and chaos in ecological systems. *Ecology* 66: 93–106.

Schaffer, W. M. 1988. Perceiving order in the chaos of nature. In M. S. Boyce, ed., *Evolution of life histories of mammals*, 313–350. New Haven, Conn.: Yale University Press.

Schoener, T. W. 1973. Population growth regulated by intraspecific competition for energy or time. *Theoretical Population Biology* 4: 56–84.

Schullery, P. 1995. A bee in every bouquet. *Yellowstone Science* 3(1): 8–14.

Settle, T. B. 1988. Galileo's use of experiment as a tool of investigation. In E. McMullin, ed., *Galileo: Man of science*, 315–337. Princeton Junction, N.J.: Scholar's Bookshelf.

Sherratt, T. N. and A. D. MacDougall. 1995. Some population consequences of variation in preference among individual predators. *Biological Journal of the Linnaean Society* 55: 93–107.

Singer, F. J., A. Harting, K. K. Symonds, and M. B. Coughenour. 1997. Density dependence, compensation, and environmental effects on elk calf mortality in Yellowstone National Park. *Journal of Wildlife Management* 61: 12–25.

Spalinger, D. E., T. A. Hanley, and C. T. Robbins. 1988. Analysis of the functional response in foraging in the Sitka black-tailed deer. *Ecology* 69: 1166–1175.

Stephens, D. W. and J. R. Krebs. 1986. *Foraging theory*. Princeton, N.J.: Princeton University Press.

Taylor, R. J. 1984. *Predation*. New York: Chapman & Hall.

Theberge, J. B. 1990. Potentials for misinterpreting impacts of wolf predation through prey:predator ratios. *Wildlife Society Bulletin* 18: 188–192.

Thompson, W. R. 1937. *Science and common sense: An Aristotelian excursion*. London: Longmans, Green.

Tuljapurkar, S. D. 1989. An uncertain life: Demography in random environments. *Theoretical Population Biology* 35: 227–294.

Vales, D. J. and J. M. Peek. 1993. Estimating the relations between hunter harvest and gray wolf predation on the Gallatin, Montana, and Sand Creek, Idaho, elk populations. In R. Cook, ed., *Ecological issues on reintroducing wolves into Yellowstone National Park*, 118–172. U.S. National Park Service, Scientific Monograph NPS/NR YELL/NRSM-93/22.

Vales, D. J. and J. M. Peek. 1995. Projecting the potential effects of wolf predation on elk and mule deer in the East Front portion of the northwest Montana wolf recovery area. In L. N. Carbyn, S. H. Fritts, and D. R. Seip, eds., *Ecology and conservation of wolves in a changing world*, 211–222. Edmonton, Alberta: Canadian Circumpolar Institute, occasional publication 35.

Volterra, V. 1926. Fluctuations in the abundance of a species considered mathematically. *Nature* 118: 558–560.

Volterra, V. 1931. *Leçons sur la théorie mathématique de la lutte pour la vie*. Paris: Gauthier-Villars.

Walters, C. J. 1986. *Adaptive management of renewable resources*. New York: Macmillan.

Walters, C. J. and C. S. Holling. 1990. Large-scale management experiments and learning by doing. *Ecology* 71: 2060–2068.

Yodzis, P. 1989. *Introduction to theoretical ecology*. New York: Harper & Row.

## Chapter 9

# Population Viability Analysis: Data Requirements and Essential Analyses

GARY C. WHITE

The biological diversity of the earth is threatened by the burgeoning human population. To prevent extinctions of species, conservationists must manage many populations in isolated habitat parcels that are smaller than desirable. An example is maintaining large-bodied predator populations in isolated, limited-area nature reserves (Clark et al. 1996).

A population has been defined as "a group of individuals of the same species occupying a defined area at the same time" (Hunter 1996:132). The viability of a population is the probability that the population will persist for some specified time. Two procedures are commonly used for evaluating the viability of a population. Population viability analysis (PVA) is the method of estimating the probability that a population of a specified size will persist for a specified length of time. The minimum viable population (MVP) is the smallest population size that will persist some specified length of time with a specified probability. In the first case, the probability of extinction is estimated, whereas in the second, the number of animals that is needed in the population to meet a specified probability of persistence is estimated. For a population that is expected to go extinct, the time to extinction is the expected time the population will persist. Both PVA and MVP require a time horizon: a specified but arbitrary time to which the probability of extinction pertains.

Definitions and criteria for viability, persistence, and extinction are arbitrary, such as a 95 percent probability of a population persisting for at least 100 years (Boyce 1992). Mace and Lande (1991) discussed criteria for extinction. Ginzburg et al. (1982) suggested the phrase "quasi-extinction risk" as the probability of a population dropping below some critical threshold, a concept also promoted by Ludwig (1996a) and Dennis et al. (1991). Schneider and Yodzis (1994) used the term *quasi-extinction* to mean a population drop such that only 20 females remain.

The usual approach for estimating persistence is to develop a probability distribution for the number of years before the model "goes extinct," or falls below a specified threshold. The percentage of the area under this distribution in which the population persists beyond a specified time period is taken as an estimate of the probability of persistence. To obtain MVP, probabilities of extinction are needed for various initial population sizes. The expected time to extinction is a misleading indicator of population viability (Ludwig 1996b) because for small populations, the probability of extinction in the immediate future is high, even though the expected time until extinction may be quite large. The skewness of the distribution of time until extinction thus makes the probability of extinction for a specified time interval a more realistic measure of population viability.

Simple stochastic models have yielded qualitative insights into population viability questions (Dennis et al. 1991). But because population growth is generally considered to be nonlinear, with nonlinear dynamics making most stochastic models intractable for analysis (Ludwig 1996b), and because catastrophes and their distribution pose even more difficult statistical problems (Ludwig 1996b), analytical methods are generally inadequate to compute these probabilities. Therefore, computer simulation is commonly used to produce numerical estimates for persistence or MVP. Analytical models lead to greater insights given the simplifying assumptions used to develop the model. However, the simplicity of analytical models precludes their use in real analyses because of the omission of important processes governing population change such as age structure and periodic breeding. Lack of data suggests the use of simple models, but lack of data really means lack of information. Lack of information suggests that no valid estimates of population persistence are possible because there is no reason to believe that unstudied populations are inherently simpler (and thus justify simple analytical models) than well-studied populations for which the inadequacy of simple analytical models is obvious. The focus of this chapter is on computer simulation models to estimate population viability via numerical techniques, where the population model includes the essential features of population change relevant to the species of interest.

The most thorough recent review of the PVA literature was provided by Boyce (1992). Shaffer (1981, 1987), Soulé (1987), Nunney and Campbell

(1993), and Remmert (1994) provided a historical perspective of how the field developed. In this chapter I discuss procedures to develop useful viability analyses. Specifically, statistical methods to estimate the variance components needed to develop a PVA, the need to incorporate individual heterogeneity into a PVA, and the need to incorporate the sampling variance of parameter estimates used in a PVA are discussed.

## ■ Qualitative Observations About Population Persistence

Qualitatively, population biologists know a considerable amount about what allows populations to persist. Some generalities about population persistence (Ruggiero et al. 1994) are as follows:

• Connected habitats are better than disjointed habitats.

• Suitable habitats in close proximity to one another are better than widely separated habitats.

• Late stages of forest development are often better than younger stages.

• Larger habitat areas are better than smaller areas.

• Populations with higher reproductive rates are more secure than those with lower reproductive rates.

• Environmental conditions that reduce carrying capacity or increase variance in the growth rates of populations decrease persistence probabilities.

This list should be taken as a general set of principles, but you should recognize that exceptions occur often. In the following section, I discuss these generalities in more detail and suggest contradictions that occur.

### GENERALITIES

Typically, recovery plans for an endangered species try to create multiple populations of the species, so that a single catastrophe will not wipe out the entire species, and increase the size of each population so that genetic, demographic, and normal environmental uncertainties are less threatening (Meffe and Carroll 1994). However, Hess (1993) argued that connected populations can have lower viability over a narrow range in the presence of a fatal disease transmitted by contact. He demonstrated the possibilities with a model, but had no data to support his case. However, the point he made seems biologically sound, and the issue can be resolved only by optimizing persistence between these two opposing forces.

Spatial variation, that is, variation in habitat quality across the landscape, affects population persistence. Typically, extinction and metapopulation theories emphasize that stochastic fluctuations in local populations cause extinction and that local extinctions generate empty habitat patches that are then available for recolonization. Metapopulation persistence depends on the balance of extinction and colonization in a static environment (Hanski 1996; Hanski et al. 1996). For many rare and declining species, Thomas (1994) argued that extinction is usually the deterministic consequence of the local environment becoming unsuitable (through habitat loss or modification, introduction of a predator, etc.); that the local environment usually remains unsuitable following local extinction, so extinctions only rarely generate empty patches of suitable habitat; and that colonization usually follows improvement of the local environment for a particular species (or long-distance transfer by humans). Thus persistence depends predominantly on whether organisms are able to track the shifting spatial mosaic of suitable environmental conditions or on maintenance of good conditions locally.

Foley (1994) used a model to agree that populations with higher reproductive rates are more persistent. However, mammals with larger body size can persist at lower densities (Silva and Downing 1994) and typically have lower annual and per capita reproductive rates. Predicted minimal density decreases as the $-0.68$ power of body mass, probably because of less variance in reproduction relative to life span in larger-bodied species.

The last item on the list—that environmental conditions that reduce carrying capacity or increase variance in the growth rates of populations decrease persistence probabilities—suggests that increased variation over time leads to lower persistence (Shaffer 1987; Lande 1988, 1993). One reason that increased temporal variation causes lowered persistence is that catastrophes such as hurricanes, fires, or floods are more likely to occur in systems with high temporal variation. Populations in the wet tropics can apparently sustain themselves at densities much lower than those in temperate climates, probably because of less environmental variation. The distinction between a catastrophe and a large temporal variance component is arbitrary, and on a continuum (Caughley 1994). Furthermore, even predictable effects can have an impact. Beissinger (1995) modeled the effects of periodic environmental fluctuations on population viability of the snail kite (*Rostrhamus sociabilis*) and suggested that this source of variation is important in persistence.

## CONTRADICTIONS

Few empirical data are available to support the generalities just mentioned, but exceptions exist. Berger (1990) addressed the issue of MVP by asking how long different-sized populations persist. He presented demographic and weather data spanning up to 70 years for 122 bighorn sheep (*Ovis canadensis*) populations in southwestern North America. His analyses revealed that 100 percent of the populations with fewer than 50 individuals went extinct within 50 years, populations with more than 100 individuals persisted for up to 70 years, and the rapid loss of populations was not likely to be caused by food shortages, severe weather, predation, or interspecific competition. Thus, 50 individuals, even in the short term of 50 years, are not a minimum viable population size for bighorn sheep. However, Krausman et al. (1993) questioned this result because they know of populations of 50 or less in Arizona that have persisted for more than 50 years.

Pimm et al. (1988) and Diamond and Pimm (1993) examined the risks of extinction of breeding land birds on 16 British islands in terms of population size and species attributes. Tracy and George (1992) extended the analysis to include attributes of the environment, as well as species characteristics, as potential determinants of the risk of extinction. Tracy and George (1992) concluded that the ability of current models to predict the risk of extinction of particular species on particular island is very limited. They suggested that models should include more specific information about the species and environment to develop useful predictions of extinction probabilities. Haila and Hanski (1993) criticized the data of Pimm et al. (1988) as not directly relating to extinctions because the small groups of birds breeding in any given year on single islands were not populations in a meaningful sense. Although this criticism may be valid, most of the "populations" that conservation biologists study are questionable. Thus results of the analysis by Tracy and George (1992) do contribute useful information because the populations they studied are representative of populations to which PVA techniques are applied. Specifically, small populations of small-bodied birds on oceanic islands (more isolated) are more likely to go extinct than are large populations of large-bodied birds on less isolated (channel) islands. However, interaction of body size with type of island (channel vs. oceanic) indicated that body size influences time to extinction differently depending on the type of island. The results of Tracy and

George (1992, 1993) support the general statements presented earlier in this chapter. As with all ecological generalities, exceptions quickly appear.

## ■ Sources of Variation Affecting Population Persistence

The persistence of a population depends on stochasticity, or variation (Dennis et al. 1991). Sources of variation, and their magnitude, determine the probability of extinction, given the population growth mechanisms specific to the species. The total variance of a series of population measurements is a function of process variation (stochasticity in the population growth process) and sampling variation (stochasticity in measuring the size of the population). Process variation is a result of demographic, temporal and spatial (environmental), and individual (phenotypic and genotypic) variation. In this section, I define these sources of variation more precisely and develop a simple mathematical model to illustrate these various sources of stochasticity, thus demonstrating how stochasticity affects persistence.

### NO VARIATION

Consider a population with no variation, one that qualifies for the simple, density-independent growth model $N_{t+1} = N_t(1 + R)$, where $N_t$ is the population size at time $t$ and $R$ is the finite rate of change in the population. This model is deterministic, and hence, so is the population. $R \geq 0$ guarantees that the population will persist, in contrast to $R < 0$, which guarantees that the population will go extinct (albeit in an infinite amount of time because a fraction of an animal is allowed in this model). $R$ can be considered to be a function of birth and death rates, so that $R = b - d$ defines the rate of change in the population as a function of birth rate ($b$) and death rate ($d$). When the birth rate exceeds or equals the death rate, the population will persist with probability 1 in this deterministic model. These examples are illustrated in figure 9.1.

### STOCHASTIC VARIATION

Let us extend this naive model by making it stochastic. I will change the parameter $R$ to be a function of two random variables. At each time $t$, I determine stochastically the number of animals to be added to the population by births and then the number to be removed by deaths. Suppose the birth rate

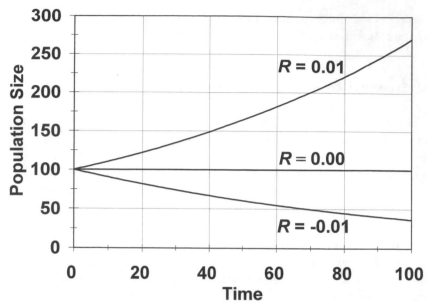

**Figure 9.1**  Deterministic model of population growth. For values of $R \geq 0$, the population persists indefinitely. For values of $R < 0$, the population will eventually go extinct in that the number of animals will approach zero.

equals the death rate, say $b = d = 0.5$. That is, on average 50 percent of the $N_t$ animals would give birth to a single individual and provide additions to the population, and 50 percent of the $N_t$ animals would die and be removed from the population. Thus the population is expected to stay constant because the number of births equals the number of deaths. A reasonable stochastic model for this process would be a binomial distribution. For the binomial model, you can think of flipping a coin twice for each animal. The first flip determines whether the animal gives birth to one new addition to the population in $N_{t+1}$ and the second flip determines whether the animal currently a member of $N_t$ remains in the population for another time interval, to be a member of $N_{t+1}$, or dies. If we start with $N_0 = 100$, what is the probability that the population will persist until $t = 100$? Three examples are shown in figure 9.2.

You might be tempted to say the probability is 1 that the population will persist until $t = 100$ because the expected value of $R$ is 0 given that the birth rate equals the death rate—that is, $E(R) = 0$, so that $E(N_{t+1}) = E(N_t)$. You would be wrong! Implementation of this model on a computer shows that the probability of persistence is 98.0 percent; that is, 2.0 percent of the time the

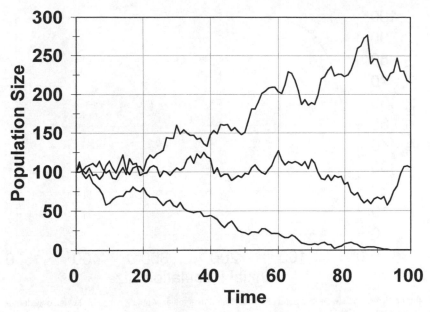

**Figure 9.2**  Three examples of the outcome of the population model with only demographic variation. The smaller population goes extinct at time 93. Birth and death probabilities are both 0.5, making the expected value of $R = 0$.

population does not persist for 100 years without $N_t$ becoming 0 for some $t$. These estimates were determined by running the population model 10,000 times and recording the number of times the simulated population went extinct before 100 years had elapsed. Lowering the initial population to $N_0 = 20$ results in persistence of only 53.2 percent of the populations, again based on 10,000 runs of the model. Setting $N_0 = 500$ improves the persistence rate to nearly 100 percent. Note that the persistence is not linear in terms of $N_0$ (figure 9.3). Initial population size has a major influence on persistence.

## DEMOGRAPHIC VARIATION

Other considerations affect persistence. The value of $R$ (the birth rate minus the death rate) is critical. $R$ can be negative (death rate exceeds birth rate) and the population can still persist for 100 years, which may seem counterintuitive. Furthermore, $R$ can be positive (birth rate exceeds death rate) and the population can still go extinct. For example, suppose $R$ is increased to 0.02 by making the birth rate 0.51 and the death rate 0.49. The persistence for $N_0 = 20$ increases

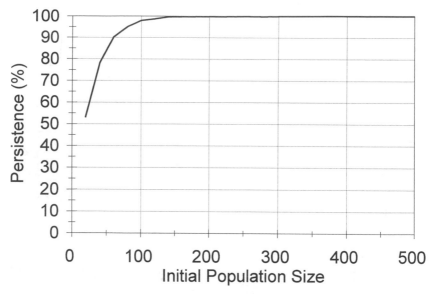

**Figure 9.3** Persistence of a population as a function of initial population size ($N_0$) when only demographic variation is incorporated into the model. Birth and death probabilities are both 0.5, making the expected value of $R = 0$. The model was run 10,000 times to estimate the percentage of runs in which the population persisted until $t = 100$.

to 84.3 percent from 53.2 percent for $R = 0$. Even though the population is expected to increase, stochasticity can still cause the population to go extinct.

The type of stochasticity illustrated by this model is known as demographic variation. I like to call this source of variation "penny-flipping variation" because the variation about the expected number of survivors parallels the variation about the observed number of heads from flipping coins. To illustrate demographic variation, suppose the probability of survival of each individual in a population is 0.8. Then on average, 80 percent of the population will survive. However, random variation precludes exactly 80 percent surviving each time this survival rate is applied. From purely bad luck on the part of the population, a much lower proportion may survive for a series of years, resulting in extinction. Because such bad luck is most likely to happen in small populations, this source of variation is particularly important for small populations, hence the name demographic variation. The impact is small for large populations. As the population size becomes large, the relative variation decreases to zero. That is, the variance of $N_{t+1}/N_t$ goes to zero as $N_t$ goes to infinity. Thus demographic variation is generally not an issue for persistence of larger populations.

To illustrate further how demographic variation operates, consider a small population with $N = 100$ and a second population with $N = 10,000$. Assume both populations have identical survival rates of 0.8. With a binomial model of the process, the probability that only 75 percent or less of the small population survives is 0.1314 for the small population, but $3.194E - 34$ for the larger population. Thus the likelihood that up to 25 percent of the small population is lost in 1 year is much higher than for the large population.

## TEMPORAL VARIATION

A feature of all population persistence models is evident in figure 9.2. That is, the variation of predicted population size increases with time. Some realizations of the stochastic process climb to very large population values after long time periods, whereas other realizations drop to zero and extinction. This result should be intuitive because as the model is projected further into the future, certainty about the projections decreases.

However, in contrast to population size, our certainty about the extinction probability increases as time increases to infinity. The probability of eventual extinction is always unity if extinction is possible. This is because the only absorbing state of the stochastic process is extinction; that is, the only population size at which there is no chance of change is zero.

Another way to decrease persistence is to increase the stochasticity in the model. One way would be to introduce temporal variation by making $b$ and $d$ random variables. Such variation would be exemplified by weather in real populations. Some years, winters are mild and survival and reproduction are high. Other years, winters are harsh and survival and reproduction are poor. To incorporate this phenomenon into our simple model, suppose that the mean birth and death rates are again 0.5, but the values of the birth rate and the death rate at a particular time $t$ are selected from a statistical distribution, say a beta distribution. That is, each year, new values of $b$ and $d$ are selected from a beta distribution.

A beta distribution is bounded by the interval 0–1 and can take on a variety of shapes. For a mean of 0.5, the distribution is symmetric about the mean, but the amount of variation can be changed by how peaked the distribution is (figure 9.4).

The beta distribution is described by two parameters, $\alpha > 0$ and $\beta > 0$. The

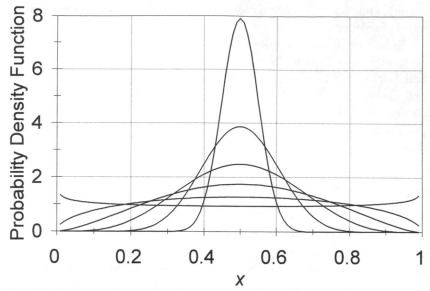

**Figure 9.4** Examples of the beta distribution, all with mean 0.5. The standard deviations proceeding from the tallest curve to the lowest curve at $x = 0.5$ are 0.05 to 0.3 in increments of 0.05.

mean of the distribution is given by $\alpha/(\alpha + \beta)$ and the variance as $\alpha\beta/[(\alpha + \beta)^2(\alpha + \beta + 1)]$, with the mode $(\alpha - 1)/(\alpha + \beta - 2)$ (mode only for $\alpha \geq 1$). Most random number generation techniques for the beta distribution require you to specify values for $\alpha$ and $\beta$. For a given mean ($\mu$) and variance ($\sigma^2$) or standard deviation ($\sigma$),

$$\alpha = \frac{\mu^2 (1 - \mu)}{\sigma^2} - \mu$$

and

$$\beta = \frac{[\sigma^2 + \mu(\mu - 1)](\mu - 1)}{\sigma^2}$$

However, the amount of variation possible is limited because the distribution is bounded on the [0, 1] interval. Thus for a mean of 0.5, the maximum variance approaches 0.25 as $\alpha$ and $\beta$ approach zero.

The standard deviations of the birth and death rates over time affect persistence because these values determine the standard deviation of $R$. The smaller the standard deviations, the more the model approaches the demographic variation case, and thus, as $N_t$ approaches infinity, the deterministic case. As the standard deviation increases, the more the variation in $N_t$, regardless of population size, and the less likely the population is to persist. Thus a standard deviation of 0.2 for both the birth and death rates results in only 28.5 percent persistence for $N_0 = 100$. Compare this to the 77.4 percent persistence achieved for a standard deviation of 0.1 (figure 9.5) or to the 98.0 percent persistence when no variation in birth and death rates occurred but demographic variation is still present.

This second source of variation in our simple model is temporal variation, that is variation in the parameters of the model across time. As the example shows, increasing temporal variation decreases persistence. The simple model illustrated assumed that no correlation existed between the birth rate and the death rate, that is, that the two rates were independent. However, in real populations there is probably a high correlation between birth rates and death rates across years. Good years with lots of high-quality resources available to the ani-

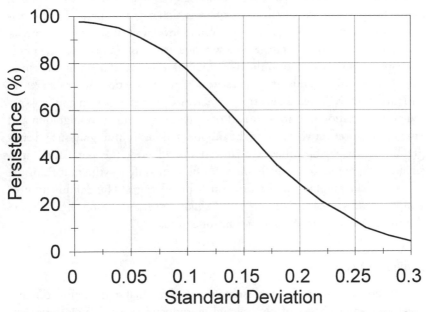

**Figure 9.5**   Persistence of a population of 100 animals at $t = 0$ to $t = 100$ years as a function of the standard deviation of birth (mean = 0.5) and death (mean = 0.5) rates (temporal variation). Demographic variation is still included in the model.

mals probably result in increased reproduction and survival, whereas bad years result in poor reproduction and high mortality. Including a negative covariance of birth and death rates (or a positive covariance between birth and survival rates) in the model results in an even bigger impact of temporal variation on persistence. That is, the bad years are really bad because of both poor reproduction and high mortality, and the good years are really good. The net effect of this negative covariance of birth and death rates is to decrease persistence.

## SPATIAL VARIATION

Spatial variation is the variation across the landscape that is normally associated with populations. Factors causing geographic variation include geologic differences that affect soil type, and thus habitat, and weather patterns (e.g., differences in rainfall across the landscape). If the immigration and emigration rates are high across the landscape, so that subpopulations are depleted because of local conditions, high spatial variation can lead to higher persistence. This is because the probability of all the subpopulations of a population being affected simultaneously by some catastrophe is low when high spatial variation exists and spatial autocorrelation is low. High positive spatial autocorrelation causes low levels of spatial variation, whereas high negative spatial autocorrelation causes high levels of spatial variation, as low levels of spatial autocorrelation generally do. In contrast, with low spatial variation (and hence high positive spatial autocorrelation), the likelihood of a bad year affecting the entire population is high. Thus, in contrast to temporal variation, where increased variation leads to lowered persistence, increased spatial variation and low spatial autocorrelation lead to increased persistence, given that immigration and emigration are effectively mixing the subpopulations. If immigration and emigration are negligible, then spatial variation divides the population into smaller subpopulations, which are more likely to suffer extinction from the effect of demographic variation on small populations. The combination of temporal and spatial variation is called *environmental* variation. Both dictate the animal's environment, one in time, one in space.

## INDIVIDUAL VARIATION

All the models examined so far assume that each animal in the population has exactly the same chance of survival and reproducing, even though these rates change with time. What happens if each animal in the population has a different rate of survival and reproduction? Differences between the individuals in the population are called individual heterogeneity, and this creates individual variation. Many studies have demonstrated individual heterogeneity of individual survival and reproductions; for example, Clutton-Brock et al. (1982) demonstrated that lifetime reproductive success of female red deer (*Cervus elaphus*) varied from 0 to 13 calves reared per female. Differences in the frequency of calf mortality between mothers accounted for a larger proportion of variance in success than differences in fecundity. Bartmann et al. (1992) demonstrated that overwinter survival of mule deer fawns was a function of the fawn's weight at the start of the winter, with larger fawns showing better survival.

Individual variation is caused by *genetic* variation, that is, differences between individuals because of their genome. Individual heterogeneity is the basis of natural selection; that is, differences between animals is what allows natural selection to operate. However, *phenotypic* variation is also possible, where individual heterogeneity is not a result of genetic variation. Animals that endure poor nutrition during their early development may never be as healthy and robust as animals that are on a higher nutritional plane, even though both are genetically identical. Animals with access to more and better resources have higher reproductive rates, as in the red deer studied by Clutton-Brock et al. (1982). Thus individual heterogeneity may result from both genetic and phenotypic variation. Lomnicki (1988) developed models of resource partitioning that result in phenotypic variation of individuals.

Another example of individual heterogeneity in reproduction was provided by Burnham et al. (1996) in northern spotted owls (*Strix occidentalis caurina*). In the case of northern spotted owls, repeated observations of reproduction across numerous individuals were used to estimate individual variation with analysis of variance procedures. The age of the female produced individual heterogeneity. This study also demonstrated temporal and spatial variation in owl fecundity rates.

Undoubtedly, natural selection plays a role in the genetic variation left in a declining population. Most populations for which we are concerned about extinction probabilities have suffered a serious decline in numbers. The genotypes remaining after a severe decline are unlikely to be a random sample of the original population (Keller et al. 1994). I expect that the genotypes persisting through a decline are the "survivors," and would have a much better chance of persisting than would a random sample from the population before the decline. Of course, this argument assumes that the processes causing the decline remain in effect, so that the same natural selection forces continue to operate.

To illustrate individual variation, start with the basic demographic variation model developed earlier in this chapter. Instead of each animal having

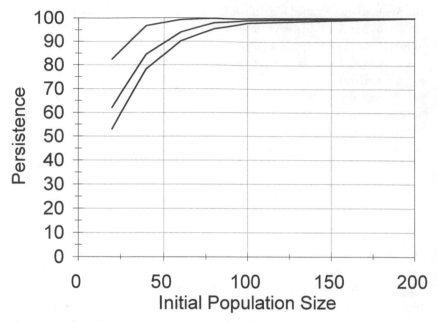

**Figure 9.6**    Effect of individual variation on population persistence. The three lines from top to bottom have standard deviations of 0.1, 0.05, and 0.01 for the birth and death rates. Compare these results with figure 9.3, where the standard deviation of individual variation is zero.

exactly a birth rate of 0.5 and a death rate of 0.5, let's select these values from beta distributions with a mean of 0.5. The birth and death rates assigned to an animal remain for its lifetime. As new animals are added to the population, they likewise are assigned lifetime birth and death rates. How does persistence of this new model compare with the results from the demographic model? The answer is in figure 9.6.

The reason that increased individual heterogeneity increases population persistence is that increased variation results in more chance that a few animals have exceptionally high reproductive potential and high survival. Therefore, these animals are unlikely to suffer mortality and be removed from the population and can be relied on to contribute new births each year. As a result, the population may remain small but will not go extinct as often. Individual heterogeneity has seldom, if ever, been included in a population viability analysis, except as genetic variation. Yet as this simple example shows, individual heterogeneity not a function of genetic variation is a very important element in maintaining viability.

**PROCESS VARIATION**

The combined effect of demographic, temporal, spatial, and individual varia-tion is called process varisation. That is, each of these sources of variation affects population processes. *Process variation* is used as a general term for the inherent stochasticity of changes in the population level. Process variation is in contrast to *sampling* variation, which is the variation contributed when biolo-gists attempt to measure population processes. That is, researchers are unable to measure the exact survival rate of a population. Rather, they observe real-izations of the process, but not the exact value. Even if the fate of every animal in the population is observed, the resulting estimate of survival is only an esti-mate of the true but unknown population survival rate. The concept of sam-pling variation is explained later in this chapter, where methods of separating sampling variation from process variation are developed.

Several lessons should be learned from this simple exercise. Persistence is a stochastic phenomenon. Even though the expected outcome for a particular model is to persist, random variation prevents this outcome from always occurring. Small populations are much more likely to go extinct than larger populations because of demographic variation. Increased temporal variation results in decreased persistence. Increased individual variation results in increased persistence.

## ■ Components of a PVA

As demonstrated earlier in this chapter, many factors affect the persistence of a population. What components are needed to provide estimates of the proba-bility that a population will go extinct, and what are the tradeoffs if not all these components are available?

• A basic population model is needed. A recognized mechanism of popula-tion regulation, density dependence, should be incorporated because no pop-ulation can grow indefinitely. "Of course, exponential growth models are strictly unrealistic on time scales necessary to explore extinction probabilities" (Boyce 1992:489). The population cannot be allowed to grow indefinitely, or persistence will be overestimated. Furthermore, as discussed later in this chap-ter, the shape of the relationship between density and survival and reproduc-tion can affect persistence, and density dependence cannot be neglected for moderate or large populations (Ludwig 1996b). Density dependence can pro-vide a stabilizing influence that increases persistence in small populations.

• Demographic variation must be incorporated in this basic model. Otherwise, estimates of persistence will be too high because the effect of demographic variation for small populations is not included in the model.

• Temporal variation must be included for the parameters of the model, including some probability of a natural catastrophe. Examples of catastrophes (for some species) are fires (e.g., Yellowstone National Park, USA, 1988), hurricanes, typhoons, earthquakes, and extreme drought or rainfall resulting in flooding. Catastrophes must be rare, or else the variation would be considered part of the normal temporal variation. However, the covariance of the parameters is also important. Good years for survival are probably also good years for reproduction. Likewise, bad years for reproduction may also lead to increased mortality. The impact of this correlation of reproduction and survival can drastically affect results. For example, the model of Stacey and Taper (1992) of acorn woodpecker population dynamics performs very differently depending on whether adult survival, juvenile survival, and reproduction are bootstrapped as a triplet or given as individual rates across the 10-year period. If the positive correlation of the survival rates and reproduction is included in the model, estimated persistence is improved.

• Spatial variation in the parameters of the model must be incorporated if the population is spatially segregated. If spatial attributes are to be modeled, then immigration and emigration parameters must be estimated, as well as dispersal distances. The difficulty of estimating spatial variation is that the covariance of the parameters must be estimated as a function of distance; that is, what is the covariance of adult survival of two subpopulations as a function of distance?

• Individual heterogeneity must be included in the model or the estimates of persistence will be too low. Individual heterogeneity requires that the basic model be extended to an individual-based model (DeAngelis and Gross 1992). As the variance of individual parameters increases in the basic model, the persistence time increases. Thus, instead of just knowing estimates of the parameters of our basic model, we also need to know the statistical distributions of these parameters across individuals. This source of variation is not mentioned in discussions of population viability analysis such as Boyce (1992), Remmert (1994), Hunter (1996), Meffe and Carroll (1994), or Shaffer (1981, 1987).

• For short-term projects, the sources of variation just mentioned may be adequate. However, if time periods of more than a few generations are projected, then genetic variation should be considered. I would expect the population to change as selection takes place. Even if no selection is operating, genetic drift is expected for small population sizes. However, the importance of genetic effects is still an issue in question; see Joopouborg and Van Groenendael (1996). Lande (1988, 1995) suggested that demographic variation or genetic effects can be lethal to a small population.

•   For long-term persistence, we must be willing to assume that the system will not change, that is, that the levels of stochasticity will not change through time, the species will not evolve through selection, and the supporting capacity of the environment (the species habitat) remains static. We must assume that natural processes such as long-term succession and climatic change do not affect persistence and that human activity will cease (given that humans have been responsible for most recent extinctions). To believe the results, we have to assume that the model and all its parameters stay the same across inordinately long time periods.

After examining this list, I am sure you will agree with Boyce (1992:482): "Collecting sufficient data to derive reliable estimates for all the parameters necessary to determine MVP is simply not practical in most cases." Of course, limitations of the data seldom slow down modelers of population dynamics. Furthermore, managers are forced to make decisions, so modelers attempt to make reasonable guesses. In the next three sections, I explore statistical methods to obtain the necessary data to develop a reasonable PVA model and suggest modeling techniques to incorporate empirical data into the persistence model.

### ■ Direct Estimation of Variance Components

The implication of the list of requirements in the previous section is that population parameters or their distributions are known without error; that is, exact parameter values are observed, not estimated. In reality, we may be fortunate and have a series of survival or reproduction estimates across time that provides information about the temporal variation of the process. However, the variance of this series is not the proper estimate of the temporal variation of the process. This is because each of our estimates includes sampling variation; that is we have only an estimate of the true parameter, not its exact value. To properly estimate the temporal variation of the series, the sampling variance of the estimates must be removed. In this section, I demonstrate a procedure to remove the sampling variance from a series of estimates to obtain an estimate of the underlying process variation (which might be temporal or spatial variation). The procedure is explained in Burnham et al. (1987).

Consider the example situation of estimating overwinter survival rates each year for 10 years from a deer population. Each year, the survival rate is different from the overall mean because of snow depth, cold weather, and other factors. Let the true but unknown overall mean be $S$. Then the survival rate for each year can be considered to be $S$ plus some deviation attributable to temporal variation, with the expected value of the $e_i$ equal to zero:

### Environmental Variation

|        | Mean | Year I         | Year I        |
|--------|------|----------------|---------------|
| 1      | $S$  | $S + e_1$      | $S_1$         |
| 2      | $S$  | $S + e_2$      | $S_2$         |
| 3      | $S$  | $S + e_3$      | $S_3$         |
| 4      | $S$  | $S + e_4$      | $S_4$         |
| 5      | $S$  | $S + e_5$      | $S_5$         |
| 6      | $S$  | $S + e_6$      | $S_6$         |
| 7      | $S$  | $S + e_7$      | $S_7$         |
| 8      | $S$  | $S + e_8$      | $S_8$         |
| 9      | $S$  | $S + e_9$      | $S_9$         |
| 10     | $S$  | $S + e_{10}$   | $S_{10}$      |
| Mean   | $S$  | $S$            | $S$           |

The true population mean $S$ is computed as $\overline{S}$:

$$\overline{S} = \frac{\sum_{i=1}^{10} S_i}{10}$$

with the variance of the $S_i$ is computed as

$$\hat{\sigma}^2 = \frac{\sum_{i=1}^{10} (S_i - \overline{S})^2}{10}$$

where the random variables $e_i$ are selected from a distribution with mean 0 and variance $\sigma^2$. In reality, we are never able to observe the annual rates because of sampling variation or demographic variation. For example, even if we observed all the members of a population, we would still not be able to say the observed survival rate was $S_i$ because of demographic variation. Consider flipping 10 coins. We know that the true probability of a head is 0.5, but we will not always observe that value exactly. If you have 11 coins; the true value is not even in the set of possible estimates. The same process operates in a population as demographic variation. Even though the true probability of survival is 0.5, we would not necessarily see exactly half of the population survive on any given year.

Hence, what we actually observe are the quantities following:

### Environmental Variation + Sampling Variation

| $I$ | Mean | Truth Year $I$ | Observed Year $I$ |
|---|---|---|---|
| 1 | $S$ | $S + e_1 + f_1$ | $\hat{S}_1$ |
| 2 | $S$ | $S + e_2 + f_2$ | $\hat{S}_2$ |
| 3 | $S$ | $S + e_3 + f_3$ | $\hat{S}_3$ |
| 4 | $S$ | $S + e_4 + f_4$ | $\hat{S}_4$ |
| 5 | $S$ | $S + e_5 + f_5$ | $\hat{S}_5$ |
| 6 | $S$ | $S + e_6 + f_6$ | $\hat{S}_6$ |
| 7 | $S$ | $S + e_7 + f_7$ | $\hat{S}_7$ |
| 8 | $S$ | $S + e_8 + f_8$ | $\hat{S}_8$ |
| 9 | $S$ | $S + e_9 + f_9$ | $\hat{S}_9$ |
| 10 | $S$ | $S + e_{10} + f_{10}$ | $\hat{S}_{10}$ |
| Mean | $S$ | $\bar{S}$ | $\bar{\hat{S}}$ |

where the $e_i$ are as before, but we also have additional variation from sampling variation, or demographic variation, or both, in the $f_i$.

The usual approach to estimating sampling variance separately from temporal variance is to take replicate observations within each year so that within-cell replicates can be used to estimate the sampling variance, whereas the between-cell variance is used to estimate the environmental variation. Years are assumed to be a random effect, and mixed-model analysis of variance procedures are used (e.g., Bennington and Thayne 1994). This approach assumes that each cell has the same sampling variance. An example of the application of a random effects model is Koenig et al. (1994). They considered year effects, species effects, and individual tree effects on acorn production by oaks in central California.

Classic analysis of variance methodology assumes that the variance within cells is constant across a variety of treatment effects. This assumption is often not true; that is, the sampling variance of a binomial distribution is a function of the binomial probability. Thus, as the probability changes across cells, so does the variance. Another common violation of this assumption is caused by the variable of interest being distributed log-normally, so that the coefficient of variation is constant across cells and the cell variance is a function of the cell mean. Furthermore, the empirical estimation of the variance from replicate measurements may not be the most efficient procedure. Therefore, the remainder of this section describes methods that can be viewed as extensions of the usual variance component analysis based on replicate measurements within cells. We examine estimators for the situation in which the within-cell variance is estimated by an estimator other than the moment estimator based on replicate observations.

Assume that we can estimate the sampling variance for each year, given a value of $\hat{S}_i$ for the year. For example, an estimate of the sampling variation for a binomial is

$$\text{vâr}(\hat{S}_i \mid S_i) = \frac{\hat{S}_i(1 - \hat{S}_i)}{n_i}$$

where $n_i$ is the number of animals monitored to see whether they survived. Then, can we estimate the variance term due to environmental variation, given that we have estimates of the sampling variance for each year?

If we assume all the sampling variances are equal, the estimate of the overall mean is still just the mean of the 10 estimates:

$$\overline{\hat{S}} = \frac{\sum\limits_{i=1}^{10} \hat{S}_i}{10}$$

with the theoretical variance being

$$\text{var}(\overline{\hat{S}}) = \frac{\sigma^2 + E[\text{var}(\hat{S} \mid S)]}{10}$$

i.e., the total variance is the sum of the environmental variance plus the expected sampling variance. This total variance can be estimated as

$$\mathrm{v\hat{a}r}(\overline{\hat{S}}) = \frac{\sum\limits_{i=1}^{10}\left(\hat{S}_i - \overline{\hat{S}}\right)^2}{10(10-1)}$$

We can estimate the expected sampling variance as the mean of the sampling variances

$$\hat{\mathrm{E}}[\mathrm{var}(\hat{S}|S)] = \frac{\sum\limits_{i=1}^{10}\mathrm{v\hat{a}r}\left(\hat{S}_i|S_i\right)}{10}$$

so that the estimate of the environmental variance is obtained by solving for $\sigma^2$

$$\hat{\sigma}^2 = \frac{\sum\limits_{i=1}^{10}\left(\hat{S}_i - \overline{\hat{S}}\right)^2}{(10-1)} - \frac{\sum\limits_{i=1}^{10}\mathrm{v\hat{a}r}\left(\hat{S}_i|S_i\right)}{10}$$

However, sampling variances are usually not all equal, so we have to weight them to obtain an unbiased estimate of $\sigma^2$. The general theory says to use a weight, $w_i$,

$$w_i = \frac{1}{\sigma^2 + \mathrm{var}\left(\hat{S}_i|S_i\right)}$$

so that by replacing $\mathrm{var}(\hat{S}_i|S_i)$ with its estimator $\mathrm{v\hat{a}r}(\hat{S}_i|S_i)$, the estimator of the weighted mean is

$$\overline{\hat{S}} = \frac{\sum\limits_{i=1}^{10} w_i \hat{S}_i}{\sum\limits_{i=1}^{10} w_i}$$

with theoretical variance (i.e., sum of the theoretical variances for each of the estimates)

$$\text{var}(\overline{\hat{S}}) = \frac{1}{\sum\limits_{i=1}^{10} w_i}$$

and empirical variance estimator

$$\hat{\text{var}}(\overline{\hat{S}}) = \frac{\sum\limits_{i=1}^{10} w_i \left( \hat{S}_i - \overline{\hat{S}} \right)^2}{\left( \sum\limits_{i=1}^{10} w_i \right)(10 - 1)}$$

When the $w_i$ are the true (but unknown) weights, we have

$$\frac{1}{\sum\limits_{i=1}^{10} w_i} = \frac{\sum\limits_{i=1}^{10} w_i \left( \hat{S}_i - \overline{\hat{S}} \right)^2}{\left( \sum\limits_{i=1}^{10} w_i \right)(10 - 1)}$$

giving the following

$$1 = \frac{\sum\limits_{i=1}^{10} w_i \left( \hat{S}_i - \overline{\hat{S}} \right)^2}{10 - 1}$$

Therefore, all we have to do is manipulate this equation with a value of $\sigma^2$ to obtain an estimator of $\sigma^2$.

To obtain a confidence interval on the estimator of $\sigma^2$, we can substitute the appropriate chi-square values in this relationship. To find the upper confidence interval value, $\hat{\sigma}^2_U$ solve the equation

$$\frac{\sum_{i=1}^{10} w_i \left( \hat{S}_i - \bar{\hat{S}} \right)^2}{10 - 1} = \frac{\chi^2_{10-1, \alpha_L}}{10 - 1}$$

and for the lower confidence interval value, $\hat{\sigma}^2_L$ solve the equation

$$\frac{\sum_{i=1}^{10} w_i \left( \hat{S}_i - \bar{\hat{S}} \right)^2}{10 - 1} = \frac{\chi^2_{10-1, \alpha_U}}{10 - 1}$$

As an example, consider the following fawn survival data from overwinter survival of mule deer fawns at the Little Hills Wildlife Area, west of Meeker, Colorado, USA.

| Year | Collared | Lived | Estimated Survival | Estimated Variance |
|------|----------|-------|--------------------|--------------------|
| 1981 | 46 | 15 | 0.3260870 | 0.0047773 |
| 1982 | 114 | 38 | 0.3333333 | 0.0019493 |
| 1983 | 118 | 5 | 0.0423729 | 0.0003439 |
| 1984 | 106 | 19 | 0.1792453 | 0.0013879 |
| 1985 | 155 | 59 | 0.3806452 | 0.0015210 |
| 1986 | 161 | 61 | 0.3788820 | 0.0014617 |
| 1987 | 116 | 15 | 0.1293103 | 0.0009706 |

The survival rates are the number of collared animals that lived divided by the total number of collared animals. For example, $\hat{S}_{1981} = 15/46 = 0.326087$ for 1981. The sampling variance associated with this estimate is computed as

$$\text{var}(\hat{S}_{1981}) = \frac{\hat{S}_{1981}\,(1 - \hat{S}_{1981})}{46}$$

which equals 0.0047773. A spreadsheet program (VARCOMP.WB1) computes the estimate of temporal process variation for 1981–87, $\hat{\sigma}^2$, as 0.0170632 ($\hat{\sigma}$ = 0.1306262), with a 95 percent confidence interval of (0.0064669, 0.0869938) for $\sigma^2$, and (0.0804167, 0.2949472) for $\sigma$. These confidence intervals represent the uncertainty of the estimate of temporal variation, that is, the sampling variation of the estimate of temporal variation.

The procedure demonstrated here is applicable to estimation of other sources of variation (e.g., spatial variation) and to variables other than survival rates, such as per capita reproduction. The method is more general than the usual analysis of variance procedures because each observation is not assumed to have the same variance, in contrast to analysis of variance, in which each cell is assumed to have the same within-cell variance.

## ■ Indirect Estimation of Variance Components

Individual heterogeneity occurs in both reproduction and survival. Estimation of individual variation in reproduction is an easier problem than estimation of individual variation in survival because some animals reproduce more than once, whereas they only die once. Bartmann et al. (1992) demonstrated that overwinter survival of mule deer fawns is related to their mass at the start of the winter. Thus one approach to modeling individual heterogeneity is to find a correlate of survival that can be measured and develop statistical models of the distribution of this correlate. Then, the distribution of the correlate can be sampled to obtain an estimate of survival for the individual. Lomnicki (1988) also suggests mass as an easily measured variable that relates to an animal's fitness.

To demonstrate this method, I use a simplification of the logistic regression model of Bartmann et al. (1992):

$$\log\left(\frac{S}{1 - S}\right) = \beta_0 + \beta_1\,\text{mass}$$

where survival ($S$) is predicted as a function of weight. Weight of fawns at the start of winter was approximately normally distributed, with mean 32 kg and standard deviation 4.2. To simulate individual heterogeneity in overwinter fawn survival, values can be drawn from this normal distribution to generate survival estimates.

This model can be expanded to incorporate temporal variation (year effects), sex effects, and area effects, as described for mule deer fawns by Bartmann et al. (1992). An example of modeling temporal variation in greater flamingos (*Phoenicopterus ruber roseus*) as a function of winter severity is provided by Cézilly et al. (1996). The approach suggested here of modeling winter severity as a random variable and estimating survival as a function of this random variable is an alternative to the variance estimation procedures of the previous section. Both provide a mechanism for injecting variation into a population viability model. The main advantage of using weather data to drive the temporal variation of the model is that considerably more weather data are available than are biological data on survival or reproductive rates.

The major drawback of the indirect estimation approach proposed in this section is that sampling variation of the functional relationship is ignored in the simulation procedure. That is, the logistic regression model includes sampling variation because its parameters are estimated from observed data. The parameter estimates of the logistic regression model include some unknown estimation error. Their direct use results in potentially biased estimates of persistence, depending on how much sampling error is present. Therefore, a "good" model relating the covariate to the biological process is needed.

■ **Bootstrap Approach**

Stacey and Taper (1992) used a bootstrap procedure to incorporate temporal variation into a model of acorn woodpecker (*Melanerpes formicivorus*) population viability. They used estimates of adult and juvenile survival and reproductive rates resulting from a 10-year field study to estimate population persistence. To incorporate the temporal variation from the 10 years of estimates, they randomly selected with replacement one estimate from the observed values to provide an estimate in the model for a year.

This procedure is known in the statistical literature as a bootstrap sampling procedure. The technique is appealing because of its simplicity. However, for estimating population viability, a considerable problem is inherent in the procedure. That is, the estimates used for bootstrapping contain sampling variation and demographic variation, as well as the environmental variation that the modeler is attempting to incorporate. To illustrate how demographic variation is included in the estimates, consider an example population of 10 animals with a constant survival rate of 0.55. Thus, the actual temporal variation is zero, yet a sequence of estimates of survival from this population suggests considerable variation. That is, the estimates of survival would have a variance of $0.55(1 - 0.55)/10 = 0.02475$ if all 10 animals had a survival probability of 0.55. Furthermore, the only observed values of survival would be 0, 0.1, . . . , 1.0. However, if the size of the population were increased to 100, you would find that the variance of the sequence of estimates becomes 0.002475, a considerable decrease. Thus randomly sampling the estimates from a population of size 10 results in considerably more variation than from a population of 100. As a result, the demographic variation from the sampled population is incorporated into the persistence model if the bootstrap approach is used.

A similar example can be used to demonstrate that sampling variation is also inherent in bootstrapping from a sample of observed estimates. Suppose a sample of 10 radiocollared animals is used to estimate survival for a population of 100,000 animals (i.e., the finite sample correction term can be ignored). The sampling variation of the estimates would be $S(1 - S)/10$, where $S$ is the true survival rate for the population assuming all animals had the same survival rate. Now if a sample of 100 radiocollared animals is taken, the sampling variation reduces to $S(1 - S)/100$. Thus randomly sampling estimates with a bootstrap procedure incorporates the sampling variation of the estimates into the persistence model. As a result of the increased variation, persistence values will be underestimated.

Therefore, I suggest not using the bootstrap approach demonstrated by Stacey and Taper (1992) if unbiased estimates of persistence are required. Persistence estimates developed with this procedure will generally be too low; that is, you will conclude that the population is more likely to go extinct than it really will. However, methods such as shrinkage estimation of variances (K. P. Burnham, personal communication 1997) may prove useful in removing sampling variance from the estimates and make the bootstrap procedure more applicable to estimating population persistence.

## ■ Basic Population Model and Density Dependence

Leslie matrix models (Leslie 1945, 1948; Usher 1966; Lefkovitch 1965; Caswell 1989; Manly 1990) are commonly used as the modeling framework for population viability models. Density dependence must be incorporated into the model; that is, basic parameters must be a function of population size.

Thus the resulting model is not a true Leslie matrix. Each iteration of the calculation also requires a temporal variance component, and making the parameters of the Leslie matrix into random variables (Burgman et al. 1993) is the standard approach but eradicates the analytical results that normally are benefits of Leslie's creative work. If multiple patches are modeled, each patch requires a spatial variance component. Demographic variation can be built into the model. Still, the resulting model doesn't resemble the elegant matrix model that Leslie originally developed.

However, use of the Leslie matrix framework ignores individual heterogeneity, and thus is likely to underestimate persistence. Incorporation of individual heterogeneity requires an individual-based model (DeAngelis and Gross 1992) and thus is conceptually different from the basic Leslie matrix approach. Individual-based models can be spatially explicit (Conroy et al. 1995; Dunning et al. 1995; Holt et al. 1995; Turner et al. 1995), providing another approach to incorporating spatial stochasticity into the model.

As suggested by Boyce (1992), Stacey and Taper (1992), and Burgman et al. (1993), density dependence is an important part of estimating a population's persistence. Lande (1993) demonstrates that the importance of environmental stochasticity and random catastrophes depends on the density-dependence mechanism operating in the population, based on the value of $K$ carrying capacity. However, how density dependence is incorporated into the model greatly affects the estimates of persistence (Pascual et al. 1997). In persistence models, as a population declines, compensation for small population size takes the form of increased birth rates and decreased death rates (density dependence) and so is a significant factor in increasing population persistence.

Consider the model

$$N_{t+1} = N_t[1 + R(t)]$$

Stacey and Taper (1992) tested two forms of density dependence with their data; the logistic form

$$R(t) = R_0\left(1 - \frac{N(t)}{K}\right)$$

and the θ-logistic form

$$R(t) = R_0\left[1 - \left(\frac{N(t)}{K}\right)^\theta\right]$$

Expressed as a difference equation, the θ-logistic model would be

$$N_{t+1} = N_t\left\{1 + R_0\left[1 - \left(\frac{N(t)}{K}\right)^\theta\right]\right\}$$

For θ = 1, the two models are identical. Although Stacey and Taper's data precluded a significant test between these models, their data did show significant correlations between adult survival and population size, suggesting that density dependence was operating in the population.

The distinction between the two models can be very important. In the first, the rate of change of the birth and death rates with population size is linear (i.e., the classic logistic population growth model). In the second, the change can be very nonlinear. As a result, the θ-logistic model can cause populations to be very persistent, or very extinction prone, depending on the shape of the function. In figure 9.7, the curve for per capita recruitment with θ = 10 results in a population with much greater persistence than the curve with θ = 0.1 because as the population size becomes small, the θ = 10 population is at peak reproduction for populations below 60, whereas peak reproduction is reached only at a population size of zero for the θ = 0.1 population.

Burgman et al. (1993) and May and Oster (1976) summarize other functional relationships to incorporate density dependence. Possibilities, expressed as a difference equations, include those by Hassell (1975), Hassell et al. (1976), and May (1976):

$$N_{t+1} = \frac{\lambda N_t}{(1 + aN_t)^b}$$

by Moran (1950) and Ricker (1954, 1975:282):

$$N_{t+1} = N_t \exp\left[r\left(1 - \frac{N_t}{K}\right)\right]$$

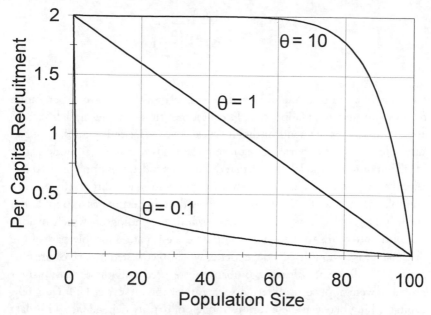

**Figure 9.7** Three examples of possible relationships of recruitment per individual to population size ($N_t$). Typical sigmoid population growth demonstrated by the logistic curve results for $\theta = 1$. A steeper curve initially results for $\theta = 0.1$, whereas a flatter curve initially results for $\theta = 10$.

by Pennycuick et al. (1968), Usher (1972), and Beddington (1974), taken from May and Oster (1976):

$$N_{t+1} = \frac{\lambda N_t}{1 + \exp[-A(1 - N_t/B)]}$$

by Beverton and Holt (1957) and Ricker (1975):

$$N_{t+1} = \frac{1}{\rho + (k/N_t)}$$

and by Maynard-Smith and Slatkin (1973):

$$N_{t+1} = \frac{R_0 N_t}{1 + (R_0 - 1)\left(\dfrac{N_t}{k}\right)^c}$$

Each of these models results in a different relationship between per capita recruitment and population size. Furthermore, these simple models can be applied to various segments of the life cycle, such as fecundity rates, neonatal survival, and adult survival, to achieve more realistic biological models. But the use of different models means that density dependence is implemented differently at a particular population level and population viability is affected. For example, Mills et al. (1996) reported widely differing estimates of population viability of grizzly bear (*Ursus arctos horribilis*) depending on which of four computer programs were used to compute the estimate. Probably part of the discrepancy is in how density dependence was implemented in each of the programs, but different functions probably were used and these relationships probably were applied to differing segments of the life cycle. Unfortunately, distinguishing between these various models of density dependence with data is not practical because of the stochasticity (noise) in observed population levels, as Pascual et al. (1997) demonstrated by fitting a collection of models to Serengeti wildebeest (*Connochaetes taurinus*) data.

Fowler (1981, 1994) argues that both theory and empirical information support the conclusion that most density-dependent change occurs at high population levels (close to the carrying capacity) for species with life history strategies typical of large mammals, such as deer ($\theta > 1$). The reverse is true for species with life history strategies typical of insects and some fishes, with $\theta < 1$.

Note that explicit estimates of carrying capacity ($K$) and its variance are not needed to incorporate density dependence into a population model, although such an approach is possible. If the functional relationships between birth and death rates and population density are available, the carrying capacity is determined by these relationships. Furthermore, if these relationships incorporate temporal and spatial variation, then the resulting model will have temporal and spatial variation in its carrying capacity, and thus stochastic density dependence.

Another example of how density dependence can operate in small populations is provided by the Allee effect (Allee 1931): The per capita birth rate declines at low densities (figure 9.8) because, for example, of the increased difficulty of finding a mate (Yodzis 1989). This is known as Allee-type behavior (of the per capita birth rate), and its effect on the per capita population growth rate, $R(t)$, is called an Allee effect. In theory, a low-density equilibrium would

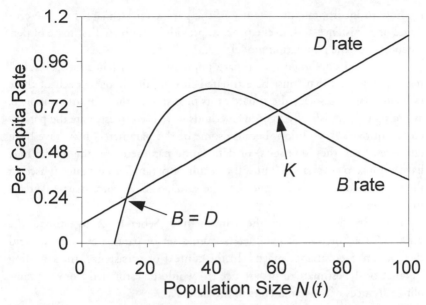

**Figure 9.8**  Example of how an Allee effect is created by a declining birth rate at low densities. Two equilibrium points exist, where the birth rate equals the death rate. The lower equilibrium may be stable in a deterministic system but could easily lead to extinction in a stochastic system.

be sustained in a deterministic equilibrium, where the birth rate equals the death rate. However, given stochasticity, the population could be driven below the low density equilibrium, and thus slide into extinction.

A second example of incorporating density dependence into a population viability analysis is provided by Armbruster and Lande (1993). They used estimates of life history parameters of elephant (*Loxodonta africana*) populations in a fluctuating environment from studies at Tsavo National Park, Kenya, to develop an age-structured, density-dependent model. Density regulation was implemented by changes in the age of first reproduction and calving interval. They modeled environmental stochasticity with drought events affecting sex- and age-specific survival.

■ **Incorporation of Parameter Uncertainty into Persistence Estimates**

In a previous section, I demonstrated how to remove the sampling variation from estimates of process variation. Unbiased estimates of process variances such as temporal and spatial variation can be achieved. In this section, I examine how to incorporate uncertainty of the parameter estimates into the estimates of persistence and, in the process, provide an unbiased estimate of persistence given the population model.

Any model developed to estimate population persistence has several to many parameters that must be estimated from available data. Each of these estimates has an associated estimate of its precision in the form of a variance, assuming that statistically rigorous methods were used to estimate the parameter from data. In addition, because some of the parameters may have been estimated from the same set(s) of data, some parameters in the model may have a nonzero covariance. Thus, the vector of parameter estimates $\hat{\theta}$ used in the model to estimate persistence has the variance–covariance matrix [vâr($\hat{\theta}$)] to measure uncertainty.

Typically, statisticians use the delta method (Seber 1982) to estimate the variance of a function of parameters from a set of parameter estimates and their variance–covariance matrix. In the context of persistence, the sampling variance of the estimate of persistence ($\hat{p}$) would be estimated from the sampling variances of the parameters in the model as

$$\mathrm{v\hat{a}r}(\hat{p}) = \frac{\delta f(\hat{\theta})^T}{\delta\theta}\, \mathrm{v\hat{a}r}(\hat{\theta})\, \frac{\delta f(\hat{\theta})}{\delta\theta}$$

where $\hat{p} = f(\hat{\theta})$. That is, the function $f$ represents the model used to estimate persistence. However, for realistically complex persistence models, the analytical calculation of partial derivatives needed in this formula is probably not feasible.

The lack of explicit analytical partial derivatives suggests that numerical methods be used. The most feasible, albeit numerically intensive, appears to be the parametric bootstrap approach (Efron and Tibshirani 1993; Urban Hjorth 1994). With a parametric bootstrap, a realization of the parameter estimates is generated based on their point estimates and sampling variance–covariance matrix using Monte Carlo methods. A multivariate normal distribution probably is used as the parametric distribution describing the set of parameter estimates, although other distributions or combinations of distributions may be more realistic biologically. Using this set of simulated values in the persistence model, persistence is estimated. This step requires a large number of simulations to properly estimate persistence with little uncertainty; typically 10,000 simulations are conducted. Then, a new set of parameter values is generated and persistence again estimated. This process is repeated for many sets of parameter estimates (at least 100, but more likely 1,000) to obtain a set of estimates of persistence. The variation of the resulting estimates of persistence is then a measure of uncertainty attributable to the variation of the parameter estimates as measured by their variance–covariance matrix. The process is diagrammed as

PARAMETRIC BOOTSTRAP LOOP (1,000 iterations):
      Select realization of parameter estimates
    MONTE CARLO LOOP (10,000 iterations):
      Tabulate percentage of model runs resulting in persistence
    END MONTE CARLO LOOP
END PARAMETRIC BOOTSTRAP LOOP

However, even more critical to our viability analysis is the fact that the mean of this set of 1,000 estimates of persistence is probably less than the estimate we obtained using our original point estimates of model parameters. More formally, the expected value of estimated persistence

$$[E(\hat{p})]$$

is less than the value of persistence predicted by our model using the point estimates of its parameters,

$$E(\hat{p}) < f[E(\hat{\theta})]$$

an example of Jensen's inequality. This difference is caused by large probabilities of early extinction for certain parameter sets that are likely, given their sampling variation (Ludwig 1996a). Therefore, to estimate persistence, the mean of the bootstrap estimates of persistence should be used, not the estimate of persistence obtained by plugging our parameter estimates directly into our population model.

Confidence intervals on persistence could be constructed using the usual ±2 SE procedure based on the set of 1,000 estimates. This confidence interval represents the variation attributable to the uncertainty of the parameter estimates used in the model. Uncertainty about the model is not included in this confidence interval because the model is assumed to be known. However, a better confidence interval will probably be achieved by sorting the 1,000 values into ascending order and using the 25th and 975th values as a 95 percent confidence interval. This procedure accounts for the likely asymmetric distribution of the estimates of persistence.

## ■ Discussion

The real problem with PVA is not the model, but obtaining the data to drive the model (Ruggiero et al. 1994). Much of the published work on PVA ignores this essential point (Thomas 1990). For example, Mangel and Tier (1994: 608–611) simplify the process to the point that they miss major issues concerning data reliability and quality of the product (estimates of persistence). Their four "facts" are as follows:

• "A population can grow, on average, exponentially and without bound and still not persist." This is because of catastrophes that will bring even a thriving population to zero.

• "There is a simple and direct method for the computation of persistence times that virtually all biologists can use." They suggest a simple model with one age class and a population ceiling that the population cannot exceed, but the ceiling does not cause density dependence effects of growth parameters. As a result, their approach to estimating persistence is likely to underestimate persistence if the ceiling is set too low because the population can never grow away from the absorbing state of extinction.

• "The shoulder of the MacArthur–Wilson model occurs with other models as well, but disappears when catastrophes are included." They suggest a slow, steady rise in persistence times as the population ceiling is increased.

• "Extinction times are approximately exponentially distributed and this means that extinctions are likely." Thus, they conclude the most likely value of a population is zero, or the mode of an exponential distribution. I believe this result is obtained because of the simplistic assumptions they have used. Realistic models that incorporate the sources of variation described in this chapter do not result in such simplistic results.

Another misguided example is Tomiuk and Loeschcke (1994). Their mathematics cover up the real problem of obtaining realistic estimates of the parameter values to use in the models. Their model emphasizes demographic variation and ignores the bigger issues of temporal variation and individual heterogeneity.

A common problem with PVA is that the sampling variation of the parameter estimates is ignored (e.g., Stacey and Taper 1993; Dennis et al. 1991). In both cases, estimates of persistence are too pessimistic because the sampling variation of the population parameters is included in the population model as if it were temporal variation. Furthermore, individual heterogeneity was left out of the model, further biasing the estimates of persistence too low.

"Most PVAs have ignored fundamentals of ecology such as habitat, focusing instead on genetics or stochastic demography" (Boyce 1992:491). For small populations (less than 50) of endangered species, such a strategy may be justified, particularly for short-term predictions. But incorporating only demographic variation results in overestimates of persistence because temporal variation has been ignored. On the other hand, the remaining survivors of an endangered species may be the individuals with high survival and reproductive rates, so the lack of individual heterogeneity may underestimate persistence.

The studies just mentioned should not lead the reader to believe that useful attempts to estimate persistence do not exist. Schneider and Yodzis (1994) developed a model of marten (*Martes americana*) population dynamics that incorporated the behavior and physiology of individual martens, spatial dynamics, and demographic and environmental stochasticity. Undoubtedly some readers would quibble with some of the assumptions and data used to build the model, but I contend that a realistic model with some of the inputs guessed (and clearly stated to be such) is a much more reasonable approach than a simplistic model that ignores important processes affecting persistence. Furthermore, such realistic models identify data needs that can be addressed with time, even though the actual estimate of persistence is of questionable value. The alternative of using simplistic and naive models leads to invalid estimates and little progress in improving the situation, with a rapid loss of credibility by the field of conservation biology.

Murphy et al. (1990) proposed two different types of PVA. For organisms with low population densities that are restricted to small geographic ranges (typical vertebrate endangered species), genetic and demographic factors should be stressed. For smaller organisms such as most endangered invertebrates, environmental uncertainty and catastrophic factors should be stressed because these organisms are generally restricted to a few small habitat patches, but are capable of reaching large population sizes within these patches. Nunney and Campbell (1993) noted that demographic models and genetic models both have resulted in similar estimates of minimum viable population size, but that the ideal spatial arrangement of reserves remains an issue.

Lande (1995) suggests that genetic mutations may affect fitness, so ignoring genetic effects results in underestimates of viability. Mutation can affect the persistence of small populations by causing inbreeding depression, by maintaining potentially adaptive genetic variation in quantitative characters, and through the erosion of fitness by accumulation of mildly detrimental mutations. Populations of 5,000 or more are required to maintain evolutionary viability. Theoretical results suggest that the risk of extinction caused by the fixation of mildly detrimental mutations may be comparable in importance to environmental stochasticity and could substantially decrease the long-term viability of populations with effective sizes as large as a few thousand (Lande 1995). If these results are correct, determining minimum viable population numbers for most endangered species is an exercise in futility because almost all of these populations are already below 5,000.

Conservation biologists would like to have rules to evaluate persistence (Boyce 1992), such as the magical Franklin–Soulé number of 500 (Franklin 1980; Soulé 1980) that is the effective population size ($N_e$) to maintain genetic variability in quantitative characters. Unfortunately, these rules lack the realism to be useful. The Franklin–Soulé number was derived from simple genetic models and hence lacks the essential features of a PVA model discussed here. Attempts with simplistic models such as Mangel and Tier (1994) and Tomiuk and Loeschcke (1994) also do not provide defensible results because of the lack of attention to the biology of the species and the stochastic environment in which the population exists. Until conservation biologists do good experimental studies to evaluate population persistence empirically, I question the usefulness of the rules and simplistic models suggested in some of the literature.

PVA can be viewed as a heuristic tool to explore the dynamics of an endangered population but not as a predictive tool. PVA could be used to identify variables to which the population may be sensitive and to investigate the relative benefits of alternative kinds of management. Some readers will argue that in this context, the absolute reliability of the model estimates of extinction probability, or time to extinction, matters much less than the extent to which risk is affected by different demographic and environmental variables. I disagree because conclusions from PVA so strongly depend on which sources of variation are included in the model and their relative magnitudes. For example, the importance of demographic variation is stressed in PVA because it happens to be the simplest source of variation to model, generally requiring only the assumption of binomial variation. Temporal variation has received less emphasis because it is a more difficult to obtain estimates of the temporal variance of population parameters. Individual heterogeneity has received no attention because this source of variation is by far the most difficult source to quantify, particularly for survival rates. The only way to draw valid inferences about the importance of various sources of stochasticity affecting a population is to have reasonably good estimates of these parameters. Simplistic PVA models based on few or no data lead to simplistic and unreliable answers. Without data, why would you expect anything else?

Until rigorous experimental work can be conducted, conservation biologists should borrow information from game species, where long-term studies have been done that will provide estimates of temporal and spatial variation and individual heterogeneity. Rules that predict temporal variation in survival as a function of weather, or individual variation in survival as a function of body characteristics, provide alternative sources of data. For at least some game species, data exist to develop such rules. Furthermore, these kinds of data probably will never be available for many endangered species; the opportunity to collect such data was lost with the decline of the population to current (threatened) levels. Thus I suggest the use of surrogate species to help meet the data needs of realistic models of persistence. Taxonomically related species may provide information, although species in the same ecological guild may also provide information on temporal and spatial variation. Note the distinction between using estimates of the temporal and spatial variation and individual heterogeneity from a related species and using estimates of survival and recruitment from a surrogate species. Estimates of survival and recruitment from a stable or increasing population would obviously be inappropriate for a species with a declining population.

## ■ Conclusion

In summary, most estimates of population viability are nearly useless because one or more of the following mistakes or omissions are made in developing a model to estimate persistence. By listing omissions, this list suggests the essential ingredients to develop a useful PVA.

• Few or no data are available to estimate basic parameters in the population model, with almost all the parameter estimates just guesses. The resulting estimate of persistence is therefore strictly a guess.

*Lesson:* To do a valid PVA, you must have data to build a realistic population model.

• The model ignores spatial variation, which increases population viability. As suggested by Stacey and Taper (1992), immigration can occasionally rescue a population from extinction.

*Lesson:* If the population is widely distributed geographically, incorporate spatial variation.

• The model uses estimates of temporal variation that are at best poor guesses. This statement assumes that the modeler understood the difference between process variation and sampling variation. Often, sampling variation is assumed to substitute for process variation; as a result, the estimates of persistence are too pessimistic. Sampling variation has nothing to do with population persistence. Estimates of population parameters must not be treated as if they are the true parameter value.

*Lesson:* Obtain reliable estimates of temporal variation and don't confuse sampling variation and temporal variation.

• The model uses demographic variation as a substitute for temporal variation in the process and ignores true temporal variation.

*Lesson:* Incorporate both demographic and temporal variation into the PVA.

• The model ignores life-long individual heterogeneity that increases population viability and assumes that all individuals endure the same identical survival and reproduction parameters. Such a naive assumption results in population viability being underestimated.

*Lesson:* Individual heterogeneity must be incorporated into a PVA model if you don't want to underestimate viability.

• The model assumes that current conditions are not changing; that is, the stochastic processes included in the model are assumed constant for the indefinite future. Loss of habitat and other environmental changes that affect these stochastic processes are ignored. Thus, as discussed by Caswell (1989), the model probably is not useful in forecasting (i.e., predicting what will happen) but is useful in projecting (i.e., predicting what will happen if conditions do not change).

*Lesson:* Recognize that your model does not predict the future; it only projects what might happen if the system doesn't change (which is unlikely).

Before you use the estimates of persistence from any population viability analysis, compare the approach used to obtain the estimate with the necessary components discussed here. If you discover omissions and errors in the approach used to obtain the estimate, recognize the worth (or lack thereof ) of the estimate of persistence. Although the estimates of persistence obtained from a PVA may have little value, the process of formulating a model and identifying missing information (i.e., parameters that are poorly estimated) may still have value in developing measures to conserve the species in question.

### Literature Cited

Allee, W. C. 1931. *Animal aggregations.* Chicago: University of Chicago Press.

Armbruster, P. and R. Lande. 1993. A population viability analysis for African elephant (*Loxodonta africana*): How big should reserves be? *Conservation Biology* 7: 602–610.

Bartmann, R. M., G. C. White, and L. H. Carpenter. 1992. Compensatory mortality in a Colorado mule deer population. *Wildlife Monographs* 121: 1–39.

Beddington, J. R. 1974. Age distribution and the stability of simple discrete time populations models. *Journal of Theoretical Biology* 47: 65–74.

Beissinger, S. R. 1995. Modeling extinction in periodic environments: Everglades water levels and Snail Kite population viability. *Ecological Applications* 5: 618–631.

Bennington, C. C. and W. V. Thayne. 1994. Use and misuse of mixed model analysis of variance in ecological studies. *Ecology* 75: 717–722.

Berger, J. 1990. Persistence of different-sized populations: An empirical assessment of rapid extinctions in bighorn sheep. *Conservation Biology* 4: 91–98.

Beverton, R. J. H. and S. J. Holt. 1957. On the dynamics of exploited fish populations. *Great Britain Ministry of Agriculture, Fisheries and Food, Fishery Investigations* (Series 2) 19: 5–533.

Boyce, M. S. 1992. Population viability analysis. *Annual Review of Ecology and Systematics* 23: 481–506.

Burgman, M. A., S. Ferson, and H. R. Akçakaya. 1993. *Risk assessment in conservation biology.* London: Chapman & Hall.

Burnham, K. P., D. R. Anderson, and G. C. White. 1996. Meta-analysis of vital rates of the northern spotted owl. *Studies in Avian Biology* 17: 92–101.

Burnham, K. P., D. R. Anderson, G. C. White, C. Brownie, and K. H. Pollock. 1987. Design and analysis experiments for fish survival experiments based on capture–recapture. *American Fisheries Society Monograph* 5: 260–278.

Caswell, H. 1989. *Matrix population models.* Sunderland, Mass.: Sinauer.

Caughley, G. 1994. Directions in conservation biology. *Journal of Animal Ecology* 63: 215–244.

Cézilly, F., A. Viallefont, V. Boy, and A. R. Johnson. 1996. Annual variation in survival and breeding probability in greater flamingos. *Ecology* 77: 1143–1150.

Clark, T. W., P. C. Paquet, and A. P. Curlee. 1996. Special section: Large carnivore conservation in the Rocky Mountains of the United States and Canada. *Conservation Biology* 10: 936–936.

Clutton-Brock, T. H., F. E. Guinness, and S. D. Albon. 1982. *Red deer: Behavior and ecology of two sexes*. Chicago: University of Chicago Press.

Conroy, M. J., Y. Cohen, F. C. James, Y. G. Matsinos, and B. A. Maurer. 1995. Parameter estimation, reliability, and model improvement for spatially explicit models of animal populations. *Ecological Applications* 5: 17–19.

DeAngelis, D. L. and L. J. Gross, eds. 1992. *Individual-based models and approaches in ecology: Populations, communities, and ecosystems*. New York: Chapman & Hall.

Dennis, B., P. L. Munholland, and J. M. Scott. 1991. Estimation of growth and extinction parameters for endangered species. *Ecological Monographs* 6: 115–143.

Diamond, J. and S. Pimm. 1993. Survival times of bird populations: A reply. *American Naturalist* 142: 1030–1035.

Dunning, J. B. Jr., D. J. Stewart, B. J. Danielson, B. R. Noon, T. Root, R. H. Lamberson, and E. E. Stevens. 1995. Spatially explicit population models: Current forms and future uses. *Ecological Applications* 5: 3–11.

Efron, B. and R. J. Tibshirani. 1993. *An introduction to the bootstrap*. New York: Chapman & Hall.

Foley, P. 1994. Predicting extinction times from environmental stochasticity and carrying capacity. *Conservation Biology* 8: 124–136.

Fowler, C. W. 1981. Density dependence as related to life history strategy. *Ecology* 62: 602–610.

Fowler, C. W. 1994. Further consideration of nonlinearity in density dependence among large mammals. *Report of the International Whaling Commission* 44: 385–391.

Franklin, I. R. 1980. Evolutionary changes in small populations. In M. E. Soulé and B. A. Wilcox, eds., *Conservation biology: An evolutionary–ecological perspective*, 135–149. Sunderland, Mass.: Sinauer.

Ginzburg, L. R., L. B. Slobodkin, K. Johnson, and A. G. Bindman. 1982. Quasiextinction probabilities as a measure of impact on population growth. *Risk Analysis* 2: 171–181.

Haila, Y. and I. K. Hanski. 1993. Birds breeding on small British islands and extinction risks. *American Naturalist* 142: 1025–1029.

Hanski, I. 1996. Metapopulation ecology. In O. E. Rhodes, Jr., R. K. Chesser, and M. H. Smith, eds., *Population dynamics in ecological space and time*, 13–43. Chicago: University of Chicago Press.

Hanski, I., A. Moilanen, and M. Gyllenberg. 1996. Minimum viable metapopulation size. *American Naturalist* 147: 527–541.

Hassell, M. P. 1975. Density-dependence in single-species populations. *Journal of Animal Ecology* 44: 283–295.

Hassell, M. P., J. H. Lawton, and R. M. May. 1976. Patterns of dynamical behaviour in single-species populations. *Journal of Animal Ecology* 45: 471–486.

Hess, G. R. 1993. Conservation corridors and contagious disease: a cautionary note. *Conservation Biology* 8: 256–262.

Holt, R. D., S. W. Pacala, T. W. Smith, and J. Liu. 1995. Linking contemporary vegetation models with spatially explicit animal population models. *Ecological Applications* 5: 20–27.

Hunter, M. L., Jr. 1996. *Fundamentals of conservation biology*. Cambridge, Mass.: Blackwell.

Joopouborg, N. and J. M. Van Groenendael. 1996. Demography, genetics, or statistics: Comments on a paper by Heschel and Paige. *Conservation Biology* 10: 1290–1291.

Keller, L. F., P. Arcese, J. N. M. Smith, W. M. Hochachka, and S. C. Stearns. 1994. Selection against inbred song sparrows during a natural population bottleneck. *Nature* 372: 356–357.

Koenig, W. D., R. L. Mumme, W. J. Carmen, and M. T. Stanback. 1994. Acorn production by oaks in central California: Variation within and among years. *Ecology* 75: 99–109.

Krausman, P. R., R. C. Etchberger, and R. M. Lee. 1993. Persistence of mountain sheep. *Conservation Biology* 7: 219.

Lande, R. 1988. Genetics and demography in biological conservation. *Science* 241: 1455–1460.

Lande, R. 1993. Risks of population extinction from demographic and environmental stochasticity and random catastrophes. *American Naturalist* 142: 911–927.

Lande, R. 1995. Mutation and conservation. *Conservation Biology* 9: 782–791.

Lefkovitch, L. P. 1965. The study of population growth in organisms grouped by stages. *Biometrics* 21: 1–18.

Leslie, P. H. 1945. On the use of matrices in certain population mathematics. *Biometrika* 33: 183–212.

Leslie, P. H. 1948. Some further notes on the use of matrices in population mathematics. *Biometrika* 35: 213–245.

Lomnicki, A. 1988. *Population ecology of individuals*. Princeton, N.J.: Princeton University Press.

Ludwig, D. 1996a. Uncertainty and the assessment of extinction probabilities. *Ecological Applications* 6: 1067–1076.

Ludwig, D. 1996b. The distribution of population survival times. *American Naturalist* 147: 506–526.

Mace, G. M. and R. Lande. 1991. Assessing extinction threats: Toward a reevaluation of IUCN threatened species categories. *Conservation Biology* 5: 148–157.

Mangel, M. and C. Tier. 1994. Four facts every conservation biologist should know about persistence. *Ecology* 75: 607–614.

Manly, B. F. J. 1990. *Stage-structured population sampling, analysis and simulation*. London: Chapman & Hall.

May, R. M. 1976. Models for single populations. In R. M. May, ed., *Theoretical ecology: Principles and applications*, 4–25. Oxford, U.K.: Blackwell Scientific.

May, R. M. and G. F. Oster. 1976. Bifurcations and dynamic complexity in simple ecological models. *American Naturalist* 110: 573–599.

Maynard-Smith, J. and M. Slatkin. 1973. The stability of predator–prey systems. *Ecology* 54: 384–391.

Meffe, G. K. and C. R. Carroll. 1994. *Principles of conservation biology*. Sunderland, Mass.: Sinauer.

Mills, L. S., S. G. Hayes, C. Baldwin, M. J. Wisdom, J. Citta, D. J. Mattson, and K. Murphy. 1996. Factors leading to different viability predictions for a grizzly bear data set. *Conservation Biology* 10: 863–873.

Moran, P. A. P. 1950. Some remarks on animal population dynamics. *Biometrics* 6: 250–258.

Murphy, D. D., K. E. Freas, and S. B. Weiss. 1990. An environment-metapopulation approach to population viability analysis for a threatened invertebrate. *Conservation Biology* 4: 41–51.

Nunney, L. and K. A. Campbell. 1993. Assessing minimum viable population size: Demography meets population genetics. *Trends in Ecology and Evolution* 8: 234–239.

Pascual, M. A., P. Kareiva, and R. Hilborn. 1997. The influence of model structure on conclusions about the viability and harvesting of Serengeti wildebeest. *Conservation Biology* 11: 966–976.

Pennycuick, C. J., R. M. Compton, and L. Beckingham. 1968. A computer model for simulating the growth of a population or of two interacting populations. *Journal of Theoretical Biology* 18: 316–329.

Pimm, S. L., H. L. Jones, and J. M. Diamond. 1988. On the risk of extinction. *American Naturalist* 132: 757–785.

Remmert, H., ed. 1994. *Minimum animal populations*. New York: Springer-Verlag.

Ricker, W. E. 1954. Stock and recruitment. *Journal of the Fisheries Research Board of Canada* 11: 559–623.

Ricker, W. E. 1975. *Computation and interpretation of biological statistics of fish populations*. Ottawa: Fisheries Research Board of Canada, bulletin 191.

Ruggiero, L. F., G. D. Hayward, and J. R. Squires. 1994. Viability analysis in biological evaluations: Concepts of population viability analysis, biological population, and ecological scale. *Conservation Biology* 8: 364–372.

Schneider, R. R., and P. Yodzis. 1994. Extinction dynamics in the American marten (*Martes americana*). *Conservation Biology* 4: 1058–1068.

Seber, G. A. F. 1982 (2d ed.). *Estimation of animal abundance and related parameters*. New York: Macmillan.

Shaffer, M. L. 1981. Minimum population size for species conservation. *BioScience* 31: 131–134.

Shaffer, M. L. 1987. Minimum viable populations: Coping with uncertainty. In M. E. Soulé, ed., *Viable populations for conservation*, 69–86. Cambridge, U.K.: Cambridge University Press.

Silva, M. and J. A. Downing. 1994. Allometric scaling of minimal mammal densities. *Conservation Biology* 8: 732–743.

Soulé, M. E. 1980. Thresholds for survival: maintaining fitness and evolutionary potential.

In M. E. Soulé and B. A. Wilcox, eds., *Conservation biology: An evolutionary-ecological perspective,* 151–170. Sunderland, Mass.: Sinauer.

Soulé, M. E. 1987. *Viable populations for conservation.* New York: Cambridge University Press.

Stacey, P. B. and M. Taper. 1992. Environmental variation and the persistence of small populations. *Ecological Applications* 2: 18–29.

Thomas, C. D. 1990. What do real population dynamics tell us about minimum viable population sizes? *Conservation Biology* 4: 324–327.

Thomas, C. 1994. Extinction, colonization, and metapopulations: Environmental tracking by rare species. *Conservation Biology* 8: 373–378.

Tomiuk, J. and V. Loeschcke. 1994. On the application of birth–death models in conservation biology. *Conservation Biology* 8: 574–576.

Tracy, C. R. and T. L. George. 1992. On the determinants of extinction. *American Naturalist* 139: 102–122.

Tracy, C. R. and T. L. George. 1993. Extinction probabilities for British island birds: A reply. *American Naturalist* 142: 1036–1037.

Turner, M. G., G. J. Arthaud, R. T. Engstrom, S. J. Hejl, J. Liu, S. Loeb, and K. McKelvey. 1995. Usefulness of spatially explicit population models in land management. *Ecological Applications* 5: 12–16.

Urban Hjorth, J. S. 1994. *Computer intensive statistical methods.* London: Chapman & Hall.

Usher, M. B. 1966. A matrix approach to the management of renewable resources, with special reference to selection forests. *Journal of Applied Ecology* 3: 355–67.

Usher, M. B. 1972. Developments in the Leslie matrix model. In J. N. R. Jeffers, ed., *Mathematical models in ecology,* 29–60. Oxford, U.K.: Blackwell.

Yodzis, P. 1989. *Introduction to theoretical ecology.* New York: Harper & Row.

## Chapter 10

# Measuring the Dynamics of Mammalian Societies: An Ecologist's Guide to Ethological Methods

DAVID W. MACDONALD, PAUL D. STEWART, PAVEL STOPKA, AND
NOBUYUKI YAMAGUCHI

Today, biologists interpret behavior within a context fortified by theories of
cognition, behavioral evolution, and games (Axelrod 1984; Findlay et al.
1989; Hemelrijk 1990; Hare 1992; de Waal 1992), and any or all of four
processes may lead to cooperation: kin selection, reciprocity and byproduct
mutualism, and even trait-group selection (reviewed by Dugatkin 1997). The
processes that fashion societies are set within an ecological context (Macdonald 1983), and a species' ecology can scarcely be interpreted without understanding its social life. As the specialties within whole-animal biology diversify
and the once close-knit family of behavioral and ecological disciplines risks
drifting apart, our purpose is to alert ecologists to the ethologist's tools for
measuring social dynamics.

### ■ Social Dynamics

"If animals live together in groups their genes must get more benefit out of the
association than they put in" (Dawkins 1989). What methods are available to
measure the negotiations—the social dynamics—in this profit and loss
account? We define a social dynamic simply as the change in social interaction
or relationship under the influence of extrinsic or intrinsic factors. Our purpose here is to show how these changes and the factors influencing them may
be measured and identified. Likely candidates include the forces of ecological
and demographic change, together with changes in the experiences and characters of group members. Ontogenic effects (individuals growing up and

changing roles) might also affect the long-term social dynamics of a group that change the demography and hence the character of the society (Geffen et al. 1996). Effects on social dynamics may be erratically stochastic or predictably circadian, seasonal, annual, or of an even longer periodicity, largely following environmental rhythms. Predictable changes in social structure may also follow as a population or group progresses in a social succession toward carrying capacity after colonization or population crashes. Against this backdrop of almost continual flux, the study of social dynamics requires the measurement of changes in behavioral parameters. These measures become the currency with which to assess predictions designed to test whether the forces of change have been isolated correctly. The concept of a group's social dynamic is a vital and often neglected foil to attempts to characterize a typical social structure. Therefore, the concept of social dynamics lies at the interface of sociobiology, ethology, and behavioral ecology and even includes aspects of complexity theory and emergent systems. This alone makes it a topic of dauntingly large scope.

Research on social behavior commonly seeks a conclusion as to whether a particular type of social interaction maximizes fitness (Krebs and Davies 1991). However, to avoid the hazards of naive interpretation, one cannot draw such a conclusion without knowing the pattern of other interactions within which the behavior in question is set. Behavioral ecologists may pick individuals for which they score an approximation of fitness against a continuum of strategies. This approach is more hazardous as the web of social interactions in which individuals of a species are enmeshed becomes more complex. Occam's razor may suggest making the simplest explanation on the basis of what you observe, but in a social network the system is seldom simple, so it is prudent to make those observations thoroughly and in a wide context before that razor can be wielded confidently.

That the social dynamics of a species are both determinants and consequences of its ecology may be clear. To a field ecologist seeking to understand any part of this loop, it may be much less obvious how to characterize a social system in replicable, enduring, and quantitative terms as a basis for modern analysis. Historically, ethology pursued its own agenda—often with captive primates—of characterizing societies by observing behavioral interactions and directionality of behaviors within groups (Hemelrijk 1990). Classic ethologists were careful to record the detail of behavior with a view to allowing comparisons between studies and between species; although modern comparative methods have brought elegance to the task of making comparisons, modern

fashions have drifted away from assembling the data classic ethologists so carefully gleaned.

## ■ Context

In a book on ecological methods, the function of this chapter is to provide a map
for ecologists through the portions of the rapids that ethologists have already
negotiated. When we launched into this task, no text existed to bring together
the wisdom of a receding era of ethology and the innovations promised by avant
garde methods; as we finished, Lehner's (1996) comprehensive second edition
*Handbook of Ethological Methods* was published. Lehner's updated text is set to
become the benchmark, so for many topics for which this chapter whets the
reader's appetite, that handbook will be the place to find the full meal.

Our treatment has five sections. First we mention three reasons why documenting social dynamics is important. Second, we tackle the question of how
to describe social dynamics in ways that provide a framework within which to
compare studies; in this, we follow Hinde (1981, 1983) in describing a hierarchical approach to the study of social dynamics. We show that single behavioral
interactions are the fundamental unit of social structure, and that it is the
changing nature of repeated interactions—relationships—that gives social
structure its potentially dynamic component. Third, we follow this framework
to discuss how behavioral parameters can be described, classified, and recorded
during social interactions. In this we explore interactions during grooming and
dominance. In the fourth section we explore methods used for gathering data
on social interaction and then, in the fifth section, we use example data to illustrate analytical techniques for elucidating relationships, and how patterns of
relationships can be combined to reveal social networks and social structure.
Because mammalian societies are ethologically complex and because our own
experience is largely with mammals, we draw our examples from that class. Furthermore, we often use examples from our own work, not because it deserves
mention more than any other, but because we know most well the lessons
learned and pitfalls encountered therein; our intention is merely to illustrate
points, not to review them compendiously (that task is more fully accomplished
by Colgan 1978; Hazlett 1977; and Lehner 1996). Throughout, we focus on
common mistakes in approach, methodological conflicts, and the use of new
technology to solve problems (and how it creates some new ones). We have
doubtless fallen short of our own prescriptions on many previous occasions,
and are aware that many methods outlined here could be improved further.

## ■ Why Study Social Dynamics?

Social dynamics merit study for three major reasons. First, the changing relationships between individuals are the building blocks from which dynamic social structures are assembled. Understanding their emergent properties, such as dominance hierarchies and social competition, is essential to tackling fundamental questions about the evolution of sociality (Pollock 1994). Second, because social dynamics are the product of interaction between individuals' ecology and behavior (Rubenstein 1993), they are relevant to predicting and managing the consequences of many human interventions for conservation or management. Third, an important motivation for understanding nonhuman societies is the light this throws on human behavior. Each of these three topics is vast, but we fleetingly mention them in turn.

### EVOLUTION OF SOCIALITY

The diverse relationships of individuals in a social network interact to create complex emergent patterns. These patterns, like the vortex that appears in an emptying sink, is not contained in the structure of a single component. Because a society represents a whole with properties different from those of its component parts, the ultimate consequences of social interactions may be remote from an observed action. This is a fact that evolution by natural selection can take in its stride but that we, as primarily linear cause-and-effect thinkers, may find hard to accommodate. It may be clear that a lion killing a zebra is behaving adaptively, but less clear whether it is adaptive when the same lion prevents a conspecific from feeding at the kill. The immediate effect is that the first lion may have more food to eat, but the ultimate effects reverberate through a stochastically unpredictable system of long-term consequences among the whole pride. Denied food or coalitionary aid by an ally of the snubbed individual at a later date, the fitness consequences for the originally possessive lion may be far from advantageous. An understanding of social dynamics offers insight into the adaptation of individual responses evolved from selection operating on them from the level of emergent systems.

### CONSERVATION APPLICATIONS

Many problems in wildlife conservation and management involve humans causing changes to animal populations or their environment. In applied work, an understanding of the dynamics of a social system is a prerequisite to predicting the effect of human activities on, for example, spatial organization, population dynamics, and dispersal. For example, attempts to control the transmission of bovine tuberculosis by killing badgers, a reservoir of the disease, clearly disrupts the society of survivors. The effects of such perturbation on social dynamics may alter the transmission of the disease, plausibly for the worse (Swinton et al. 1997). A similar case may be argued regarding rabies control (Macdonald 1995). Translocation of elephants without regard for the social structure that provides adolescent discipline has led to problem animals in some African parks (McKnight 1995). Tuyttens and Macdonald (in press) review some consequences of behavioral disruption for wildlife management. Population control has been shown to affect the rate and pattern of dispersal (Clout and Efford 1984), home range size (Berger and Cunningham 1995), territoriality, mating system (Jouventin and Cornet 1980), and the nature of social interactions (Lott 1991) in a variety of species.

## UNDERSTANDING OURSELVES

So much is similar in the basic biology of vertebrates, and so universal are the processes of evolution, that an understanding of nonhuman sociality is likely to illuminate human society. This point was hitherto neglected, but stressed by Tinbergen in the foreword to Kruuk's (1972:xi–xiii) book, in which he concluded, "It is therefore imperative for the healthy development of human biology that studies of primates be supplemented by work on animal species that have evolved adaptations to the same way of life as ancestral man." Following Wilson's (1975) *Sociobiology,* it has been widely and sometimes controversially discussed. Clearly, two routes come to mind as fruitful sources of this insight: looking at the societies of species most similar to our current condition (including some hypersocial aspects that put us in circumstances to which we have not yet had time to evolve) and focusing on those currently entering evolutionary phases through which we have already passed. The first approach has prompted (or at least its promise has funded) much primatological research. The closeness of this parallel might be diluted if, as Hinde (1981) suggested, the societies of humans differ from those of other animals in that social structure in nonhumans is determined primarily by the sum of the interactions of its component individuals, whereas in human groups a structure is more often imposed from above by government or tradition in the form of Dawkins's (1989) memes. Hinde's dichotomy may imply that the imposition of structure can cause stresses in human social systems when natural roles conflict with assigned roles. On the other hand, one could take the view that the dichotomy is not profound because the constraints of governmental ideology are loosely parallel to those imposed on all species by ecological factors such as resource dispersion. If so, a different understanding of social responses to the imposition of external constraints might be revealed by species more recently launched onto a trajectory of sociability. Examples we explore in this chapter include badgers and farm cats living as groups in agricultural settings. Certainly, badger groups show weight reduction, higher incidences of wounding, and lower reproductive success per breeding individual as group size increases (Woodroffe and Macdonald 1995a). For badgers, group living may be a social innovation facilitated by the development of agriculture; individuals may be evolving towards capitalizing on this newly imposed structure (by manipulation, support, interdependence of roles, and other factors), but for the moment the stress is showing.

## ■ How to Describe Social Dynamics

It does not detract from the excitement of behavioral, ecological, and sociobiological insights to note that recent enthusiasm for these topics (much stimulated by Wilson 1975) has been characterized by a plethora of short, snappy papers with a clear adaptive punchline and a concomitant neglect of the empirical foundations of ethology. Historically, this arises because behavioral ecology and sociobiology were pioneered to offer ultimate functional explanations, whereas ethology embraced adaptive significance and evolution along with mechanisms and ontogeny (Tinbergen 1963). This vogue has led to the widespread abandonment of the ethological aspirations of the late 1970s, epitomized by Hinde's (1981, 1983) careful use of terminology and hierarchical classification to ensure compatibility between studies used for comparative work. At its purest, this traditional ethological approach placed greater emphasis on the facility of later reinterpretation of results than on the quest for a desired result. In contrast, there is an invasive tendency to treat hard data and description as disposable assets sacrificed to analytical elegance and discussion. In a science in its infancy (such as social biology) this brings the risk that future research may be doomed to repeat previous field work solely to attempt reinterpretation of undisputed results.

### ACTION, INTERACTION, AND RELATIONSHIPS

Adapting Hinde's (1983) classification, the basic units of social exchange between individual primates are *action* and *interaction*. Actions are directed toward the environment, and are often important in understanding a previous or subsequent interaction with individuals. Interactions (A attacks B) are hard to interpret without records of actions (A accidentally drops fruit, B picks up fruit).

Above action and interaction in the hierarchy of social dynamics are social relationships. Relationships are quantified by the rates, frequencies, and patterning of component interactions and may be described in terms of the diversity of interactions, the degree of reciprocity or complementarity, relative frequency and pattern of interactions, synchronicity, and multidimensional qualities. In principle, relationships may be stationary or transitionary. The former do not change with prior experience (intrinsic development) or conditions (extrinsic modification) and are at most rare and perhaps nonexistent in mammalian societies. Generalizations about relationships can be sought in various ways. Dyads may be assigned to predefined categories such as age, sex, kinship, or even personality (Faver et al. 1986). Personality, in this context, is a consistent moderator of interactions; for example, a shy animal tends to act differently from a bold animal under the same circumstances (Stevenson-Hinde 1983). Block model methods allow subgroupings to be isolated from sociomatrices. For example, Iacobucci (1990) compared 13 methods for recovering subgroup structure from dyadic interaction data. In general, block models are based on structural equivalence of sociomatrices (see "Analysis of Observational Data"). Some individuals behave similarly in respect to their age, status, or sex. Using block models, these relationships between individuals can be extracted and further studied, for example, using tests for reciprocity and interchange of behaviors (Hemelrijk and Ek 1991) or using more detailed methods based on time structure of the processes (Haccou and Meelis 1992). In their study of capybara mating systems, for example, Herrera and Macdonald (1993) disentangled the effects of dominance on mating success. In that example, dominant males secured more matings than any other individual, but fewer matings than subordinate males as a class; this arose because while the dominant was busy driving off one subordinate, another sought quickly to mate with the female.

## SOCIAL NETWORKS

The sum of social relationships may be compiled in a matrix of dyadic interactions to produce a social network (Pearl and Schulman 1983). Analysis of sociomatrices assumes stationarity, which, as we have noted, is effectively nonexistent. The solution is to divide sociomatrices into appropriately defined periods that approximate stationarity (see "The Bout"). This might involve consideration of, for example, "the first 50 interactions," "the second 50 interactions," and so forth, or "wet season interaction" versus "dry season interaction," or "simultaneous presence of dominant" versus "absence of dominant." Nested analysis is a common way to achieve this goal.

## SOCIAL STRUCTURE, FROM SURFACE TO DEEP

A social network provides a snapshot of one facet of society. By analogy, one analysis of a social network is akin to the view through one window into a large and labyrinthine house; the view through all windows gives the structure. It is therefore necessary to compile several networks that describe different facets of one society. This task may be made harder because different sociometric variables (e.g., grooming, aggression, play) may not follow similar patterns of stationarity; aggression may covary with age and presence of dominant, whereas grooming may not. Notwithstanding these complexities, a society's structure can be described in terms of these networks. Indeed, there are layers of completeness to this description. The structure that prevails may vary on a circadian basis, or seasonally or annually; it may also be characteristic of a species' society in only one habitat or set of environmental conditions. At its most fundamental, elements of social structure may characterize all populations of a species. Therefore, social structure might usefully be categorized on a continuum from surface structures to deep structures. The study of social dynamics seeks to describe and explain the patterning of transitions and stability of social structures. An important goal of evolutionary biology is to identify the rules, derived from a variety of empirical and theoretical sources, that are thought to guide an individual's decisions in a social context—Axelrod's (1984) seminal question of whether to cooperate. An accurate description of structure is clearly a prerequisite to a sensible exploration of these rules. Operationally, the point at which a thorough description of social structure is complete is probably the first point at which it is legitimate to consider exchanging data language (grooms, fights) for theory language (alliance formation, competition). These interpretative substitutions are topics for the discussion section of a paper, whereas in the results section data should be presented without such interpretation.

Exploring exhaustively the layers of structure in animal societies is a major undertaking. Not least because a major objective of studying social dynamics is to contribute to the solution of practical conservation problems, it is inevitable that such studies may sometimes have to be undertaken quickly. The obvious shortcut, within the framework of existing theory, is to use selected revealing behavior as a guide to an overall system. As a loose analogy, this is akin to using the seating arrangement at a formal dinner as a guide to the social role of the guests in contexts beyond the dinner. Naturally, such shortcuts necessitate validation.

## ■ Behavioral Parameters

### THE BOUT

Even to a casual observer it is obvious that most behavior patterns occur in bouts; that is, they do not occur randomly in time. Although the existence of bouts may be obvious, how best to define them is not. A review of major textbooks and many papers reveals that many prefer to avoid this issue. The most common usages define bouts as a repetitive occurrence of the same behavioral act (states or events) or a short sequence of behavioral actions that occur in some functional pattern (Lehner 1996). States usually have durations, and states with extremely short durations are called events.

The difficulties of defining the hierarchy of bouts, states, and events is illustrated by allogrooming by a mouse (figure 10.1). This comprises a series of actions (nibbles) of short duration; an uninterrupted string of these nibbling actions might make up a bout of grooming.

However, a student of the detail of mouse grooming will see that these strings of nibbles may sometimes transfer from one body region to another (e.g., from head to neck to flanks or back), and for some purposes it may be helpful to distinguish bouts at this finer scale (a bout of head grooming distinct from one of flank grooming). The problem is that the best definition of a bout depends on the purpose of the analysis in which it will be used. There is a hierarchy of bouts within bouts, as depicted in figure 10.1. Depending on the scale of resolution required, even a short sequence of nibbles at one patch on the flank might be distinguished from another bout of grooming at the next patch of fur. Ultimately, each nibble could be defined as a state, punctuated by another state (shifting the head a fraction to grasp the next tuft of fur). At a given level of resolution it may be helpful to define states from which bouts are built up, but often, under closer scrutiny, a state will emerge to have a structure that could itself represent a bout (rather than one nibble, or one sweep of the paws, while grooming). In some contexts this wracking down of the microscope to reveal more and more detail may seem merely a quest to

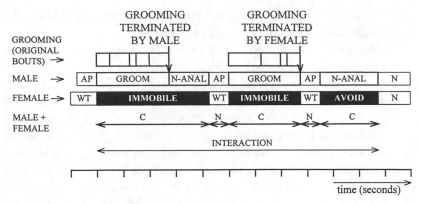

**Figure 10.1**  Data on the interactions between male and female wood mice, illustrating the arbitrariness in defining a bout. In practice, the best definition of a bout depends on the purpose of the analysis. The top row illustrates that grooming (of the female by the male) is split into a series of brief nibbles partitioned by fleeting changes of position. Each period of nibbling might be defined as an event or as a bout. If each nibble is an event, then the sequence of them might make up a bout of grooming. Bouts might also be defined in terms of contact between the two mice, and in that case the period during which the male was first grooming and then nasoanal sniffing the females constitutes one bout of contact, during which the female was immobile. Finally, the entire period of male–female interaction might be defined as a bout. This scheme is based on real data on the behavior of wood mice. Depending on the model under analysis, grooming bouts could be distinguished as one continuous process or as a series of original grooming bouts in which a mouse shifts between body positions. Even an interaction can be considered a bout if several criteria are fulfilled. C = contact behavior, N = noncontact behavior, AP = approach, WT = wait, N-anal = nasoanal contact.

bring into view the number of angels perched on the pinhead, but an important point nonetheless emerges: the unit of much behavioral analysis is the bout, and the usefulness of a definition of a bout is affected by the level of magnification at which the analysis is being undertaken. Bouts must be defined very carefully because their definition will have far-reaching statistical consequences for any analysis in which they are involved and because of their role as an indicator of motivation and neural processes.

From the mathematical point of view, when a behavior is modeled it is easiest to keep definitions simple, so Haccou and Meelis (1995:7) define a bout as a "time interval during which a certain act is performed. A bout length is the duration of such a time interval." In the calculation of transition matrices, transitions to the same act are impossible, so diagonal elements in the matrix are treated as zero (the notion of transitions from one bout of behavior to another of the same behavior presupposes that the observer can distinguish a bout from an interval between two consecutive bouts).

The complexities of defining bouts seem particularly difficult to accommodate in studies of mammals, for which one action pattern often appears to slur into another. For example, although the task of defining a bout of drinking by a fox (made up of a string of lapping events) is feasible, the parallel quest of distinguishing the gulps of a rat or primate (which drink continuously) might be dubious. In comparison it would be straightforward, in a parallel study of a bird, to identify consecutive pecks and the interval between them, which can be analyzed using log–survivor plots (Machlis 1977). The awkward truth is that the convenient hierarchy of bout, state, and event is often arbitrary and often concerns a continuum of lengths (clearly revealed by techniques such as the Noldus videotape analysis system, which measures even the durations of events).

To continue with the example of allogrooming wood mice (figure 10.1) and opting for Haccou and Meelis's (1995) definition of a bout as either the duration of a state or the interval between two states, it can be straightforward to recognize bouts. Wood mice, for example, often switch between two states during exploration: scanning while walking and rearing up or scanning while immobile. In this example, the behavioral elements *scan, rear up,* and *walk* each have durations and are therefore states. For example, the duration of scanning is a bout. During scanning, however, wood mice often turn their heads back for only a tenth of a second; *turn-head-back* is an event (Stopka and Macdonald, 1998). In this case, one bout may include scanning interrupted by several turn-head-back events. Sometimes, however, it is difficult to distinguish between bouts, although methods are available to do so (Langton et al. 1995; Sibley et al. 1990). When behavior is studied in sequences (i.e., continuous records), the existence of bouts can be confirmed because the distribution of their lengths follows an exponential distribution as long as successive bouts comply with the assumptions of a first-order continuous-time Markov chain model (Haccou and Meelis 1992). In practice, if bout durations do not follow an exponential distribution, there are two possible explanations. First, the observer incorrectly recognized the bouts and therefore measured the wrong thing, perhaps because some bouts were only partially observed through insufficient time for observation (Bressers et al. 1991). Second, the first-order Markovian assumption is not upheld. In the latter case, there are again two possibilities. First, bout lengths may exhibit dependency (between successive bouts or every second or third bout). In this case it is always better to use a semi-Markovian model (Haccou and Meelis 1992). Second, and more abstruse, there may be second- or third-order Markov models, which according to Haccou and Meelis (1992) are rare and intractable in ethology. In general, bouts approximate to a first-order Markov process and therefore can be distinguished on the basis of testing their distribution or fitting a nonlinear curve to the logarithm of observed frequencies of gap length. Because most behavioral studies are based on parameters such as latency, duration, and time intervals, bout length is a very important parameter to study at the outset.

## STATIONARITY

Obviously, if circumstances change over a series of bouts it is confusing or misleading to lump them for analysis. Consequently, it is important to identify stationarity, during which things do not change (formally, "a process with transition probabilities independent of time"; Haccou and Meelis 1992:193). Plots of transition frequencies from one act to another in several consecutive periods can be used to judge whether the behavioral process is approximately stationary. Haccou and Meelis (1995) emphasize that nonstationarity can mask treatment effects and make the results of an analysis ambiguous. Quite apart from the mathematical implications of inhomogeneity in the data, the existence of the motivational change that causes nonstationarity may be the very object of study, so it is important to identify it from an ethological standpoint. The complexities of stationarity reverberate through ethological methodology. Bekoff (1977), for example, is skeptical that the concept of stationarity can be applied in a social context and therefore concludes that Markovian analysis should be avoided.

## THE ETHOGRAM

Early in any study a researcher must classify the behavior patterns to be documented. This classification is an ethogram, and constitutes a dictionary of the researcher's language; without an ethogram, meaning is not fixed. Ethograms have become unpopular because (like dictionaries) they take up large amounts of space and are dull to read. However, they are vital reference material for those wishing to assess a study's conclusions critically. As a corollary, it is not uncommon for the brunt of reviewers' comments now to fall on statistics rather than data acquisition, reflecting the same shift in emphasis away from data, and onto interpretation, as the hard currency of behavioral science. By analogy, consider the shift from real objects to paper representations in financial systems: 50 years on, would you rather find a stash of gold or war bonds?

We propose that the Internet, with its capacity to archive and disseminate large quantities of multimedia information, will revolutionize the use and usefulness of ethograms. For example, in parallel to a written publication on badger vocalizations (Wong et al. 1999), we have created a multimedia vocal ethogram and another (Stewart et al. 1997) based on digital video.

Ethograms are often constructed in a pretrial period during which behavior appropriate to the study is chosen and described. However, short studies are unlikely to detect a complete catalog of relevant behavior and are likely to miss social behavior characteristic of context-sensitive dyadic interactions. Fagen (1978) tentatively proposed that a sample of 50,000 acts or more might generally be necessary to estimate the repertoire of a typical carnivore or primate species. Fagen's estimate was based on the biologically general type–token relationship. This relationship was shown by May (1975) to be a general statistical property of all catalogs in which an infinite number of objects belong to a finite number of categories. It prescribes that the logarithm of the number of types in an inventory depends approximately linearly on the logarithm of the total number of acts in the catalog (Fagen and Goldman 1977). In practice, our badger study confirmed the point: Some antagonistic behaviors (coordinated attack on the rump of an animal after a partner has grasped its head) that we predicted might occur from their occasional appearance in ritualized play between cubs were seen for the first time in earnest between adults only after thousands of hours of direct observation. Although rare in badgers, these coordinated attacks gave an important insight into a previously undocumented realm of cooperative aggression.

What are the components of an ethogram? Ideally, each element might be purely descriptive, free of any imputed function. A heroic attempt at this purity of description was Golani's (1976; see also Schleidt et al. 1984) use of balletic choreographic scores to quantify the spatial and sequential organization of body part movement, together with its qualities such as speed and force and the degrees of variation tolerated within categories. It was intended to help solve what Golani (1992) called a blind spot in the behavioral sciences: the need for a universal language to describe animal movement. Despite potential for universality of description, such quantitative choreography has proved impractical for most field studies. Also, focusing on the minutiae of postures may be too detailed for this purpose. A realistic option, facilitated by the Internet, is the creation of an archive of film clips and spectrographic catalogs within a taxonomic library. However, such a "content and quality" ethogram is still denuded of context. Is a bird pecking a conspecific attempting fighting, grooming, or feeding? The answer may seem obvious from the rate of pecking, its strength, and the reaction of the other bird, but logging these contextual clues is time-consuming. This is why many ethograms incorporate a shorthand notation that uses not only purely descriptive language but also proximate function language, derived from context and likely result. The demons of tautology and teleology lurk in such language.

## BEWARE TELEOLOGY

Omnipresent dangers in the study of social behavior are the closely allied traps of teleology and unwitting anthropomorphism. Teleology, or the doctrine of final causes, infers purpose in nature (e.g., to infer the existence of a creator from the works of creation). The teleological conviction that mind and will are the cause of all things in nature is not within the scope of scientific method (Romanes 1881). In the context of animal behavior, teleological explanation would name and account for a behavior by its presumed ultimate effect (e.g., appeasement, submission, or punishment) and not by its proximate causes or physical appearance. It may be convenient to label as "punishment" the category of attack launched by a dominant meerkat on the only member of her group not to join in a fight with territorial trespassers. The danger lies in (inadvertently) interpreting the functional nuance of this convenient label as the proximate cause of the attacker's behavior. We know only that one individual did not join the fight and that the dominant member of its clan then initiated an attack on it (joined by all its group-mates). It is a matter of interpretation whether this attack functioned as punishment and a matter of speculation whether the attacking meerkat had punishment (or anything else) in mind when it attacked; it is certainly unwarranted to conclude that the putatively punitive attack was launched with the ultimate purpose of increasing the fitness of the attacker (although that may well be its consequence).

Interpretation based on context again harbors the pitfalls of premature use of proximate function language. It would be folly, for example, to categorize as "supportive" an instance of a large female grooming subordinate kin but to label as "repressive" the grooming by the same female of nonkin subordinates. In such a case, proximate function language would have prematurely slipped all the way to ultimate function language. Clearly, the risk is that the very hypothesis used to make this interpretation may later be said to be supported by the observation; in this hypothetical case, the erroneously circular conclusion would be that kin selection theory is supported by the observation that females are supportive of kin and repressive to nonkin. In fact, such theories can be tested only by following actions through to fitness consequences—a very difficult task. In the progress of a given scientific inquiry, there is no virtue in premature exchange of data language for theory language.

The teleological trap arises because in humans, conscious action (and even empathy to other people's unconscious actions) often involves mental plans and objectives that stimulate or explain the performance of behaviors. But consciousness has not been proved in other species and to confuse proximate factors with ultimate factors in this way may stifle correct interpretation. As Kennedy (1986:23–24) noted, "Teleological terms such as 'searching' . . . really describe the animal's presumed state of mind. . . . Trying to find an objective substitute for a teleological term will always pay off in research because it forces a mental break-out and a closer look at the components of the behaviour actually observed, components which the handy teleological term leaves unnoticed or at best unformulated in the back of the mind." Kennedy concluded that *scanning* would be a more useful term for searching. The worst cases of teleology may use such "catchy" labels that proximate and even ultimate function are obscured in the mind of the reader (if not the writer) because of the human innuendo of the word. *Rape,* used as a label for resisted matings, has been widely considered a case in point.

Used correctly, "mock anthropomorphism" can be a very valuable heuristic tool to guess the function of behavior, using our own mental processes to explain behavior in an ultimate fitness context (Mitchell et al. 1997). Such hypotheses can be tested, but even if they are supported we must beware of unwittingly crossing the threshold of assuming that the animal has used the same mental processes when deciding on its own actions. The temptation of teleology—to impose our mental model on others—may be greatest when dealing with other primates. Tinbergen (1963:413–414) foresaw this when he wrote, "Teleology may be a stumbling block to causal analysis in its less obvious forms. . . . The more complex the behaviour systems we deal with, the more dangerous this can be." These strictures against teleology do not imply that ethologists diminish animals to the level of Descartean *machina anima;* rather, the objective is to be mindful of Lloyd Morgan's canon that "In no case may we interpret an action as the outcome of the exercise of a higher psychical faculty, if it can be interpreted as the outcome of the exercise of one which stands lower in the psychological scale" (Kennedy 1986). This is a methodological rather than ideological stance, which may be more likely to be correct than the converse assessment. As Dawkins (1989:95) put it, a pragmatic aid in navigating this tricky terrain is to advance "always reassuring ourselves that we could translate our sloppy language into respectable terms if we wanted to."

## CLASSIFICATIONS OF BEHAVIORAL INTERACTIONS

How, in practice, can we impartially assign positive or negative implications to behavioral interactions such as proximity or grooming that are ambivalent, without recourse to the very theories we seek to support? One solution lies in sequence analysis, using unequivocal behavior patterns as anchors; high-intensity attacks are certainly detrimental to the receiver's fitness, whereas allowing mating is usually beneficial to fitness. If transition probabilities revealed grooming as a predictor of attack, then it might usefully be classed as aggressive, but if grooming often precedes acceptance of mating, it is probably amicable. Firm grounding of a behavior in context may justify an interpretative classification. Often, however, it is impossible to establish a uniform link between ambivalent behaviors (e.g., grooming) and anchors (e.g., coalitionary aid or attack) until the whole social network of interactions within the group has been described. Descriptions based on ultimate function should be eschewed until such links are established.

A very long list of behavior patterns or relational types could be described and analyzed at each level of the structural hierarchy of social dynamics. Here, we consider just one example at each of three levels: grooming as an interaction, dominance as a relationship, and the concept of "social group" as a structure.

### Grooming

Allogrooming can be abundant, observable, and clearly bidirectional or unidirectional, a combination of attributes that have made it perhaps the most frequently studied social interaction. Despite the common assumption that it is amicable or beneficent, it may also cue either dominance or submission in a ritualized context. The style of grooming, be it simultaneous, alternating, or routinely one-sided, is often largely species typical and even a small sample may give important clues to social relations in a wider context. Grooming has been the focus of many primatological studies (reviewed by Goosen 1987), among which it appears to have more to do with social bonding than with hygiene. For example, there is a significant correlation between time spent grooming and group size but not body size (Dunbar 1988). The idea that primates compete for grooming access in a social setting has also been very influential (Seyfarth 1983).

To unravel more complex intricacies requires more probing analytical techniques, of which an especially rigorous example is Hemelrijk and Ek's

(1991) study of reciprocity and interchange of grooming and agonistic "support" during conflict in captive chimpanzees. Reciprocity was defined as one act being exchanged for the same thing (grooming for grooming), whereas interchange was defined as involving two different kinds of acts being bartered (i.e., grooming for support). They created an actor matrix (who initiates to whom) and a receiver matrix (who is recipient from whom) and used a correlation procedure, the $K_r$ test, to examine links between the two matrices while taking account of individual variation in tendency to direct acts. They also attempted to retain stationarity in the data by distinguishing between periods when an alpha male was clearly established or when alpha status was in dispute. Hemelrijk (1990) had shown that grooming and support were both independently correlated with dominance rank, so whenever reciprocity or interchange was part of a triangle of significant correlations, they partialed out the third variable to see whether the remaining two retained significance. Between the females, for example, a significant interchange between grooming and support received did not depend on dominance rank. Females groomed individuals that had supported them regardless of rank, indicating a social bond. However, the reverse was not true in that support was not given in return for grooming, so it was not possible to "buy" support with grooming. Patterns of relationships were different between the sexes.

Cheney (1992) used the distribution of grooming among individuals within a group to examine whether adversity, in the form of rivalry between groups, had a uniting effect within the group; the expectation of this hypothesis was a more egalitarian spread of grooming within the group when competition with neighbors was intense. This expectation was not upheld, perhaps because neglected individuals would suffer a greater loss of fitness by shirking in their support of their own group than by attempting to trade this support for better treatment, even assuming a lack of punishment for failure to cooperate (Clutton-Brock and Parker 1995). Even when intergroup competition was strong, marked hierarchies in grooming favoritism characterized intragroup relations. This study used the distribution of grooming to generate data with which to infer intragroup cohesion and equality. The ultimate functional explanations given depend directly on the validity of that assumption. Our acceptance of such hypotheses may be influenced more by our intuitive anthropomorphic bias for their seeming correctness than by our purely objective assessment of the supporting evidence. This is the dilemma of being a social mammal attempting to describe the societies of other mammals.

The site to which grooming is directed may have social implications. Among long-tailed macaques (*Macaca fascicularis*), for example, low-ranking females expose the chest, face, and belly less often to higher-ranking females than vice versa. Male–male dyads almost never groomed the face, chest, and belly. Similarly the face, chest, and belly were underrepresented in grooming between individuals that groomed each other less frequently. One interpretation is that recipients of grooming expose less vulnerable parts of their body to, and avoid eye contact with, individuals they perceive as potentially dangerous (Moser et al. 1991; Borries 1992). Another analysis of the form, distribution, and context of social grooming was undertaken in two groups of Tonkean macaques (*Macaca tonkeana*) by Thierry et al. (1990). They found that kinship and dominance role had no effect on the form or distribution of social grooming among adult females, the most common class of individuals to be observed grooming.

The caveat that no one method of disentangling social relationships is universally appropriate has been stressed by Dunbar (1976). For example, in a study of grooming in three species of macaques, Schino et al. (1988) found that whereas frequency and total duration of grooming were highly correlated, there was low correlation between its total duration and mean duration, and frequency and mean duration were not correlated. A closer inspection of the data also revealed that even this general equivalence between measures of frequency and total duration did not always hold within specific grooming relationships. In general, frequency is a measure of initiation and mean duration is a measure of continuation. In Schino et al.'s case, both measures were needed to interpret adequately the macaques' society.

Not surprisingly, one important stimulus for grooming in rodents is soiled fur (Geyer and Kornet 1982). However, because grooming can be undertaken alone or mutually, it has social implications transcending cleanliness. Stopka and Macdonald (in press) used a similar approach to Hemelrijk and Ek's (1991) analysis of grooming in captive chimpanzees to examine patterns of reciprocity in grooming among wood mice (*Apodemus sylvaticus*). A powerful method for investigating this involved row-wise matrix correlation and the Mantel test (de Vries et al. 1993). Allogrooming was less common than autogrooming and was most commonly directed by male wood mice to females. Having described a social network in grooming, Markov chain analysis then revealed that the termination rate of female contact behavior (expressed as her tendency to flee the male's attentions) depended largely on the male's tendency to terminate allogrooming. The pattern of transition rates from one behavior to another revealed that the most common transition was from allogrooming of the female by the male to male nasoanal contact with the female. This indicates the motivation underlying the male's tendency to allogroom females:

males groom a female in order to gain the opportunity to sniff her nasoanal region and, ultimately, to decrease her tendency to flee, thereby allowing them the opportunity to mate. Females are most likely to mate with males that groom them the most. This constitutes an unusual example of the market effect (Noe et al. 1991) in a short-lived mammal, with females essentially selling sex in exchange for grooming.

Allogrooming is the clearest and most common form of cooperation among badgers. It appears to be a valuable supplement to self-grooming because it is focused on the parts of a badger (back of neck, shoulders, and upper rump) that are most difficult for an individual to reach for itself (Stewart 1997; Stewart and Macdonald, unpublished data). The proportion of grooming directed to each segment of the body differs, but is complementary between self-grooming and allogrooming (figure 10.2).

Allogrooming cooperation is maintained in this case by the simple expedient of ensuring that as much grooming is given as received. Adult badgers groom one another simultaneously using a responsive rule set (Axelrod and Hamilton 1981) that dictates that grooming can be initiated generously, but rapidly withdrawn (at a mean of just 1.2 seconds) if the recipient does not respond in kind. If the recipient reciprocates, then the two badgers groom each

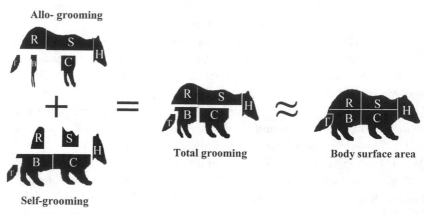

**Figure 10.2** Silhouettes represent the proportion of time, per square centimeter, spent grooming different portions of the body surface. It shows that although self-grooming or allogrooming alone is strongly biased toward particular body regions, a combination of self-grooming and allogrooming produces a fairly even body coverage. This supports a generally utilitarian function for badger grooming. The slight overrepresentation of the shoulder is the result of that being the most common site of grooming initiation during greeting, indicating an additional social function or a necessary social etiquette during such cooperative relations. R = rump, S = side, H = head, T = tail, B = belly, C = chest.

other simultaneously until one individual defects. The other will then retaliate to this defection, typically in less than half a second, by terminating allogrooming. No elaborate scorekeeping or partner recognition is required to guard against cheating in this cooperative system. (see figure 10.3a).

We hypothesize that the bartering of allogrooming is so direct because in the putatively primitive society of the badger, individuals spend most of their waking hours foraging separately from one another (Kruuk 1978). As a consequence, there are few opportunities to pay back beneficence in other currencies such as coalitionary aid. Perhaps more importantly, there are also few opportunities for badgers to exert manipulative pressure and extort grooming from other individuals by threat of negative reciprocity. An exception may help prove the rule: During the breeding season, when larger badgers can despotically control breeding opportunities around the sett, smaller individuals sometimes abandoned the "You scratch my back, I'll scratch yours" strategy and groomed larger individuals even without reciprocation (figure 10.3c). The only other exception to the direct reciprocation rule was found when mothers groomed cubs that were too young to reciprocate grooming, a behavior that doubtless accrues fitness benefits in terms of cub survival for the mother (figure 10.3b).

## Dominance

Terms such as *dominance* are used to describe predictable aspects of repeated dyadic interactions (i.e., relationships). The term *dominance* is a valuable concept made controversial largely by a proliferation of definitions; Drews (1993), having reviewed 13 definitions, concluded that dominance is characterized by a consistent outcome in favor of the same individual within a dyad, within which the opponent invariably yields rather than escalating the interaction. This excludes many uses of dominance and, in particular, does not allow the winner of a single agonistic interaction to be defined as dominant. A dominance hierarchy may be produced by ordering the dominance relationships between dyads. There may be a problem in determining what constitutes yielding as an outcome. If a limiting commodity is involved, then yielding may be allowing the "winner" deferential access to the commodity. Where no commodity is involved, a working definition may be used. For example, Chase (1982) considered an overall yielding response to occur if a hen chicken delivered any combination of three strong aggressive contact acts when there was more than a 30-minute interval after the third action during which the receiver of the actions did not attack the initiator. A certain degree of subjective intu-

# a

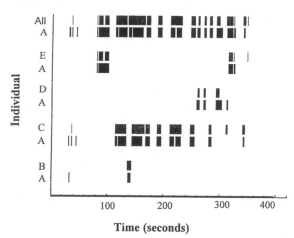

# b

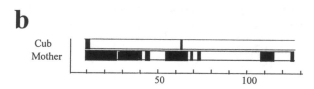

# c

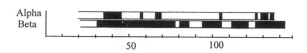

**Time (seconds)**

**Figure 10.3** Barplots of badger allogrooming behavior: (a) Typical pattern of simultaneously reciprocal allogrooming bouts for one individual (A) with other individuals (B, C, D, and E) and all combined (All) in a group-grooming huddle. (The grooming interactions between B, C, D, and E, are not shown. (b) Allogrooming interactions of a mother and her cub, showing the characteristically one-way beneficence. (c) Grooming of one adult male to a large adult female, one of the small minority of observations in which allogrooming between adults is not simultaneously reciprocal.

ition may go into such definitions; for example, fox cubs may launch quite ferocious attacks on their mother when food is under dispute, but the lack of reaction and effective supplanting of the mother at the food would not be interpreted by many researchers as a sign that the cubs are meaningfully dominant to the mother.

Dominance is often assumed to depend on body size, as a proxy for combative prowess, and this may generally be true (Lindstedt et al. 1986; Hansson 1992). However, this may sometimes be too simple an interpretation (Barbault 1988). For example, Berdoy et al. (1994) showed that among male Norway rats (*Rattus norvegicus*) age (itself broadly correlated with weight) was a better predictor of dominance than was weight. Age of male is also an important predictor of mate choice by female spotted hyenas (*Crocuta crocuta;* Hofer and East 1993). Advanced age, obviously a necessary corollary of good survival, may also be a better measure of fitness than current musclepower.

There is a tendency to assume too readily that all societies are arranged in a straightforward dominance hierarchy. Our own work with badgers revealed some of the complexities of elucidating and interpreting the dominance concept. We were interested in feeding interactions because the pattern of food availability and its manner of exploitation are believed to be central to badger social organization (Woodroffe and Macdonald 1995b). We investigated feeding dominance by establishing artificial food stations in the field (Stewart 1997). This experimental approach was deemed necessary because although dominance interactions may be key components of a social system, they may also be too rarely expressed to investigate with statistical rigor in a purely natural context. If a longer period is spent accumulating data, the requirement of stationarity may be breached and dominance relations may change during the observation period. Artificial provisioning experiments allow wild social groups with settled relationships to serve as subjects while improving the number, quality, and rate of observations. There are clear perils to the approach, however; ethically we had to ensure that injurious levels of aggression were not provoked (Cuthill 1991), and scientifically the general relevance of the experimental protocol had to be verified in an unmanipulated context (Wrangham 1974; Dunbar 1988). For this reason we maintained surveillance and control over the experiments using a live infrared video link and pursued further related observations in different social contexts.

We found that when badgers were presented with a single food source requiring contest competition for access, there was little direct evidence of default yielding to certain challengers and hence strict dominance relations: a feeding badger generally escalated aggression to some degree against any challenger. However, the study revealed predictors of contest success, so that we could predict relative supremacy in a dyad. These predictors of supremacy included length and weight and, to a lesser degree, age. These attributes alone were not sufficient to predict the direction of supremacy in a dyad, however, because the role of challenged feeding individual also proved to be associated with contest success. And even when an individual occupied the same role against the same adversary, we still could not fully predict the contest winner; subsequent rechallenges often had a different outcome, indicating that some motivational factor such as hunger can also play a decisive role. Hence, in the badger's case it did not prove helpful to attempt construction of a dominance hierarchy, linear or otherwise, from the provisioned patch data. A measure of relative supremacy based on an individual's characteristics and role in a contest proved to be more heuristically valuable.

To further complicate matters, it became clear that if more food sources were provisioned at one site, so that contest competition was no longer necessary for feeding access, scramble (Milinski and Parker 1991) became the preferred method of competition. This cast doubt on the relevance of our predictors of relative supremacy derived from contest competition because most natural badger foods are presented in a way that does allow such scramble competition. Interestingly, however, the behavioral patterns of contest we observed during feeding competition (e.g., side-to-side flank barging) were also seen in male–male competition for estrous females. Analysis of those data revealed that weight, length, and age as well as prior ownership were also predictors of copulation access, leading us to believe that these assets may indeed have general relevance in predicting the outcome of competitive dyadic interactions among similarly motivated badgers.

*Social Groups*

In terms of their constituent individuals, a group may be transient, as in herds of wildebeest, or nearly permanent, as in packs of wolves. Some may even move daily between these extremes; in the dry season, stable groups of capybaras may coalesce at water holes into transient herds (Macdonald 1981; Herrera and Macdonald 1989). Many groups can be defined spatially, in terms of the proximity that members maintain to each other. Within territorial species, cohabiting occupants constitute a spatial group, and may be brought together by limited or variably available resources or by sociological advantage (Macdonald and Carr 1989). Membership of spatial groups can be assigned using spatial criteria, within which indices of association can be derived from dyadic proximity interactions; Macdonald et al. (1987) report systematic patterns in the frequency with which members of a farm cat colony sit adjacent to other individuals. Among these cats, and among the aforementioned capybaras, the social status of some low-status individuals is defined by their spatial peripheralization. As Martin and Bateson (1993) noted, extra meaning can be given to such criteria if it can be shown that there is some aspect of synchrony or complementarity between the actions of individuals in each defined group or dyad relative to outgroup individuals. Clutton-Brock et al. (1982) used a behavior similarity index to quantify such complementarity. At a broader scale there may even be physiological synchrony within groups or between clusters of groups, as within packs of dwarf mongooses (Creel et al. 1991) or between packs of Ethiopian wolves (Sillero-Zubiri et al. 1998).

Spatially coherent groups can include a variety of types. Brown's (1975) schema, adapted by Lehner (1996), is useful in that context. It identified five broad categories of group: kin groups (families and extended families), mating groups (pairs, harems, leks, and spawning groups), colonial groups (e.g., nesting sea birds), survival groups (groups drawn to each other for reasons of foraging, herding, or huddling) and aggregation groups (formed by physical factors acting on animals to force them into one geographical location).

Although membership of a spatial group is a frugal expression of nonrandom associations, a social group should exhibit some persistent convergence of behavior or unified purpose that results directly from the association and interaction of the constituent network of individuals. That is not to say there will be no competition, antagonism, or discord within the group. It may be endemic. But to differentiate a social group (e.g., a pride of lions) from an aggregation (brown bears at a salmon run may be an example), there must be some element of cooperation (positive interaction) to offset the disharmony. Some populations of badgers are organized into groups that cohabit within the same territory, develop social relationships, and include close kin, yet for a long time there was scant evidence that they accrued any sociological benefit from their association. However as more has been learned of the social system of badgers, more potential advantages of social living have been found. These include winter huddling to conserve energy (Roper 1992), sporadic alloparental behavior (Woodroffe 1993), and occasional examples of coalitionary aid (Stewart 1997). We have even observed young individuals apparently using the main sett as an information center and trailing older adults to new feeding grounds after weaning. The important point to emerge was that none of these advantages appeared sufficient to explain social grouping, particularly because at the same time disadvantages of group living have been identified (Rogers et al. 1997; Woodroffe and Macdonald 1995b). Rather, these social advantages appear to be secondary benefits derived from, but not fully explaining, their group living habit. On one hand, this spatial congruity is distinct from the aggregation of jackals or bears at a carcass or salmon leap, and perhaps even distinct from the community of antagonistic monogamous pairs of maras (*Dolichotis patagonum*) at a communal breeding den (Taber and Macdonald 1992). On the other hand, the badger spatial group is equally distinct from the social group of African wild dogs, whose cooperative hunting or collective defense of prey leads sociologically to per capita advantage (Creel 1997; see also Packer and Caro 1997). Furthermore, the jackals meeting at a carcass may come from many different territories, but they may nonetheless meet frequently and have well-established social relationships. Indeed, as Macdonald and Courtenay (1996) showed, neighboring territorial canids may even be bonded by familial ties (see also Evans et al. 1989 for genetic evidence of links between adjoining groups of badgers). It is therefore difficult to formulate a precise definition, and social group risks becoming a nebulous concept. A social group constitutes a network of variously close affiliations: closely connected subclusters of individuals comprising subgroups, and groups connected by primary links between core individuals defined as supergroups. Clusters preferentially and mutually exchanging services, support, or aid may be said to display friendships (Smuts 1985), and where clusters act in concert against others we may define them as coalitions. Of course, the benefits an individual can accrue through associations are not always mutually beneficial, and Kraft et al. (1994) designed a diving-for-food experiment to illustrate theft in rat society. Rats were trained to dive into water in order to obtain food from another compartment. All individuals learned to dive; however, those that were able to steal effectively from their diving companions often opted for this less arduous means of securing food.

Macdonald et al. (1987) distinguished a hierarchy of questions to be tackled in quantifying cat sociality, and these could be adapted to a simple description of any society. First, on the basis of their individual comings and goings, what are the probabilities of a given dyad of cats being available to each other for interaction? Second, when they are simultaneously present, what proximity do they maintain? Third, what is the overall frequency of interactions between them, and how does that translate to a rate per unit time in association? Fourth, what are the qualities of interaction? Fifth, what are the directions of flow from initiator to recipient for each type of interaction?

First, indices of association raise the question of what level of proximity constitutes being together, and the answer may vary between species from physical to visual or olfactory contact (Martin and Bateson 1993). Clutton-Brock et al. (1982) classified red deer within 50 m of each other as in the same party, whereas farm cats generally position themselves within 1 m of the nearest neighbor (figure 10.4).

Indices of association can be calculated in which $x$ is the number of observation periods during which A and B are observed together, $ya$ the number of observation periods during which only A is observed, $yb$ is the number of observation periods during which only B is observed, and $yab$ is the number of observation periods during which both A and B are observed but they are not together (see figure 10.5). In some studies, as Ginsberg and Young (1992) pointed out, the proportion of individual A's time spent with B $[x/(x + ya + yab)]$ and the proportion of individual B's time spent with A $[x/(x + yb + yab)]$ may differ only because of a viewing bias concerning $ya$ or $yb$ (see also Cairns and Schwager 1987). However, in the case of the farm cats the value $x + ya + yab$ differs from $x + yb + yab$ because individuals were genuinely present at the resource center at different frequencies (figure 10.5).

This leads to a problem with the simplest association index $[x/(x + ya + yb + yab)]$ because A's association with B is very strong, whereas B's association

**Figure 10.4**  What constitutes proximity between individuals differs between species and may take account of physical contact, sight, sound, and smell. Farm cats within a matrilineal group may cluster closely, but with individuals positioned in different annuli with respect to the most central females, whereas two red deer stags might be interacting intimately over a distance much wider than that separating entire societies of cats.

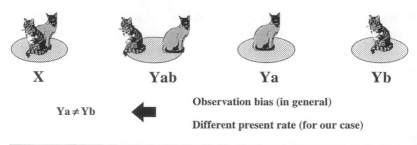

$$Ya \neq Yb \quad \blacktriangleleft \quad \text{Observation bias (in general)}$$

**Different present rate (for our case)**

| | Total scan | Present | Together | % together | Association index |
|---|---|---|---|---|---|
| Cat-A | | 40 | | 20% | |
| | 40 | | 8 | | 0.2 |
| Cat-B | | 10 | | 80% | |

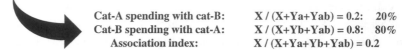

Cat-A spending with cat-B:    X / (X+Ya+Yab) = 0.2:    20%
Cat-B spending with cat-A:    X / (X+Yb+Yab) = 0.8:    80%
Association index:    X / (X+Ya+Yb+Yab) = 0.2

**Figure 10.5** Considerations in scoring indices of association are illustrated by the comings and going of farm cats at a resource center. Observations may be biased by the observer's success in seeing them or by genuine differences in individuals' attendance records, but the time for which the two cats are simultaneously present defines the opportunity for interacting in the calculation of the example association index.

with A is not necessarily so, and vice versa. For example, if cat A is at the resource center 40 times out of 40 scan censuses and cat B is there 10 times out of the same 40 scan censuses, and they actually sit together during 8 scans, then according to the simple association index, their association index is $[8/(8 + 30 + 0 + 2)] = 8/40 = 0.2$. However, looking at this another way, it is also true that while at the resource center cat A sits with cat B only 20 percent of its time, whereas cat B spends 80 percent of its time sitting with cat A. This asymmetry is accommodated if we calculate an association index for each cat rather than for each pair of cats. Thus cat A's association with cat B is $[x/(x + ya + yab)] = 0.2$, whereas cat B's association with cat A is $[x/(x + yb + yab)] = 0.8$.

Kerby and Macdonald (1988) and Macdonald et al. (in press) took scan samples of presence and proximity of cats at 30-min intervals at three colonies (small, medium, and large memberships), and made ad libitum records of the social interactions of focal cats. Data were summarized for each cat each month, and statistical analyses used these individual summary scores to avoid problems associated with pooling data (Machlis et al. 1985; Leger and Didrichsons 1994). Some generalizations spanned all colonies: the sexes did not differ significantly in their average monthly rate of presence, whereas age classes did; juvenile cats had the highest average monthly rate of presence, and adult males the lowest. Furthermore, each sex–age class displayed a favorite nearest-neighbor distance in terms of its tendency to maintain a given level of proximity to other cats (figure 10.6).

At the large colony, when kittens were within 1 m of an adult female, this was their mother on 71.1 percent and 58.3 percent of occasions for male and female kittens, respectively, whereas adult males tended to be within 1 m of male juveniles most frequently and of male kittens least frequently. Despite such generalizations, there were also significant differences in the pattern of proximity between the colonies. For example, at the large colony, adult males were significantly more frequently within 1 m of adult females, but at the medium-sized colony no such tendency emerged. A further corollary of spatial structure was that certain females occupied spatially peripheral positions, whereas others lived centrally, and the latter had significantly higher reproductive success (figure 10.7).

Taking into account indices of association, it is possible to explore the frequency, rate, and quality at which cats interact. Take three cats, cat A spending twice as much time in the company of cat B as of cat C (figure 10.8). If A rubs on B and C 40 and 8 times respectively, and attacks them 10 and 2 times respectively, then A rubs on and attacks B at only 2.5 times the rate at which it rubs on or attacks C, although A rubs and attacks B five times as often as C. The quality of A's relationship with B and C is the same in that 80 percent of its interactions with both are rubbing and 20 percent are attacking.

In reality, at Kerby and Macdonald's (1988) large colony, individual central males averaged 602 (±254, SE) initiations relative to 502 (±142) for peripheral males, but the latter were present so seldom that their hourly rate of interaction when present averaged 6.9 (±1.5), in comparison to the central males at 2.3 (±0.7). The situation was opposite for females: Peripheral females interacted at low frequency, and when their presence was taken into account, they interacted at an even lower rate. Furthermore, in the context of grooming, Macdonald et al. (1987) showed that whereas the flow of licking differed significantly between pairs of interacting cats, it was invariably symmetric. In marked contrast, the flow of rubbing, though also differing between dyads, correlated with aggression but not grooming and was generally highly asymmetric. They concluded that in the society of female cats, rubbing as a measure of status flowed centripetally toward central dominant individuals; grooming, which was exchanged reciprocally, was a measure of social (and often genetic) alliances, but unrelated to status (figure 10.9).

## A    Proximity to the nearest neighbour

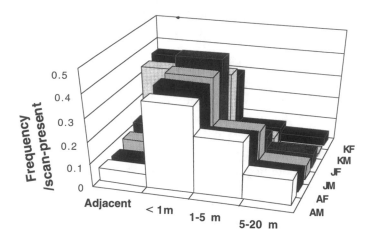

## B    Adjacent neighbour of each age-sex class

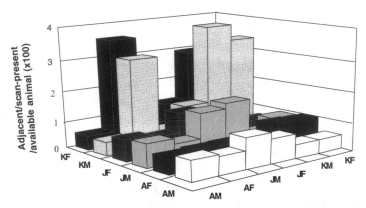

**Figure 10.6** Exploration of patterns of spatial proximity can reveal some unexpected structure to what appears superficially to be a random gathering. In the case of farm cats, these graphs, based on G. Kerby's field study in Macdonald et al. (in press), reveal (A) that different age–sex classes of individuals positioned themselves at significantly different distances to their nearest neighbors and (B) that these positionings differed significantly depending on the age–sex class to which that neighbor belonged. AF = adult female, AM = adult male, JF = juvenile female, JM = juvenile male, KF = female kitten, KM = male kitten.

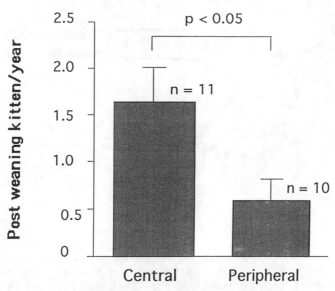

**Figure 10.7**  An ultimate goal of many ethological descriptions is to translate indices of social behavior into evolutionary consequences. In the case of Kerby and Macdonald's (1988) and Macdonald et al's (in press) farm cat study, the spatial arrangement of female cats around the farmyard resource center was correlated with the number of kittens raised annually per female.

Macdonald et al.'s (1987, in press) and Kerby and Macdonald's (1988) cat study shows how even simple quantification can reveal unexpected layers of structure in unsophisticated mammalian societies.

In summary, to define social groups and to describe social dynamics one must describe interactions that may be positive, tolerant, or negative in terms of their consequences for those involved. However, the choice of interaction type, or of the way of quantifying it, may radically affect the researcher's interpretation of the outcome. This should not surprise us. An analogy with human social dynamics shows us that very different patterns of interaction appear if we view exchange of such commodities as money, conversation, or affection. And the picture changes yet again if we look at frequency rather than quantity or quality of the exchange. The most pernicious problem lies in correct interpretation of the context of the interchange. Is the individual to which the most money is observed to be given beloved kin, despised extortionist, or scarcely known shopkeeper?

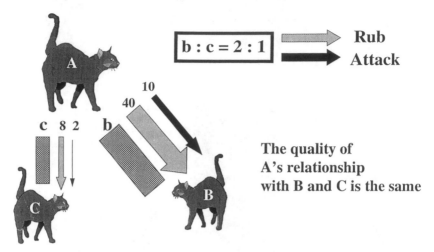

**Figure 10.8**   The same observations of social interactions can be expressed in three ways: as the total flux of a given behavior pattern, as a rate, or as a proportion. In this case, the total frequency is greater between A and B than between A and C (b > c), but qualitatively the components of their relationship are the same.

## ■ Methods for Behavioral Measurement

Although our goal is not to provide an encyclopedic guide to the formidable practical problems faced by the field biologist, we briefly review several methodologies relevant to measuring social dynamics.

### IDENTIFYING THE INDIVIDUAL

From the variety of ways of identifying individuals reviewed by Stonehouse (1978), it is generally preferable to use natural characteristics. This necessitates validation of identification skills and it is imperative that behavior is not inadvertently used as a cue to identity. For scientific as well as ethical reasons, it is also important to be alert to any influences of a marking technique on subsequent behavior. For example, in an analysis of a new fur-clipping technique used to mark badgers for individual identification, Stewart and Macdonald (1997) assessed the effect of the clips on body condition using a matched-samples test. Our concern was that the technique could cause thermoregulatory disadvantage, and no significant effect of clipping on condition was found. Furthermore, marked individuals were capable of attaining high status in their group, as measured among males by copulation frequency during female

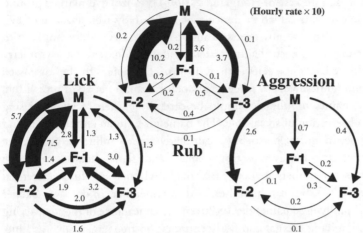

# A Asymmetry of interaction

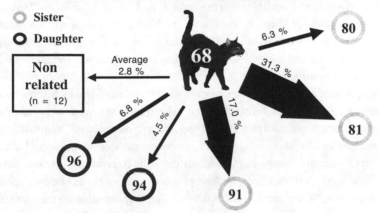

# B Aggressive interactions of adult female 68 toward her adult female relatives

**Figure 10.9** (A) The flow of rubbing between a group of four cats studied by Macdonald et al. (1987) was much less than that of grooming, and the two differed in that rubbing relationships tended to be highly asymmetric whereas grooming ones were symmetric. Furthermore, whereas the flow of grooming tended to mirror indices of association, those of rubbing and aggression did not, but the latter two were correlated with each other but not with grooming. (B) Superimposed on these patterns, relationships were modified by kinship and age; for example, female 68 (in Kerby and Macdonald's [1988] large colony) interacted more with her sisters than her adult daughters, and more with these 5 close kin than with the 12 other females available to her.

estrus. It is logically impossible to be certain that clip marks had no effect on behavior, but as in the case of Hoogland's (1995) Nyanzol dye-marked ground squirrels, we could detect no evidence that it did. However, absence of evidence is not evidence of absence, so caution is necessary. For example, bird rings or ear tags have been demonstrated to affect judgments of asymmetry and mate choice in some species (Burley 1988). This topic is further discussed by Bekoff and Jamieson (1997) (see also Moehrenschlager et al. in press). In a similar way, names of animals should be carefully chosen to avoid biasing behavioral observation of social role. The name given to an animal, be it that of a colleague, friend, or personality, can inadvertently influence all but the most objective observer. Even ascribing "male" and "female" names can cause recording biases. Numbers are better (although "A1" may still find deferential recording to "C5"), but can be rather hard to remember. In our own badger study we primarily used names derived from the appearance of the identifying fur clip mark, giving asexual and fairly connotation-free names such as Line, Dot, Dash, and Corners.

### SAMPLING AND RECORDING RULES

Sampling rules dictate which subjects to watch and when. Martin and Bateson (1993) distinguish ad libitum, focal, scan, and behavior sampling. Recording rules specify how the behavior is to be recorded. The two options are continuous recording and time sampling. Time sampling may be undertaken on an instantaneous or one–zero basis.

The qualities of each sampling rule are easily envisaged through the eyes of contrasting onlookers at a football match. The typical spectator engages in ad libitum sampling and samples everything. He or she is likely to record the goals and any spectacular fouls, with a bias toward the fouls committed by certain star players and those committed on the ball. In contrast, a talent scout might be a focal sampler, observing one player at a time. He or she may gain a poor impression of the game as a whole, but a clear impression of the qualities of particular players. Of course, the scout must be alert to biases such as the stage of the match at which a given player becomes the focal subject (it would be inappropriate to compare one player observed in the first 5 min with another observed in the final 5 min). A team manager might opt for scan sampling, censusing the behavior of each player in turn, at rapid intervals, to get an overall impression of how the team functions as a unit. Finally, a gambler, interested only in the outcome of the game, may undertake behavior sampling, recording only goals. By focal or scan sampling one could miss these events.

Turning to recording rules, a continuous record contains the times at which each behavior started and ended, and thereby reveals true frequencies and durations. Instantaneous sampling characterizes the current behavior at predefined intervals. One–zero sampling records whether a behavior happened in one particular interval of defined length. The most common combinations are focal sampling with a continuous record, scan sampling with an instantaneous record, and focal sampling with a one–zero record.

## AD LIBITUM SAMPLING

An obvious bias in ad libitum sampling would arise if a researcher's presumptions led to particular individuals being disproportionately represented in the sample, thereby distorting their importance. This effect has been credited with the overestimation of male roles in the societies of some primates and may subsequently have been countered by contrasting biases (Lindquist and Bernstein 1987; Morell 1993). Furthermore, individuals or behavior patterns that are less conspicuous (because of short duration, subtlety, or association with obscured locations) tend to be underrepresented by ad libitum sampling. For example, Bernstein (1991) demonstrated that using ad libitum sampling of rhesus monkeys (*Macaca mulatta*), the frequencies of contact aggression and avoidance were underestimated whereas that of chasing was overestimated. A related and insidious bias may affect the initial choice of study group. Sharman and Dunbar (1982) noted that studies of baboon behavior tend to involve larger than average groups. This erodes the generalizability of the results because the behavior of individuals was systematically influenced by the size of the group. In the same vein, perhaps the bias of primatologists' expectations for their own taxon has lead to primates being described as having peculiarly complex social lives (Rowell and Rowell 1993). However, other taxa also emerge to have intricate societies, an obvious example being the Carnivora (Creel and Macdonald 1996). Despite this family of biases, ad libitum sampling can be useful during the naturalistic observation phase of pilot studies and is valuable as an adjunct to more rigorous protocols that might otherwise exclude important events that fell beyond their scope.

## FOCAL SAMPLING

Because the whole attention of one observer is likely to be absorbed by one focal subject, focal sampling risks losing valuable contextual and simultaneously comparable information. The temptation to follow one focal individual for longer than is necessary reduces the effective data set. Having observed a focal chimp for 300 hours, Kawanaka (1996) discovered that subsets of 25 hours reliably and consistently represented the same time budget (see also Arnoldmeeks and McGlone 1986). Excessive recording of particular focal individuals also worsens the task of finding periods of relative stationarity. Cycling through a series of focal individuals and then checking an individual against itself in an earlier period may circumvent this. Video recording can sometimes relax these limitations. There may be a general warning in the finding of Arnoldmeeks and McGlone (1986) that focal animal sampling of one individual among a litter of young pigs provided a good measure of the time budgets of other piglets in the same period for all but social behavior.

## TIME SAMPLING

### One–Zero Sampling

One–zero sampling consistently overestimates duration and gives unreliable information on frequency (Dunbar 1976; Altmann 1974). This can be partly overcome by reducing the recording interval to periods shorter than the bout length of the behavioral categories, but the technique then often becomes more cumbersome than continuous focal sampling. Therefore, this once common sampling method has largely been abandoned. Using a cumulative Poisson process, it is possible to assess, for a given interval length, the probability that the necessary conditions will not be met for accurate frequency counts and unbiased duration estimates in one–zero sampling (Suen and Ary 1984; 1986a). Suen and Ary (1986b) also identified a post hoc correction procedure that produces results with negligible systematic errors in one–zero duration estimates. This correction procedure requires that more than five 0 scores lie between two consecutive but not adjacent 1 scores (see also Quera 1990).

Despite its general fall from favor, one–zero sampling has some potential virtues. It gives a measure of the probability that an observer will see at least one example of the response in the specified time interval. Bernstein (1991) argues that this probability may be relevant to an animal's perception: For example, when an individual is deciding whether to spend the next 5 min in the vicinity of another individual, the probability of attack may be more important than the mean duration or frequency of attacks. Martin and Bateson (1993) also note that the onset, duration, and frequency of some behavior, such as play, are hard to represent with continuous sampling and hard to score during true instantaneous sampling. The judgment that play has at least occurred in a previously defined interval is more likely to be accurate. Finally, some behavior, such as copulation, may occur often in one short interval and then not be seen for many subsequent intervals. One–zero scan sampling reveals how long a researcher is likely to have to watch the animals to see an instance of such behavior.

## Instantaneous Scan Sampling

This technique scores what some or all individuals are doing at predetermined time intervals. It produces a reliable estimate of the total durations of activities through a dimensionless measure of the proportion of occasions during which the behavior is encountered (Hansson et al. 1993). Instantaneous scans may give poor information on frequency, although this shortcoming is minimized if the sample interval is small enough to catch even the shortest intervening period between recurrence (Odagiri and Matsuzawa 1994). However, as in the case of short one–zero measures, such short bouts become more inconvenient than focal sampling. Furthermore, the need for statistical independence of the data points requires that sample intervals not be too short. For example, if 10 successive scans all fell within a single long bout of a particular behavior, then they constitute a sample of 1, not 10. Such dependence would require that the data be transformed to a proportion on an interval scale for comparison rather than being used as discrete categorical scores (and therefore could not be analyzed in an interaction sociomatrix). As a rule, Bernstein (1991) suggests that if the probability of scoring the same continuous state twice in succession is less than 0.05, then the two scans can be treated as independent. In practice this means that the interval between scans should exceed the mean duration of a state plus 1.96 standard deviations. Even long intervals between scans can provide a basis for accurate estimates of mean time budgets, but the increase in variance caused by longer sampling intervals may swamp analyses of covariation and produce type II errors (Poysa 1991).

The pros and cons of each approach, and the diverse aims of investigations, mean not only that no particular method is universally superior, but that even for one study an eclectic approach to sampling is likely to optimize data acquisition. A mix of all occurrence sampling, scan sampling, and focal individual instantaneous sampling is possible. The results of all three are likely to be enhanced if decisions on the durations of intervals and focal sessions are judged from preliminary analysis of a continuous focal sample (e.g., made by using a portable video camera in the field).

## TECHNIQUES FOR BEHAVIORAL MEASUREMENT

Simultaneous and continuous focal sampling of all individuals in a group might be the most accurate and informative combination of sampling and recording, but circumstances may render this combination unattainable. Technology is at hand to help. Computer-based event-recording software provides an automated string of actor, behavior (time), and modifiers (such as recipient of interaction and quality of interaction). Currently, memorizing the keyboard is the way to avoid taking one's eyes off the subjects, but voice recognition software may soon allow input of data spoken in the correct syntax. Various software packages already assist behavioral recording, and the most sophisticated, such as Noldus Observer, offer a variety of tools for manipulation and basic statistical analysis of the data and an easy Windows interface (Albonetti et al. 1992; Noldus Information Technology 1995). Noldus also performs sequential and nested analysis and allows simultaneous input from an alternative digitized source (e.g., heartbeat or external temperature).

Two caveats concern widespread adoption of sophisticated software. First, it may encourage a lazy uniformity in data acquisition that stifles innovation in sampling and recording rules or analyses. Second, a researcher's new-found ability to record a plethora of behavior and contexts may exceed his or her ability to frame hypotheses and manipulate and analyze the data.

There are circumstances under which video recording is an economical alternative to direct observation and offers some unique advantages. The video record can be played and replayed, allowing multiple passes over complex sequences at a resolution of up to 50 frames per second. With a microphone, complex vocalizations can be spectrographically analyzed in context (Wong et al. 1999). The video and sound sequences may be digitized and stored using a PC and video capture board; this allows efficient indexed archiving of raw data and use in electronically distributed ethograms. Video can also be time-coded and used with event recording packages to facilitate accurate data entry or even allow automatic data entry for movement, interaction, and some simple postural categories. Video can be used to calculate interobserver and intraobserver reliability, including the drift in ethogram categorization which often occurs in long studies. Limited use of video in pilot studies can allow the disparity between different behavior sampling techniques and a true continuous record to be assessed.

In automatic field use, video allows easy habituation for many species (including facility to use infrared light to provide illumination beyond the visual range of most mammals). Continuous surveillance of multiple focal sites becomes a possibility for even a single researcher, with the equipment operating in conditions that would be too arduous for the researcher to endure for protracted periods. Against these undoubted benefits video equipment can still be initially costly, with problems of equipment reliability, accessibility, and security. Video also has a more limited field of vision than a human observer, and because it requires transcription after the event, it may increase the duration of studies. Stewart et al. (1997) describe the use and construction of automatic video surveillance units, and Wratten (1994) provides a general background to video techniques.

Finally, although the naturalist's preference may be to observe social behavior directly or with video, this may be uneconomical or unfeasible. Examples of indirect measures include the frequencies of bite wounds (Woodroffe and Macdonald 1995a), chewing of ear tags (White and Harris 1994), and hunting patterns from carcass retrieval (Johnsingh 1983). Radiotracking has revolutionized the study of elusive mammals. Although initially used principally to plot movements, the use of radiotracking to supplement classic fieldcraft and as an aid to observation has long been advocated (Macdonald and Amlaner 1980). It is also possible to infer features of social behavior from the dynamic interaction between the movements of radiotracked individuals (Macdonald et al. 1980; Doncaster 1990). The practicalities of radiotracking are reviewed in Kenward (1987) and analytical considerations in White and Garrot (1990). An increasing number of software packages for analyzing tracking data are available, such as Wildtrak (Todd 1992) and Ranges V.

## ■ Analysis of Observational Data

Context must always have been important in human interpretation of the behavior of other animals. Even our own aphorisms (blood's thicker than water, do as you would be done by, and so on) reveal an intuition that presages modern conceptual insights regarding the forces fashioning animal societies, including our own. Today, biologists interpret behavior within a context fortified by theories of cognition, games, trait-group selection, kin selection, reciprocity, and byproduct mutualism. The traditional approach to observing animals, with roots embedded in the Aristotelian idea of ascribing properties to animals, could yield detailed ethograms that demonstrated the defining behavioral uniqueness of each species. Interspecific comparisons of these behavioral lexicons shed light on phylogenies even before the days of phylogenetic regression (Tinbergen 1951). The quest for more robust tools with which to elucidate the shared properties between species has, in the last 20 years, contributed to the blurring of boundaries between studies of ecology, ethology, and socio-biology. Today, the basic questions posed by those concerned with the contextual interpretation of animal behavior still echo Niko Tinbergen's (1963) four factors: causation, development, survival value, and evolution. The classic sequence starts with observation to document the behavioral repertoire in context, and this forms a basis for experimental study of causation, development, and survival value and an attempt to reconstruct the evolution of particular units of behavior (McFarland 1981). The approach rests on parsimony and reductionism. Of course, there are different forms of reductionism, and the biologist's variant, as described by Konrad Lorenz in *The Foundations of Ethology* (1981), may deviate from the physicist's method of general reduction.

Here, we consider some limitations of the rationales behind four concepts widely used in the study of social dynamics: statistical rationality, matrix facilities, lag sequential and nested analysis, and concept of uncertainty measures (Markov chain analysis).

## STATISTICAL RATIONALITY

A society is the product of social flux among its members and between them and outsiders. Therefore, for many purposes the description of a society requires quantification of the interactive components of behavior between members of, for example, a reproductive unit, group, colony, or population; these may be continuously or instantaneously recorded. In the analysis of observational data, insufficient attention to methodology can result in a rift between statistical significance and biological relevance, whereas the correspondence between these two can be fostered by attention to sampling tactics and statistical sensitivity. An inadequately sensitive statistical method or model may fail to reveal the properties of an interactive behavior, even if an appropriate sampling method was used.

The familiarity of two types of error has rendered neither rare. Type I error involves mistaken rejection of a null hypothesis (i.e., a false positive); type II error is failure to reject a false null hypothesis (i.e., ignorance) (Lindgren 1968). As a general statement, scientific validity depends partly on the sensitivity of the tests involved (decreasing the probability of type I errors) and on the number of hypotheses tested (the accumulating building blocks of the argument adding to its objectivity and decreasing the probability of type II errors). However, there is a tradeoff between these kinds of error; decreasing type I error can increase the risk of type II error (Shrader-Frachette and Mc-Coy 1992). There is also a risk that even when the hypothesis has been tightly phrased, a test may pick up "departures" from $H_0$ that result from "noise" rather than real differences. Matloff (1991) describes how confidence interval analysis may partly solve these problems. An advanced account of these topics is given in Thompson (1992) and Krishnaiah and Rao (1988).

## MATRIX FACILITIES: ANALYZING SEQUENTIAL DATA

A matrix is any rectangular array of numbers (e.g., frequencies, number of transitions). If the array has $r$ rows and $s$ columns it is called an $r$ by $s$ ($r \times s$) matrix. Thus for example,

$$
A (3 \times 3) \begin{vmatrix} a11 & a12 & a13 \\ a21 & a22 & a23 \\ a31 & a32 & a33 \end{vmatrix} \text{ and } B (2 \times 3) \begin{vmatrix} a11 & a12 & a13 \\ a21 & a22 & a23 \end{vmatrix}
$$

where each row can represent the actor entry and columns the receiver; that is, the value $a11$ represents frequency (or other value) of relations between the row descriptor ($r1$) and the column descriptor ($s1$).

Our task here is not to explain the mathematical principles of matrices (see Roberts 1992), but to describe their uses in behavioral sciences. Interaction frequencies among individuals, presented as a sociometric matrix, are often the basis for analysis of social dynamics. Individuals are classed as actors or initiators in the rows ($r$) of the matrix and as receivers in the columns ($s$).

Other sociometric matrices include distance and association matrices; in these, each cell contains a symmetric (dis)similarity measure for the pair of animals indicated by the row and the column of the cell. The cells of a transition matrix contain the frequencies with which the behavior indicated by the row (the preceding behavior) is followed by the behavior indicated by the column (the succeeding behavior) (de Vries et al. 1993).

Relationships are often presented in terms of the dyadic interactions between two individuals of the group. Frequencies of interactions within the dyad during defined time intervals are then written in sociometric matrices. These types of sociometric matrices are often used to identify the dominance hierarchy among individuals based on such key behavior patterns as initiating aggression. One individual is defined as a dominant (the winner), the other as a subordinate (the loser). The strength of the relationship is expressed in terms of linearity (Appleby 1983). Han de Vries (1995) developed an improved test of linearity in dominance hierarchies containing unknown or tied relationships. This test is based on Landau's linearity index, but takes the unknown and tied relationships into account. An unknown relationship commonly arises because two members of a dyad simply were not seen to interact. This can occur for at least four reasons: sampling may be inappropriate, linearity of the dominance hierarchy may be an artifact of experimental circumstances (e.g., dispersal cannot occur in captivity), the formation of coalitions may lead to mutual protection or mutual avoidance, and linearity may not be a feature of the group's social structure (ultimately because the ecological circumstances offer no selective advantage to linear dominance with regard to, for example, resource holding potential). However, papers acknowledging the first two categories of problem are inevitably few.

The problems of elucidating a hierarchy and revealing its consequences are illustrated by wood mice (*Apodemus sylvaticus*). Although some aspects of this species' behavior can be studied in the wild (Tew and Macdonald 1994), their small body size and large ranges thwart many studies and make them strong candidates for observation in arenas (Plesner Jensen 1993) or seminatural enclosures (Bovet 1972a, 1972b). Bovet (1972b) found that male wood mice of different statuses tended to be active at different times. Therefore, they rarely met, which thwarts study of their encounters. Such problems can be partly circumvented by the use of sociometric matrices. Because it illustrates the use of matrices, we will go step by step through Stopka and Macdonald's (in press) sociometric study of wood mice. Based on videotaped interactions of nine mice observed over 3 months, the frequencies of transition between fight or chase and avoidance for all individuals in the colony were ranked (table 10.1).

Each of these matrices can be analyzed for linearity in the relationships between the mice (a task swiftly performed by options in the package MATMAN). In the case of avoidance (table 10.1), this provides a statistic (Kendall's linearity index $K = 0.775$), which indicates that the null hypothesis that avoidance relationships are randomly distributed can be rejected in favor of the alternative that they are linearly ordered ($\chi^2 = 43.76$, $df = 20.16$, $p = 0.005$). Similarly for allogrooming ($K = 0.592$), there is significant linearity ($\chi^2 = 34.96$, $df = 20.16$, $p = 0.03$). However, linearity in avoidance relationships is stronger ($p = 0.005$) than that in allogrooming relationships ($p = 0.03$), raising the possibility that linearity in grooming merely reflects that in avoidance. The statistical tool to test this hypothesis is the $Kr$ row-wise matrix correlation test (de Vries et al. 1993; de Vries 1995). When applied to the avoidance and allogrooming matrices (in which the rows and columns are identically ordered), this reveals a

**Table 10.1 Matrix of Frequencies of Transitions from Fight to Avoidance (top) and Frequencies of Allogrooming Behavior (bottom) Among Nine Wood Mice**

| | | Receiver (avoid) | | | | | | | | |
|---|---|---|---|---|---|---|---|---|---|---|
| | | *M*1 | *M*2 | *M*3 | *F*1 | *F*2 | *M*4 | *F*3 | *F*4 | *M*5 |
| | *M*1 | * | 52 | 180 | 9 | 18 | 192 | 11 | 66 | 26 |
| | *M*2 | 8 | * | 22 | 0 | 6 | 15 | 3 | 11 | 0 |
| | *M*3 | 10 | 15 | * | 6 | 10 | 18 | 4 | 10 | 16 |
| | *F*1 | 0 | 0 | 0 | * | 1 | 0 | 0 | 1 | 2 |
| Actor (fight) | *F*2 | 1 | 1 | 6 | 0 | * | 6 | 2 | 8 | 7 |
| | *M*4 | 2 | 4 | 4 | 1 | 3 | * | 4 | 4 | 4 |
| | *F*3 | 0 | 2 | 3 | 0 | 1 | 0 | * | 0 | 3 |
| | *F*4 | 5 | 3 | 5 | 0 | 4 | 3 | 0 | * | 1 |
| | *M*5 | 0 | 0 | 1 | 0 | 0 | 0 | 0 | 0 | * |

| | | Receiver | | | | | | | | |
|---|---|---|---|---|---|---|---|---|---|---|
| | | *M*1 | *M*2 | *M*3 | *M*4 | *M*5 | *F*1 | *F*2 | *F*3 | *F*4 |
| | *M*1 | * | 1 | 0 | 3 | 0 | 62 | 15 | 124 | 13 |
| | *M*2 | 0 | * | 0 | 0 | 0 | 6 | 3 | 8 | 6 |
| | *M*3 | 0 | 0 | * | | 0 | 19 | 4 | 2 | 14 |
| | *M*4 | 1 | 0 | 0 | * | 0 | 4 | 1 | 8 | 4 |
| Actor | *M*5 | 0 | 0 | 0 | 0 | * | 0 | 0 | 0 | 0 |
| (groom) | *F*1 | 6 | 1 | 2 | 2 | 0 | * | 0 | 1 | 3 |
| | *F*2 | 0 | 0 | 3 | 0 | 0 | 0 | * | 0 | 0 |
| | *F*3 | 14 | 0 | 1 | 1 | 0 | 0 | 0 | * | 1 |
| | *F*4 | 1 | 1 | 0 | 0 | 0 | 1 | 0 | 0 | * |

From Stopka and Macdonald (submitted, b).

strong negative correlation ($p$-right = 0.941). Therefore, the dominance (avoidance) and grooming matrices exist independently. Tendencies to allogroom and avoid differ between males and females; in practice, lower-ranking males are rarely groomed and they avoid dominant males and are avoided by females, whereas females are the main recipients of grooming, which is mainly initiated to them by dominant males. This pattern led us to conclude that in order to obtain matings, male wood mice have to groom females. This hypothesis is supported by the positive row-wise correlations ($p < 0.001$), which reveal that males that are dominant and groom females most secure most copulations.

## LAG SEQUENTIAL AND NESTED ANALYSIS

Sequential analysis facilitates study of the temporal structure of sequences of events. A common variant is lag sequential analysis, which involves calculating frequencies of transitions between pairs of events within a certain lag in a time series. The first event of each pair is called the criterion event (also called antecedent, X-event, or given event) and the second event is the target event (also called the consequent or Y-event). Lag sequential analysis allows you to answer questions such as "How many times is the (criterion) event Head Up followed by the (target) event Stare?" The time window in which the transition from criterion to target event occurs is either a state lag or a time lag. In a state-lag sequential analysis, a transition is counted from a criterion event to the first target event following this criterion event. In time-lag sequential analysis, a transition is counted from a criterion event to the target event occurring within a specified time window following the criterion event. All other events that are not defined as either criterion or target events are ignored in the analysis. Lag-sequential analysis differs profoundly from other approaches (such as log-linear models), which compare estimated expected frequencies with observed frequencies because these do not have a sequential property. The only way to analyze changes of behavior through time is in terms of transitions between the antecedent and the consequent behavior. Lag sequential analysis is currently the only method that applies a contextual approach: Instead of analyzing the frequencies of different behavior over a certain period of time, lag sequential analysis uses the frequencies of behavioral transitions (Roberts 1992).

However, the results of lag sequential analysis are rarely expressed in terms of the absolute frequencies of transition from one event to the others; instead, they are often written in probabilistic terms as transition rates. The transition rates, written in transition matrices, can be calculated from the frequencies of transitions between preceding and consequent behavior using equation 1.6 in Haccou and Meelis (1992), which is formally written as a maximum likelihood estimator (MLE)—equation 10.1 or 10.2—when duration of an act is constant:

$$\alpha_{AB} = \frac{N_{AB}}{N_A} \times \frac{1}{x_A} \qquad (10.1)$$

or

$$\alpha_{AB} = \frac{N_{AB}}{N_{a}} \qquad (10.2)$$

where $\alpha_{A,B}$ is the estimator of transition rates, $N_{A,B}$ is the number of transitions from the state A to state B, $N_A$ is the number of states A, and $\bar{x}_A$ is the mean bout length of category A. MLE values are written in the state-space representation (also known as flow diagrams), that is, a set of states and transition rates between them.

In corroborative analyses at least two methods are currently used to describe behavioral processes: multiple-matrix analysis and nested analysis.

Multiple-matrix analysis involves the construction of several matrices, one each for a variety of measurable parameters of social flux (e.g., dominance hierarchy, mutual grooming, approach and avoidance, acts of support, copulations, and paternity). The pattern of flow between individuals for each behavior pattern can then be compared between matrices using, for example, row-wise matrix correlation (association) or linearity indices (de Vries 1993). This approach provides answers to questions such as "Are dominant males groomed more frequently than subordinate males, or do females given protection by other females reciprocate this support?" This can involve construction of several matrices and row-wise correlations between them.

Nested analysis, on the other hand, involves defining questions within matrices. Subsets (nested categories) of data are selected within the overall matrix. For example, a researcher might ask, "Is there a tendency for juvenile males to be more successful in contests with rivals when positioned close to their mothers?" To answer this, the researcher would analyze the outcomes of fights when key individuals were adjacent to, or distant from, their mothers.

Both multiple matrix and lag sequential or nested analyses may involve the manipulation of very large data sets, and software is available to make this task more manageable. PC programs such as the Observer event recorder lag sequential analysis (Noldus 1994) and the MATMAN program (de Vries et al. 1993) are designated for recording and analyzing such data.

## SEARCHING FOR A BEHAVIORAL PATTERN (MARKOV CHAIN)

It is useful when analyzing sequential data to characterize the chance of one behavior following another. If exclusive acts (events or states) in a sequence are independent of each other, then the distribution of bout lengths will follow an exponential distribution; such a sequence is said to have Markovian properties. Such a process is completely characterized by a set of transition rates, which are the probabilities per unit of time of switching from one state to another. By definition, acts within such a sequence are independent of each other and of time. The termination rate of an act is constant (i.e., independent of the past) only if the duration of the bouts follows an exponential distribution. This fact allows the inference of social interaction because dependencies between the terminations of different individuals would imply that they exchange signals (Haccou and Meelis 1992). Indeed, most sequences are not Markovian because the probability that most behaviors will occur depends on earlier events in the sequence and other confounding variables. Generally, for a given behavior, the probability of occurrence of any future behavior in the sequence is altered by additional knowledge concerning the past (Haccou and Meelis 1992). The program UNCERT (Hailman and Hailman 1993) analyzes serial dependencies in sequential events using the method of Markov chains and expressed in terms of the uncertainty that one event will follow another. For example, Hailman and Dzelzkalnz (1974) used Markov chain analyses to reveal that tail-wagging by mallard duck (*Anas platyrhynchos*) punctuated a long sequence of behavior patterns, indicating that a display is about to start (loosely analogous to the role of a capital letter beginning a sentence).

## PREDICTABILITY OF BEHAVIOR

Measures of uncertainty can be a valuable tool in assessing the predictability of responses of individuals to others. Again, we explore an example from our work on wood mice (Stopka and Macdonald, unpublished data). For example, all combinations of five male and four female wood mice kept under video surveillance in an enclosure with abundant nest boxes sometimes denned communally, although there was no thermoregulatory need to huddle. It was therefore surprising that cohabitants were often agonistic to each other. One of the most common categories of interaction was approach and nose-to-nose contact, categorized by Bovet (1972a, 1972b) as amicable, but lag sequential analysis revealed that although this (nose-to-nose) behavior pattern was indeed sometimes a precursor to flank-to-flank or nasoanal contact, on other occasions it was followed by attacking or chasing. Therefore, nose-to-nose encounters are defined by their context within a sequence. Sequential analysis revealed that nose-to-nose contact was sometimes the prelude to an amicable encounter and sometimes to an aggressive one (figure 10.10).

The crucial question is how to recognize mixed categories. Haccou and

Meelis (1992) recommend: performing a global inspection of bar plots to ensure stationarity, testing the bout lengths of all acts in the ethogram for exponentiality (e.g., Darling's test), using cumulative bout length plots and the likelihood ratio change point test to reveal deviations from exponentiality attributable to lack of stationarity, and devising a rational basis (in the case of nose-to-nose in wood mice, lag sequential analysis) on which to split problematic categories into new robust categories. Other methods of splitting the behavior according to the bouts are based on fitting a nonlinear curve to the logarithm of the observed frequencies of gap lengths (Sibly et al. 1990) and on the use of likelihood ratio tests in helping to determine whether the data occur in bouts (Langton et al. 1995).

However, it is important that the statistical procedure does not obscure biological insight, so the new, split categories must make sense. For example, it is clearly sensible to split the instances of follow-B (individual B) that were preceded by fight from those preceded by copulation or intromission because both new categories of follow (A, B) have different contextual functions and, it turns out, different bout lengths. Splitting of the category nose-to-nose contact did not make sense because it was equally likely to precede a fight or nasoanal contact. However, on further exploration, the probability that nose-to-nose contact would lead to one or other of these categories depended heavily on the sex of interactant. Such sex-dependent sequences may even resolve

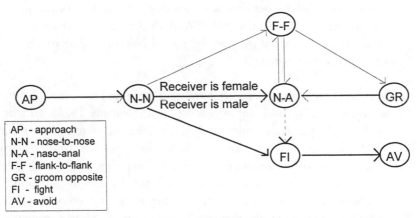

**Figure 10.10** The flow diagram (state-space representation) of the sex-dependent nose-to-nose interaction. Lines between behavioral elements represent the tendency of switching from one behavioral element to another, based on a maximum-likelihood estimator.

puzzles in video surveillance by allowing inference of the sex of an unmarked individual.

## SEQUENCES THROUGH THE MIST

All animals are social, even if only to the extent of engaging in sexual reproduction and otherwise avoiding each other. Understanding the rules that govern social dynamics in general is a step to understanding our own lives (Axelrod 1984). Everything happens in context, so it is unsurprising that we advocate a contextual approach to studying society. Social relationships are the consequences of entwining sets of events and states, and our aim has been to explore how these may be disentangled through analyzing sequences of changes in time and space. Although some of these analytical approaches may seem daunting through their unfamiliarity, they concern a phenomenon—sequentiality—that not only is at the root of life in the patterning of nucleic acids, but also suffuses every aspect of our daily experience in strings of words and sentences and stories. One story can be told in different ways for different purposes, just as an individual butterfly is expressed differently as larva, pupa, and imago. A metamorphosis in storytelling shows how the perceptions of Robert Henryson, a fifteenth-century Scottish poet, might, for the purposes of measuring the dynamics of mammalian society, be stripped to their essentials. In his *Moral Fables,* Henryson penned these words:

> This country mouse lay flattened on the ground, fearing every minute that she would be killed; for her heart was pounding with strokes of fear, and feverishly she trembled hand and foot. And when her sister found her in such straits, for the sake of pity first she grieved a bit, then comforted her with words as sweet as honey.

It is common for stories to be peppered with anthropomorphism, and one class of these, akin to mock anthropomorphism (Kennedy 1986), enriches prose with insight. Certainly, understanding is not the sole prerogative of those who wield the colorless pen of late-twentieth-century science. On the other hand, another class of anthropomorphism is blind to life's processes and, in its fancifulness, more often than not corrupts literature by diminishing understanding. Stripping out these distractions is a step toward understanding, and interpretation of the essence that remains hinges on the meaning of sequences, whether they be words or behaviors. In the contextual study of animal behavior the narration that precedes or succeeds a given action has equal weight to the action itself. As a step toward such contextual analysis, we might wonder whether Henryson had this in mind:

> One rural female wood mouse (*Apodemus sylvaticus*) adopted a dorsoventrally flattened posture, exhibiting, in an unspecified but seemingly fearful context, tachycardia and elevated body temperature, and associated tremulous movements of the fore- and hind-paws. These signs appeared to be stress-induced. A conspecific female approached and initiated prolonged nonreciprocal grooming, perhaps with the consequence of diminishing the signs of putative stress. Subsequent molecular analyses confirmed the hypothesis that these individuals were siblings.

The only merit of the aridity of this version is that it exposes the states and events that can usefully be defined in an ethogram and exposes the sequences of those that will become grist for analysis:

(I) ETHOGRAM: *crouch:* dorsoventral flattened posture with head close to the (i.e., modifier) ground; *shake:* tremulous movement of body (no modifier), or shaking fore-paw(s) (modifier 1) or shaking hind-paw (modifier 2); *approach:* one individual moves toward another; *groom opposite:* one individual grooms the other on the head, or neck, or back, or flank (modifier 1).

(II) SEQUENCE: BEHAVIOR: A-crouch $\rightarrow$ A-shake $\rightarrow$ B-approach-A $\rightarrow$ B-nose-to-nose-A $\rightarrow$ B-groom-A.

MODIFIER 1: ground paw

MODIFIER 2: foot

TIME: $t_0 \rightarrow t_1 \rightarrow t_2 \rightarrow t_3 \rightarrow t_4 \rightarrow \ldots t_n$;

DURATION: $(t_1 - t_0) (t_2 - t_1) (t_3 - t_2) (t_4 - t_3) (t_n - t_4)$

SUBJECTS: 2 female wood mice (*Apodemus sylvaticus*)

Clearly, there are many ways of describing an observation. Henryson's original and our coded version may be quests for different sorts of understanding. The point is that having designed an ethogram and having chosen the correct method of observation, we can distill an observed set of events or states into a pattern that facilitates greater precision of interpretation and decreased risk of fanciful anthropomorphism.

### Acknowledgments

We are grateful to M. Dawkins, R. Dunbar, D. Johnson, and two anonymous reviewers for comments on an earlier draft. We also thank Luigi Boitani and Todd Fuller for their invitation to participate in the conference that spawned this book and for their forbearance as editors.

### Literature Cited

Albonetti, M. E., J. Lazarus, D. W. Dickins, A. Whybrow, R. Schonheiter, J. Prenter, R. Elwood, J. D. Newman, F. O. Odberg, L. Kaiser, M. H. Pham-Delegue, V. Verguelen, J. C. Fentress, and M. L. Bocia. 1992. Software multiple review. The Observer: Software for behavioural research, version 2.0. *Ethology, Ecology and Evolution* 4: 401–414.

Altmann, J. 1974. Observational study of behaviour: Sampling methods. *Behaviour* 49: 227–267.

Appleby, M. C. 1983. The probability of linearity in hierarchies. *Animal Behaviour* 32: 600–608.

Arnoldmeeks, C. and J. J. McGlone. 1986. Validating techniques to sample behavior of confined, young-pigs. *Applied Animal Behaviour Science* 16: 149–155.

Axelrod, R. 1984. *The evolution of co-operation*. New York: Basic Books.

Axelrod, R. and W. D. Hamilton. 1981. The evolution of cooperation. *Science* 211: 1390–1396.

Barbault, R. 1988. Body size, ecological constraints, and the evolution of life-history strategies. In M. K. Hecht, B. Wallace, and G. T. Prance, eds., *Evolutionary biology,* 261–285. New York: Plenum.

Bekoff, M. 1977. Quantitative studies of three areas of classical ethology: Social dominance, behavioural taxonomy, and behavioural variability. In B. Hazlett, ed., *Quantitative methods in the study of animal behavior,* 1–46. New York: Academic Press.

Bekoff, M. and D. Jamieson. 1997. Ethics and the study of carnivores: Doing science while respecting animals. In J. L. Gittleman, ed., *Carnivore behavior, ecology and evolution,* vol. 2, 15–45. Ithaca, N.Y.: Cornell University Press.

Berdoy, M., P. Smith, and D. W. Macdonald. 1994. Stability of social status in wild rats: Age and the role of settled dominance. *Behaviour* 132: 193–212.

Berger, J. and C. Cunningham. 1995. Predation, sensitivity and sex: Why female black rhinoceroses outlive males. *Behavioural Ecology* 6: 57–64.

Bernstein, I. S. 1991. An empirical comparison of focal and ad-libitum scoring with commentary on instantaneous scans, all occurrence and one-zero techniques. *Animal Behaviour* 42: 721–728.

Borries, C. 1992. Grooming site preferences in female langurs (*Presbytis entellus*). *International Journal of Primatology* 13: 19–32.

Bovet, J. 1972a. On the social behaviour in a stable group of long-tailed field mice (*Apodemus sylvaticus*). I. An interpretation of defensive postures. *Behaviour* 41: 43–54.

Bovet, J. 1972b. On the social behaviour in a stable group of long-tailed field mice (*Apodemus sylvaticus*). II. Its relations with distribution of daily activity. *Behaviour* 41: 55–67.

Bressers, M., E. Meelis, P. Haccou, and M. Krok. 1991. When did it really start or stop: The impact of censored observations on the analysis of durations. *Behavioural Processes* 23: 1–20.

Brown, J. L. 1975. *The evolution of behaviour*. New York: Norton.

Burley, N. T. 1988. Wild zebra finches have band colour preferences. *Animal Behaviour* 36: 1235–1237.

Cairns, S. J. and S. J. Schwager. 1987. A comparison of association indices. *Animal Behaviour* 35: 1454–1469.

Chase, I. D. 1982. Dynamics of hierarchy formation: The sequential development of dominance relationships. *Behaviour* 80: 218–240.

Cheney, D. L. 1992. Intragroup cohesion and intergroup hostility: the relation between grooming distributions and intergroup competition among female primates. *Behavioural Ecology* 3: 334–345.

Clout, M. N. and M. G. Efford. 1984. Sex differences in the dispersal and settlement of brushtail possums (*Trichurus vulpecula*). *Journal of Animal Ecology* 53: 737–749.

Clutton-Brock, T. H., F. E. Guinness, and S. D. Albon. 1982. *Red deer: Behaviour and ecology of two sexes*. Chicago: University of Chicago Press.

Clutton-Brock, T. H. and G. A. Parker. 1995. Punishment in animal societies. *Nature* 373: 209–216.

Colgan, P. 1978. *Quantitative ethology*. New York: Wiley.

Creel, S. 1997. Cooperative hunting and group size: Assumptions and currencies. *Animal Behaviour* 54: 1319–1324.

Creel, S. and D. W. Macdonald. 1996. Sociality, group size, and reproductive suppression among carnivores. *Advances in the Study of Behaviors* 24: 203–257.

Creel, S., S. L. Monfort, D. E. Wildt, and P. M. Waser. 1991. Spontaneous lactation is an adaptive result of pseudopregnancy. *Nature* 351: 660–662.

Cuthill, I. 1991. Field experiments in animal behaviour: Methods and ethics. *Animal Behaviour* 42: 1007–1014.

Dawkins, R. 1976. *The selfish gene*. Oxford, U.K.: Oxford University Press.

Dawkins, R. 1989 (2d ed.). *The selfish gene*. Oxford, U.K.: Oxford University Press.

Doncaster, C. P. 1990. Non-parametric estimates of interaction from radio-tracking data. *Journal of Theoretical Biology* 143: 431–443.

Drews, C. 1993. The concept and definition of dominance in animal behaviour. *Behaviour* 125: 283–313.

Dugatkin, L. A. 1997. *Cooperation among animals: An evolutionary perspective*. Oxford, U.K.: Oxford University Press.

Dunbar, R. I. M. 1976. Some aspects of research design and their implications in the observational study of behaviour. *Behaviour* 58: 78–98.

Dunbar, R. 1988. *Primate social systems*. London: Croom Helm.

Evans, P. G. H., D. W. Macdonald, and C. L. Cheeseman. 1989. Social structure of the Eurasian badger (*Meles meles*): Genetic evidence. *Journal of Zoology* 218: 587–595.

Fagen, R. M. 1978. Information measures: Statistical confidence limits and inference. *Journal of Theoretical Biology* 73: 61–79.

Fagen, R. M. and R. N. Goldman. 1977. Behavioural catalogue analysis methods. *Animal Behaviour* 25: 261–274.

Faver, J., M. Mendl, and P. Bateson. 1986. A method for rating individual distinctiveness of domestic cats. *Animal Behaviour* 34: 1016–1025.

Findlay, C. S., R. I. C. Hansell, and C. J. Lumsden. 1989. Behavioural evolution and biocultural games: Oblique and horizontal cultural transmission. *Journal of Theoretical Biology* 137: 245–269.

Geffen, E., M. E. Gompper, J. L. Gittleman, H. K. Luh, D. W. Macdonald, and R. K. Wayne. 1996. Size, life-history traits, and social organization in the Canidae: A reevaluation. *American Naturalist* 147: 140–160.

Geyer, L. A. and C. A. Kornet. 1982. Auto- and allo-grooming in pine voles (*Microtus pinetorum*) and meadow voles (*Microtus pennsylvanicus*). *Physiology and Behavior* 28: 409–412.

Ginsberg, J. R. and T. P. Young. 1992. Measuring association between individuals or groups in behavioural studies. *Animal Behaviour* 44: 377–379.

Golani, I. 1976. Homeostatic motor processes in mammalian interactions: a choreography of display. In P. P. G. Bateson and P. H. Klopfer, eds., *Perspectives in ethology*, vol. 2, 69–134. New York: Plenum.

Golani, I. 1992. A mobility gradient in the organisation of vertebrate movement: The perception of movement through symbolic movement. *Behavioural and Brain Sciences* 15: 249–266.

Goosen, C. 1987. Social grooming in primates. In G. Mitchell and J. Erwin, eds., *Comparative primate biology*, Vol. 2: *Behaviour, cognition, and motivation*, 107–131. New York: Alan R. Liss.

Haccou, P. and E. Meelis. 1992. *Statistical analysis of behavioural data*. Oxford, U.K.: Oxford University Press.

Haccou, P. and E. Meelis. 1995 (2d ed.). *Statistical analysis of behavioural data*. Oxford, U.K.: Oxford University Press.

Hailman, J. P. and J. J. I. Dzelzkalns. 1974. Mallard tail-wagging: Punctuation for animal communication? *American Naturalist* 108: 236–238.

Hailman, E. D. and J. P. Hailman. 1993. UNCERT software (beta version) user's guide. http://www.cisab.indiana.edu/CSASAB/index.html (jhailman@macc.wisc.edu).

Hansson, L. 1992. Fitness and life history correlates of weight variations in small mammals. *Oikos* 64: 479–484.

Hansson, M. B., L. J. Bledsoe, B. C. Kirkevold, C. J., Casson, and J. W. Nightingale. 1993.

Behavioural budgets of captive sea otter mother–pup pairs during pup development. *Zoo Biology* 12: 459–477.

Hare, J. F. 1992. Colony member discrimination by juvenile Columbian ground squirrels (*Spermophilus columbianus*). *Ethology* 92: 301–315.

Harris, S. and P. C. L. White. 1992. Is reduced affiliative rather than increased agonistic behaviour associated with dispersal in red foxes? *Animal Behaviour* 44: 1085–1089.

Hazlett, B. A. 1977. *Quantitative methods in the study of animal behaviour.* New York: Academic Press.

Hemelrijk, C. K. 1990. Models and tests of reciprocity, unidirectionality and other social interaction patterns at group level. *Animal Behaviour* 39: 1013–1029.

Hemelrijk, C. K. and A. Ek. 1991. Reciprocity and interchange of grooming and support in captive chimpanzees. *Animal Behaviour* 41: 923–935.

Herrera, E. A. and D. W. Macdonald. 1989. Resource utilisation and territoriality in group-living capybaras (*Hydrochoerus hydrochaeris*). *Journal of Animal Ecology* 58: 667–679.

Herrera, E. A. and D. W. Macdonald. 1993. Aggression, dominance, and mating success among capybara males (*Hydrochoerus hydrochaeris*). *Behavioural Ecology* 4: 114–119.

Hinde, R. A. H. 1981. Social relationships. In D. McFarland, ed., *The Oxford companion to animal behaviour,* 521–527. Oxford, U.K.: Oxford University Press.

Hinde, R. A., ed. 1983. *Primate social relationships. An integrated approach.* Oxford, U.K.: Blackwell Scientific.

Hofer, H. and M. East. 1993. Loud calling in a female-dominated mammalian society. II Behavioural contexts and functions of whooping of spotted hyaenas, *Crocuta crocuta. Animal Behaviour* 42: 651–669.

Hoogland, J. L. 1995. *The black tailed prairie dog: Social life of a burrowing mammal.* Chicago: University of Chicago Press.

Iacobucci, D. 1990. Derivation of subgroups from dyadic interactions. *Psychological Bulletin* 107: 114–132.

Johnsingh, A. J. T. 1983. Large mammalian prey-predators in Bandipur. *Journal of the Bombay Natural History Society* 80: 1–53.

Jouventin, P. and A. Cornet. 1980. The sociobiology of pinnipeds. *Advances in the Study of Behavior* 11: 121–141.

Kawanaka, K. 1996. Observation time and sampling intervals for measuring behavior and interactions of chimpanzees in the wild. *Primates* 37: 185–196.

Kennedy, J. S. 1986. Some current issues in orientation to odour sources. In T. L. Payne, M. C. Birch, and C. E. J. Kennedy, eds., *Mechanisms in insect olfaction,* 11–25. Oxford, U.K.: Clarendon.

Kenward, R. 1987. *Wildlife radio-tagging.* London: Academic Press.

Kerby, G. and D. W. Macdonald. 1988. Cat society and the consequences of colony size. In D. Turner and P. Bateson, eds., *The domestic cat: The biology of its behaviour,* 67–81. Cambridge, U.K.: Cambridge University Press.

Kraft, B., C. Colin, and P. Peignot. 1994. Diving-for-food: A new model to assess social roles in a group of laboratory rats. *Ethology* 96: 11–23.

Krebs, J. R., and N. B. Davies. 1991 (3d ed.). *Behavioural ecology*. Oxford, U.K.: Blackwell Scientific.

Krishnaiah, P. R. and T. J. Rao. 1988. *The handbook of statistics*. Amsterdam: Elsevier.

Kruuk, H. 1972. *The spotted hyena*. Chicago: Chicago University Press.

Kruuk, H. 1978. Foraging and spatial organisation of the badger (*Meles meles*). *Behavioural Ecology and Sociobiology* 4: 75–89.

Langton, S. D., D. Collet, and M. R. Sibly. 1995. Splitting behaviour into bouts; a maximum likelihood approach. *Behaviour* 132: 9–10.

Ledger, D. W. and I. A. Didrichsons. 1994. An assessment of data pooling and some alternatives. *Animal Behaviour* 48: 823–832.

Lehner, P. N. 1996 (2d ed.). *Handbook of ethological methods*. Cambridge, U.K.: Cambridge University Press.

Lindgren, B. W. 1968. *Statistical theory*. London: Macmillan.

Lindquist, T. and I. S. Bernstein. 1987. A feminist perspective on dominance relationships in primates. *American Journal of Primatology* 12: 355.

Lindstedt, S. L., B. J. Miller, and S. W. Buskirk. 1986. Home range, time, and body size in mammals. *Ecology* 67: 413–418.

Lorenz, K. 1981. *The foundations of ethology*. New York: Springer-Verlag.

Lott, D. F. 1991. *Intraspecific variation in the social systems of wild vertebrates*. Cambridge, U.K.: Cambridge University Press.

Macdonald, D. W. 1983. The ecology of carnivore social behaviour. *Nature* 301: 379–384.

Macdonald, D. W. 1981. Dwindling resources and the social behaviour of capybaras (*Hydrochoerus hydrochaeris*). *Journal of Zoology* 194: 371–391.

Macdonald, D. W. 1995. Wildlife rabies: The implications for Britain. Unresolved questions for the control of wildlife rabies: Social perturbation and interspecific interactions. In *Rabies in a changing world*, 33–48. Cheltenham: British Small Animal Veterinary Association.

Macdonald, D. W. and C. J. Amlaner, Jr. 1980. A practical guide to radio tracking. In C. J. Amlaner and D. W. Macdonald, eds., *A handbook on biotelemetry and radiotracking*, 143–159. Oxford, U.K.: Pergamon.

Macdonald, D. W., P. J. Apps, G. Carr, and G. Kerby. 1987. Social behaviour, nursing coalitions and infanticide in a colony of farm cats. *Advances in Ethology* 28: 1–66.

Macdonald, D. W., F. G. Ball, and N. G. Hough. 1980. The evaluation of home range size and configuration using radio-tracking data. In D. W. Macdonald and C. J. Amlaner, eds., *A handbook on biotelemetry and radio tracking*, 143–159, 405–426. Oxford, U.K.: Pergamon.

Macdonald, D. W. and G. M. Carr. 1989. Food security and the rewards of tolerance. In V. Standen and R. A. Foley, eds., *Comparative socioecology: The behavioural ecology of humans and other mammals*, 75–99. Oxford, U.K.: Blackwell Scientific.

Macdonald, D. W. and O. Courtenay. 1996. Enduring social relationships in a population of crab-eating zorros, *Cerdocyon thous*, in Amazonian Brazil. *Journal of Zoology* 239: 329–355.

Macdonald, D. W., N. Yamaguchi, and G. Kerby. In press (2d ed.). Group-living in the domestic cat: Its sociobiology and epidemiology. In D. Turner and P. Bateson, eds., *The domestic cat: The biology of its behaviour.* Cambridge, U.K.: Cambridge University Press.

Machlis, L. 1977. An analysis of the temporal patterning of pecking in chicks. *Behaviour* 63: 1–70.

Machlis, L., P. W. D. Dood, and J. C. Fentress. 1985. The pooling fallacy: Problems arising when individuals contribute more than one observation to the data set. *Zeitschrift für Tierpsychologie* 68: 201–214.

Martin, P. and P. Bateson, eds. 1993 (2d ed.). *Measuring behaviour: An introductory guide.* Cambridge, U.K.: Cambridge University Press.

Matloff, N. S. 1991. Statistical hypothesis testing: Problems and alternatives. *Environmental Entomology* 20: 1246–1250.

May, R. M. 1975. Patterns of species abundance and diversity. In M. L. Cody and J. Diamond., eds., *Ecology and evolution of communities,* 81–120. Cambridge, Mass.: Harvard University Press.

McFarland, D. J. 1981. *The Oxford companion to animal behaviour.* Oxford, U.K.: Oxford University Press.

McKnight, B. L. 1995. Behavioural ecology of hand reared elephants (*Loxodonta africana*) in Tsavo East National Park, Kenya. *African Journal of Ecology* 33: 242–256.

Milinski, M. and G. A. Parker. 1991 (3d ed.). Competition for Resources. In J. R. Krebs and N. B. Davies, eds., *Behavioural ecology: An evolutionary approach,* 137–168. Oxford, U.K.: Blackwell Scientific.

Mitchell, B. W., N. S. Thompson, and H. L. Miles, eds. 1997. *Anthropomorphism, anecdotes and animals.* New York: State University of New York Press.

Moehrenschlager, A., D. W. Macdonald, and C. Moehrenschlager. In press. The effects of trapping and handling swift foxes. In L. Carbyn and M. Sovoda, eds., *Swift fox biology.* Edmonton, Alberta: Canadian Circumpolar Institute.

Morell, V. 1993. Primatology: Seeing nature through the lens of gender. *Science* 260: 428–429.

Moser, R., M. Cords, and H. Kummer. 1991. Social influences on grooming site preferences among captive long-tailed macaques. *International Journal of Primatology* 12: 217–230.

Noe, R., C. P. van Schaik, and J. A. R. A. M. van Hoof. 1991. The market effect: An explanation for pay-off asymmetries among collaborating animals. *Ethology* 87: 97–118.

Noldus Information Technology. 1994. *The Observer, support package for video tape analysis. Reference manual,* Version 3.0. Wageningen, the Netherlands: Noldus Information Technology.

Noldus Information Technology. 1995. *The Observer, base package for Windows. Reference manual,* Version 3.0. Wageningen, the Netherlands: Noldus Information Technology.

Odagiri, K. and Y. Matsuzawa 1994. Application of an instantaneous sampling method to lamb's play behaviour. *Japanese Journal of Animal Psychology* 43: 9–16.

Packer, C. and T. Caro. 1997. Foraging costs in social carnivores. *Animal Behaviour* 54: 1317–1324.

Pearl, M. C. and S. R. Schulman. 1983. Techniques for the analysis of social structure in animal societies. *Advances in the Study of Behavior* 13: 107–146.

Plesner Jensen, S. 1993. Temporal changes in food preferences of wood mice (*Apodemus sylvaticus* L.). *Oecologia* 94: 76–82.

Pollock, G. B. 1994. Social competition or correlated strategy? *Evolutionary Ecology* 8: 221–229.

Poysa, H. 1991. Measuring time budgets with instantaneous sampling: A cautionary note. *Animal Behaviour* 42: 317–318.

Quera, V. 1990. A generalised technique to estimate frequency and duration in time sampling. *Behavioral Assessment* 12: 409–424.

Roberts, B. A. 1992. *Sequential data in biological experiments. An introduction for research workers.* London: Chapman & Hall.

Rogers, L. M., C. L. Chesseman, and S. Langton. 1997. Body weight as an indication of density-dependent population regulation in badgers (*Meles meles*) at Woodchester Park, Gloucestershire. *Journal of Zoology* 242: 597–604.

Romanes, G. J. 1881. *Nature,* Oct. 27, 604.

Roper, T. J. 1992. Badger (*Meles meles*) sett architecture, internal environment and function. *Mammal Review* 22: 43–53.

Rowell, T. E. and C. A. Rowell. 1993. The social organization of feral *Ovis aries* ram groups in the pre-rut period. *Ethology* 95: 213–232.

Rubenstein, D. I. 1993. The ecology of female social behaviour in horses, zebras and asses. *Physiology and Ecology Japan* 29: 13–28.

Schino, G., F. Aureli, and A. Troisi. 1988. Equivalence between measures of allogrooming: an empirical-comparison in 3 species of macaques. *Folia Primatologica* 51: 214–219.

Schleidt, W. M., G. Yakalis, M. Donnelly, and J. McGarry. 1984. A proposal for a standard ethogram, exemplified by an ethogram of the bluebreasted quail (*Coturnix chinensis*). *Zeitschrift für Tierpsychologie* 64: 193–220.

Seyfarth, R. M. 1983. Grooming and social competition in primates. In R. A. Hinde, ed., *Primate social relationships,* 182–190. Oxford, U.K.: Blackwell Scientific.

Sharman, M. and R. I. M. Dunbar. 1982. Observer bias in selection of study group in baboon field studies. *Primates* 23: 567–573.

Shrader-Frachette, K. S. and E. D. McCoy. 1992. Statistics, costs and rationality in ecological inference. *Trends in Ecology and Evolution* 7: 96–99.

Sibley, R. M., H. M. R. Nott, and D. J. Fletcher. 1990. Splitting behaviour into bouts. *Animal Behaviour* 39: 63–69.

Sillero-Zubiri, C., P. J. Johnson, and D. W. Macdonald. 1998. A hypothesis for breeding synchrony in Ethiopian wolves (*Canis simensis*) *Journal of Mammalogy* 79: 853–858.

Smuts, B. 1985. *Sex and friendship in baboons.* New York: Aldine.

Stevenson-Hinde, J. 1983. individual characteristics and the social situation. In R. A. Hinde, ed., *Primate social relationships,* 28–35. Oxford, U.K.: Blackwell Scientific.

Stewart, P. D. 1997. *The social behaviour of the European badger (Meles meles)*. Doctoral thesis. Oxford, U.K.: Oxford University Press.

Stewart, P. D., S. A. Ellwood, and D. W. Macdonald. 1997. Video surveillance of wildlife: An introduction from experience with the European badger (*Meles meles*). *Mammal Review* 27: 185–204.

Stewart, P. D. and D. W. Macdonald. 1997. Age, sex, and condition as predictors of moult and the efficacy of a novel fur clip technique for individual marking of the European badger (*Meles meles*). *Journal of Zoology* 241: 543–550.

Stonehouse, B. 1978. *Animal marking: Recognition marking of animals in research*. Baltimore, Md.: University Park Press.

Stopka, P. and D. W. Macdonald. 1998. Signal interchange during mating in the wood mouse (*Apodemus sylvaticus*): The concept of active and passive signalling. *Behaviour* 135: 231–249.

Stopka, P. and D. W. Macdonald. 1999. The market effect in the wood mouse, *Apodemus sylveticus:* Selling information on reproductive status. *Ethology* 105: 913–1008.

Suen, H. K. and D. Ary. 1984. Variables influencing one–zero and instantaneous time sampling outcomes. *Primates* 25: 89–94.

Suen, H. K. and D. Ary. 1986a. Poisson cumulative probabilities of systematic errors in single-subject and multiple subject time sampling. *Behavioural Assessment* 8: 155–169.

Suen, H. K. and D. Ary. 1986b. A post hoc correction procedure for systematic errors in time sampling duration estimates. *Journal of Psychopathology and Behavioral Assessment* 8: 31–38.

Swinton, J., F. Tuyttens, D. W. Macdonald, D. J. Nokes, C. L. Cheeseman, and R. Clifton-Hadley. 1997. A comparison of fertility control and lethal control of bovine tuberculosis in badgers: The impact of perturbation induced transmission. *Philosophical Transactions of the Royal Society of London* B352: 619–631.

Taber, A. B. and D. W. Macdonald. 1992. Communal breeding in the mara *Dolichotis patagonium*. *Journal of Zoology* 227: 439–452.

Tew, T. E. and D. W. Macdonald. 1994. Dynamics of space use and male vigour amongst wood mice, *Apodemus sylvaticus,* in the cereal ecosystem. *Behavioral Ecology and Sociobiology* 34: 337–345.

Thierry, B., C. Gauthier, and P. Peignot. 1990. Social grooming in Tonkean macaques (*Macaca tonkeana*). *International Journal of Primatology* 11: 357–375.

Thompson, S. K. 1992. *Sampling*. New York: Wiley.

Tinbergen, N. 1951. *The study of instinct*. Oxford, U.K.: Oxford University Press.

Tinbergen, N. 1963. On aims and methods in ethology. *Zeitschrift für Tierpsychologie* 20: 410–433.

Todd, I. A. 1992. *Wildtrak: Non-parametric home range analysis for the Macintosh*. Oxford, U.K.: Isis Innovations.

Tuyttens, F. A. M. and D. W. Macdonald. In press. Social perturbation caused by population control: Implications for the management and conservation of wildlife. In M.

Gosling, W. Sutherland, and M. Avery, eds., *Behaviour and conservation.* Cambridge, U.K.: Cambridge University Press.

de Vries, H., W. J. Netto, and L. H. Hanegraaf. 1993. MATMAN: A program for the analysis of sociometric matrices and behavioural transition matrices. *Behaviour* 125: 157–175.

de Vries, H. 1993. The rowwise correlation between two proximity matrices and the partial rowwise correlation. *Psychometrica* 58: 53–69.

de Vries, H. 1995. An improved test of linearity in dominance hierarchies containing unknown or tied relationship. *Animal Behaviour* 50: 1375–1389.

de Waal, F. B. M. 1992. Coalitions as part of reciprocal relations in the Arnhem chimpanzee colony. In A. H. Harcourt and F. B. M. de Waal, eds., *Coalitions and alliances in humans and other animals,* 233–258. Oxford, U.K.: Oxford University Press.

White, G. C. and R. A. Garrott. 1990. *Analysis of wildlife radio-tracking data.* London: Academic Press.

White, P. C. L. and S. Harris. 1994. Encounters between red foxes (*Vulpes vulpes*): Implications for territory maintenance, social cohesion and dispersal. *Journal of Animal Ecology* 63: 315–327.

Wilson, E. O. 1975. *Sociobiology.* Cambridge, Mass.: Belknap.

Wong, J., P. D. Stewart, and D. W. Macdonald. 1999. Vocal repertoire in the European badger (*Meles meles*): Structure, context, and function. *Journal of Mammalogy* 80: 570–588.

Woodroffe, R. 1993. Alloparental behaviour in the European badger. *Animal Behaviour* 46: 413–415.

Woodroffe, R. and D. W. Macdonald. 1993. Badger sociality: Models of spatial grouping. *Symposium of the Zoological Society of London* 65: 145–169.

Woodroffe, R. and D. W. Macdonald. 1995a. Costs of breeding status in the European badger (*Meles meles*). *Journal of Zoology* 235: 237–245.

Woodroffe, R. and D. W. Macdonald. 1995b. Female–female competition in European badgers (*Meles meles*): Effects on breeding success. *Journal of Animal Ecology* 64: 12–20.

Wrangham, R. W. 1974. Artificial feeding of chimpanzees and baboons in their natural habitat. *Animal Behaviour* 22: 83–93.

Wratten, S. D., ed. 1994. *Video techniques in animal ecology and behaviour.* London: Chapman & Hall.

## Chapter 11

# Modeling Species Distribution with GIS

Fabio Corsi, Jan de Leeuw, and Andrew Skidmore

From the variety of checklists, atlases, and field guides available around the world it is easy to understand that distribution ranges are pieces of information that are seldom absent in a comprehensive description of species. Their uses range from a better understanding of the species biology, to simple inventory assessment of a geographic region, to the definition of specific management actions. In the latter case, knowledge of the area in which a species occurs is fundamental for the implementation of adequate conservation strategies. Conservation is concerned mostly with fragmentation or reduction of the distribution as an indication of population viability (Maurer 1994), given that, for any species, range dimension is considered to be correlated to population size (Gaston 1994; Mace 1994).

Unfortunately, animals move and this poses problems in mapping their occurrence. Traditional methods used to store information on species distributions are generally poor (Stoms and Estes 1993). Distributions have been described by drawing polygons on a map (the "blotch") to represent, with varying approximations, a species' ranges (Gaston 1991; Miller 1994). The accuracy of the polygons relies on the empirical knowledge of specialists and encloses the area in which the species is considered likely to occur, although the probability level associated with this "likelihood" is seldom specified. A more sophisticated approach divides the study area into subunits (e.g., administrative units, equal-size mesh grid), with each subunit associated with information on the presence or absence of the species. In this case the distribution range of a species is defined by the total of all subunits in which presence is confirmed; however, blank areas are ambiguous as to whether the species is absent or no records were available (Scott et al. 1993).

New approaches tend to overcome the concept of distribution range and move toward one of area of occupancy.[1] This concept is particularly useful for conservation action and has therefore been included in the new IUCN*Red List* criteria (IUCN1995). In this chapter we outline the basis of identifying distributions that represent a step toward the definition of a real area of occupancy.

For example, imagine a biologist who needs to find zebras. Intuitively, the odds of finding zebras in Scandinavia are very low, but moving to Kenya greatly increases the odds. This process is based on very basic assumptions such as that zebras live in warm places, say, with an average annual temperature of 13–28°C. Obviously our observer won't expect to find zebras in every place on Earth that has an average annual temperature of 13–28°C; there are many other ecological requirements, along with other reasons, such as historical constraints (see Morrison et al. 1992 for a review) and species behavioral patterns (Walters 1992), that contribute to define the distribution of the zebra. Nevertheless, if our biologist extends the same process, taking into account the preferred ranges of values of various environmental variables, the probability of finding the species in the areas in which these preferences are simultaneously satisfied increases.

If the aim of our researcher is to map the areas in which the species is most likely to be found rather than to find an individual, the entire process can be seen as a way of describing the species' presence in terms of correlated environmental variables. And if inexpensive and broadly acquired environmental data (e.g., vegetation index maps derived from satellite data) are used to define species probability of presence, then maps of species distribution can be produced quickly and efficiently.

To provide a formal approach to species distribution modeling, the process can be divided into two phases. The first phase assesses the species' preferred ranges of values for the environmental variables taken into account, and the second identifies all locations in which these preferred ranges of values are fulfilled. The first phase is generally called habitat suitability index (HSI) analysis, habitat evaluation procedures (HEP) (Williams 1988; Duncan et al. 1995), or, more generally, species–environment relationship analysis. The second, which involves the true distribution model, has seen its potential greatly enhanced in the last 10 years by the increasing use of geographic information systems (GIS), which can extrapolate the results of the first phase to large portions of territory.

The power of GIS resides in its ability to handle large amounts of spatial data, making analysis of spatial relationships possible. This increases the number of variables that can be considered in an analysis and the spatial extent to which the analysis can be carried out (Burrough 1986; Haslett 1990).

Thus GIS provides a means for addressing the multidimensional nature of the species–environment relationship (Shaw and Atkinson 1990) and the need to integrate large portions of land (eventually the entire biosphere) into the analysis (Sanderson et al. 1979; Klopatek et al. 1983; Flather and King 1992; Maurer 1994) to produce robust conservation oriented models.

This chapter is a review of models and methods used in GIS-based species distribution models; it is based on a literature review carried out on GEOBASE[2] with the following keywords: *GIS, remote sensing* (RS), *wildlife, habitat,* and *distribution.* The 82 papers collected were classified according to the main tool used (GIS or RS), the modeling approach, the analysis technique, the discussion of the assumptions, and the presence of a validation section. At the same time, information was gathered on the use of the term *habitat,* the number of variables used for modeling, and the kind of output produced.

Far from being comprehensive, the review was the starting point for a tentative classification of GIS distribution models that is presented in this chapter; at the same time, it allowed us to focus attention on some issues that we consider among the most important for correct use of GIS in species distribution modeling. In fact, although it offers powerful tools for spatial analysis, GIS has been largely misused and still lacks a clear framework to enable users to exploit its potential fully.

These issues range from unspecified objectives in the process of model building to the lack of adequate support for the assumptions underlying the models themselves. A large part of the chapter is devoted to the problem of validation, which we believe is crucial throughout the process of model building but is very seldom taken into account.

Before discussing these issues, we address the problem of terminology inconsistencies, which has a much broader extent in ecology than the specific realm of species distribution modeling. The problem emerges from our review and is probably caused, in this context, by misleading use of the same term in the different disciplines that have come to coexist under the wide umbrella of GIS.

## ■ Terminology

Multidisciplinary fields of science are very appealing because they bring together people with different experience and backgrounds whose constructive exchange of ideas may generate new solutions. In fact, many solutions that have been successfully developed and used in one field of science may, with minor changes, be used in other fields. The very nature of GIS makes it essential that specialists in different scientific disciplines contribute to the general effort of setting up and maintaining common data sets.

One drawback is that in the early phases of tool development (such as GIS), people who master the new tool tend to become generalists, invading other fields of science without having the necessary specific background. This may cause problems both in the solutions provided, which generally tend to be too simplistic, and in terminology, because the same term or concept can be used with slightly different meanings in different disciplines. This is the case, for instance, with use of the concept of *scale*. For the cartographer, large scale pertains to the domain of detailed studies covering small portions of the earth's surface (Butler et al. 1986), whereas for the ecologist large scale means an approach that covers regional or even wider areas (Edwards et al. 1994). Obviously this derives from the fact that cartographers use *scale* to mean the ratio between a unit measure on the map and the corresponding measure on the earth's surface, whereas the ecologist uses it in the sense of proportion or extent. For example, the relationship between the geographic scale and the extension of ecological studies supplied by Estes and Mooneyhan (1994) highlights that large scale in ecology is often associated with small geographic scale:

Site = 1:10,000 or larger

Local = 1:10,000 to 1:50,000

National or regional = 1:50,000 to 1:250,000

Continental = 1:250,000 to 1:1,000,000

Global = 1:1,000,000 or smaller

In ecology it would be better to use the adjectives *fine* or *broad* (Levin 1992), which places the term *scale* more in the context of its second meaning.

If the confusion arising from the two uses of *large scale* seems trivial (at least from the ecologists' point of view), we believe that the different uses that have been made of the word *habitat* give rise to major misunderstandings and thus need to be clarified (Hall et al. 1997).

## ■ Habitat Definitions and Use

The term *habitat*[3] forms a core concept in wildlife management and the distribution of plant and animal species. The fact that the actual sense in which it is used is rarely specified suggests that its meaning is taken for granted. However, Merriam-Webster's dictionary (1981) provides two different definitions and Morrison et al. (1992) observed that use of the word *habitat* remains far from unambiguous. The latter distinguished two different meanings: one concept that relates to units of land homogeneous with respect to environmental conditions and a second concept according to which habitat is a property of species.

Our literature review provided us with a variety of definitions and uses of the term *habitat* that are wider than the dichotomy suggested by Morrison et al. (1992). We arranged these various meanings according to two criteria: whether the term relates to biota (either species and or communities) or to land, and whether it relates to Cartesian (e.g., location, such as a position defined by a northing and easting) or environmental space (e.g., the environmental envelope defined by factors such as precipitation, temperature, and land cover) (table 11.1).

Although the classification in table 11.1 allows us to partition the different definitions of habitat we have traced, in reality this partition is rather hazy. For instance, definitions range from the place where a species lives (Begon et al. 1990; Merriam-Webster 1981; Odum 1971; Krebs 1985), which is a totally Cartesian space–related concept, to the environment in which it lives (Collin 1988; Moore 1967; Merriam-Webster 1981; Whittaker et al. 1973). In this last case habitat is seen as a portion of the environmental space. At both extremes of the range of definitions, the slight differences in the terms used allows us to define a continuous trend between the Cartesian and the environmental concept, which is further supported considering a few definitions that combine the Cartesian and the environmental space (Morrison et al. 1992; Mayhew and Penny 1992). These last authors define habitat as the area that has specific environmental conditions that allow the survival of a species. Note that all of these definitions relate habitat to a species and some describe it as a property of an organism.

With a similar range of definitions, another group relates habitat to both species and communities. For instance, Zonneveld (1995:26), in accordance with a Cartesian concept, defined it as "the concrete living place of an organism or community." Others relate it to both Cartesian and environmental space, defining it as the place in which an organism or a community lives, including the surrounding environmental conditions (Encyclopaedia Britannica 1994; Yapp 1922).

All of the definitions cited so far defined habitat in terms of biota. Zonneveld (1995) remarked that the term *habitat* may be used only when specifying a species (or community). Yet *habitat* has been used as an attribute of land.

**Table 11.1    Classification Scheme of the Term *Habitat***

|  | Biota | | Land |
|---|---|---|---|
|  | *Species* | *Species and Communities* |  |
| Cartesian space | Begon et al. (1990) Krebs (1985) Odum (1971) Merriam-Webster (1981) | Zonneveld (1995) |  |
| Cartesian space and environment | Morrison et al. (1992) Mayhew and Penny (1992) | Encyclopaedia Britannica (1994) Yapp (1922) | Stelfox and Ironside (1982) Kerr (1986) USFWS (1980a, 1980b) Herr and Queen (1993) |
| Environment | Collin (1988) Merriam-Webster (1981) Whittaker et al. (1973) Moore (1967) |  |  |

The various meanings of *habitat* are grouped according to whether the term relates to biota (species or species and communities) or land and whether it relates to Cartesian space, environmental space, or both.

Riparian habitat, for instance, is a specific environment, with no relation to biota. Use of *habitat* in this sense is widespread in the ecological literature (e.g., old-forest habitat, Lehmkuhl and Raphael [1993], or woodland habitat, Begon et al. [1990]). The concept predominates in ecology applied to land management such as habitat mapping (Stelfox and Ironside 1982; Kerr 1986), habitat evaluation (USFWS 1980a, 1980b; Herr and Queen 1993), and habitat suitability modeling (USFWS 1981). A similar meaning of *habitat* is used in a review of habitat-based methods for biological impact assessment (Atkinson 1985). Although it has been used very often in this sense, we were unable to find a single definition. A closely related concept, the habitat type, which is used in habitat mapping, has been defined as "an area, delineated by a biologist, that has consistent abiotic and biotic attributes such as dominant or subdominant vegetation" (Jones 1986:23). Daubenmire (1976) noted that this meaning of *habitat type* corresponds to the land unit concept (Walker et al. 1986; Zonneveld 1989). In articles dealing with habitat evaluation, the term is used in a similar sense.

The use of an ambiguous term leads to confusion in communication between scientists. The ambiguity of *habitat* is also observed within the same publication. Lehmkuhl and Raphael (1993), for instance, simultaneously used "old-forest habitat" and "owl habitat." Even ecological textbooks are not free from ambiguity. Begon et al. (1990:853) defined habitat as "the place where a micro-organism, plant or animal species lives," suggesting that they consider habitat a property of a species. However, when outlining the difference between niche and habitat, they later described habitat in terms of a land unit (Begon et al. 1990:78): "a woodland habitat for example may provide niches for warblers, oak trees, spiders and myriad of other species." Confusion arises with respect to habitat evaluation as well. When defined as a property of a species, unsuitable habitat does not exist because habitat is habitable by definition. In this case some land may be classified as habitat and all of this is suitable. When defined as a land property, all land is habitat, whether suitable or unsuitable, for a specific species.

Why is the term *habitat* used in these various senses? The word originates from *habitare,* to inhabit. According to Merriam-Webster (1981) the term was originally used in old natural histories as the initial word in the Latin descriptions of species of fauna and flora. The description generally included the environment in which the species lives. This leads to the conclusion that habitat was originally considered a species-specific property. It is interesting to note that the definitions we traced originated both from ecology and geography, suggesting that the confusion was not the result of separate developments in two fields of science.

At some time habitat started to be used as a land-related concept, most likely in conjunction with habitat mapping. A possible explanation for the change is given by Kerr (1986), who remarked that mapping habitat[4] individually for each species would be an impossible job. He argued that a map displaying habitat types and describing the occurrence of species in each type would be more useful to the land manager. This suggests that the land-related habitat concept arose because it was considered more convenient to map habitat types rather than the habitat of individual species.

We suggest that there was a second reason for the popularity of habitat type maps. In general the distribution of species is affected by more than one environmental factor. Until a decade ago it was virtually impossible to display more than one environmental factor on a single map. The habitat type, defined as a mappable unit of land "homogeneous" with respect to vegetation and environmental factors, circumvented this problem and was the basis of the land system (land concept) maps developed in the 1980s (Walker et al. 1986; Zonneveld 1989). However, it is based on the assumption that environmental factors show an interdependent change throughout the landscape and that the environmental factors are constant within the "homogeneous" area. Thus to a certain extent the land unit meaning of the term *habitat* arose as a way to overcome operational difficulties in species distribution mapping. Nevertheless, given that the variation of one environmental factor affecting the distribution of a species often tends to be independent of the other environmental factors, homogeneity is seldom the case, so there is seldom a true relationship between species and habitat types.

The advent of GIS has made it possible to store the variation of environmental factors independently and subsequently integrate these independent environmental surfaces into a map displaying the suitability of land as a habitat for a specific species.

The first examples of such GIS-based habitat mapping were published in the second half of the 1980s (e.g., Hodgson et al. 1988). Since then there has been a steady increase of the number of GIS-based habitat models (figure 11.1). The increase illustrates a move away from the general habitat-type mapping applicable for multiple species toward more realistic species-specific habitat maps.

At the same time, the habitat type loses its usefulness because of the decreasing need to classify land in homogeneous categories. In other words, species-specific habitat mapping is increasingly incorporating independent environmental databases processed using information on the preferences of the species concerned. In view of the anticipated move toward species-specific habitat models, we prefer to use the original species-related concept of habitat instead of a land-related concept; to avoid confusion, in this chapter we will use the terms *species–environment relationships* and *ecological requirements* instead of the terms *species habitat* and *habitat requirements*.

## ■ General Structure of GIS-Based Models

The rationale behind the GIS approach to species distribution modeling is straightforward: the database contains a large number of data sets (layers), each of which describes the distribution of a given measurable and mappable environmental variable. The ecological requirements of the species are defined

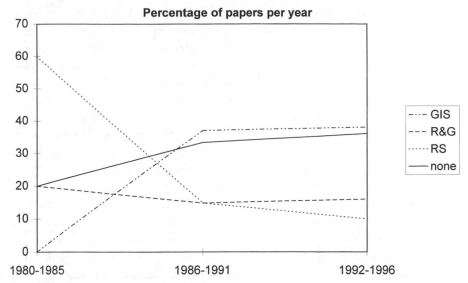

**Figure 11.1**   Percentage of the papers dealing with habitat modeling using no spatial information, RS, GIS, and a combination of RS and GIS for three periods (1980–1985, 1986–1991, and 1992–1996).

according to the available layers. The combination of these layers and the subsequent identification of the areas that meet the species' requirements identify the species' distribution range, either actual (if there is evidence of presence) or potential (if the species has never been observed in that area).

This basic scheme can be implemented using different approaches. A few classifications based on different criteria have been attempted. For example, Stoms et al. (1992) classified models based on the conceptual method used to define the species–environment relationship, whereas Norton and Possingham (1993) based their classification on the result of the model and its applicability for conservation. Accordingly, Stoms et al. (1992) classified GIS species distribution models into two main groups—deductive and inductive—whereas Norton and Possingham (1993) gave a more extensive categorization of modeling approaches.

We have tried to define logical frameworks that can be used to classify species distribution models based on the major steps that must be followed to build them. To this end, we find the deductive–inductive categorization the most suitable starting point because it focuses attention on the definition of the species–environment relationship, which is the key point for the implementation of distribution models.

The deductive approach uses known species' ecological requirements to extrapolate suitable areas from the environmental variable layers available in the GIS database. In fact, analysis of the species–environment relationship is relegated to the synthesizing capabilities and wide experience of one or more specialists who decide, to the best of their knowledge, which environmental conditions are the most favorable for the existence of the species. Once the preferences are identified, generally some sort of logical (Breininger et al. 1991; Jensen et al. 1992) or arithmetic map overlay operation (Donovan et al. 1987; Congalton et al. 1993) is used to merge the different GIS environmental layers to yield the combined effect of all environmental variables.

When the species–environment relationships are not known a priori, the inductive approach is used to derive the ecological requirements of the species from locations in which the species occurs. A species' ecological signature can be derived from the characterization of these locations. Then, with a process that is very similar to the one used in deductive modeling but is generally more objectively driven by the type of analysis used to derive the signature, it is used to extrapolate the distribution model (Pereira and Itami 1991; Aspinall and Matthews 1994).

In figure 11.2 we summarize the data flow of GIS-based species distribution models for both the deductive and the inductive approaches. Whereas in the deductive approach GIS data layers enter the analysis only to create the distribution model, in the inductive approach they are used both to extrapolate the species–environment relationship and the distribution model. Along with the data flow, the steps that need validation are also evidenced in the figure. Validation is addressed in more detail later in this chapter, but it is interesting to note here that validation procedures are needed at many different stages in the flow diagram.

Both inductive and deductive models can be further classified according to the kind of analysis performed to derive the species–environment relationship. Essentially these can be subdivided into two main categories: the descriptive and the analytical. Models pertaining to the first category use either the specialists' a priori knowledge (deductive–descriptive) or the simple overlay of known location of the species with the associated environmental variable layers (inductive–descriptive) to define the species–environment relationship. Descriptive models generally are based on very few environmental variable layers, most often just a single layer. They tend to describe presence and absence in a deterministic way; each value or class of the environmental variable is associated with presence or with absence (e.g., the species is known to live in savanna with an annual mean temperature of 15–20°C, so savanna polygons

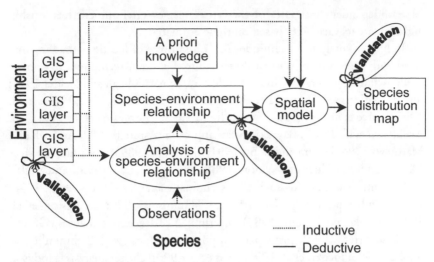

**Figure 11.2**  General data flow of the two main categories of GIS species distribution models identified in this chapter.

falling within the adequate temperature range are to be included as suitable environments). No attempt is made to define confidence intervals to the individual estimate, nor is any information provided on the relative importance of one variable over another (e.g., vegetation types vs. temperature). Moreover, no estimate of the degree of association or its variability is provided with the relationship.

On the other hand, models that fall into the analytical group introduce variability in the sense that advice from different specialists is combined to define species–environment relationships, thus introducing variability in terms of different opinions of the experts (deductive–analytical), or that the species observation data are analyzed in a way that takes into account the range of acceptability of all environmental variables measured, their confidence limits, and their correlation. Both the deductive–analytical and the inductive–analytical approaches tend to estimate the relative importance of the different environmental layers considered in the analysis, thus moving toward an objective combination of environmental variable layers.

Examples of deductive–analytical models are based on techniques such as multi-criteria decision-making (MCDM) (Pereira and Duckstein 1993), Delphi (Crance 1987), and nominal group technique (NGT) (Allen et al. 1987). Generally speaking, these techniques use the advice of more than one specialist as independent estimates of the "true" species–environment relationship and evaluate its variability based on these estimates.

Inductive–analytical techniques rely on samples of locations that are analyzed with some sort of statistical procedure. Different techniques have been used, including generalized linear models (GLMS; McCullagh and Nelder 1988; for applications see Akçakaya et al. 1995; Bozek and Rahel 1992; Pausas et al. 1995; Pearce et al. 1994; Pereira and Itami 1991; Thomasma et al. 1991; Van Apeldoorn et al. 1994), Bayes theorem approach (Aspinall 1992; Aspinall and Matthews 1994; Pereira and Itami 1991; Skidmore 1989a), classification trees (Walker 1990; Walker and Moore 1988; Skidmore et al. 1996), and multivariate statistical methods such as discriminant analysis (Dubuc et al. 1990; Flather and King 1992; Haworth and Thompson 1990; Livingston et al. 1990; Verbyla and Litvaitis 1989), discriminant barycentric analysis (Genard and Lescourret, 1992), principal component analysis (PCA) (Lehmkuhl and Raphael 1993; Picozzi et al. 1992; Ross et al. 1993), cluster analysis (Hodgson et al. 1987), and Mahalanobis distance (Clark et al. 1993; Knick and Dyer 1997; Corsi et al. 1999).

Models that use simple univariate statistics, such as ANOVA, Pearson rank correlation, and Bonferroni, pertain to a different subgroup because these analyses do not generally allow for definition of the relative importance of the environmental variables.

Further differences should be outlined for models that rely on the interpolation of density or census estimates to extrapolate distribution patterns. Although we have included these models in the inductive–analytical group, the geostatistical approach (Steffens 1992) on which they are generally based suggests putting them into a slightly different subgroup.

Finally, another means of classifying GIS distribution models can be based on their outputs. Essentially, these can be distinguished as categorical–discrete models and probabilistic–continuous models. Most often the products of the first type of models are polygon maps in which each polygon is classified according to a presence–absence criterion or a nominal category (e.g., frequent, scarce, absent). The products of the second type of model are continuous surfaces of an index that describes species presence in terms of the relative importance of any given location with respect to all the others. Indices that have been used are the suitability index (Akçakaya et al. 1995; Pereira and Itami 1991), probability of presence (Agee et al. 1989; Skidmore 1989a; Aspinall 1992; Clark et al. 1993; Walker 1990), ecological distances from "optimum" conditions (Corsi et al. 1999), and species densities (Palmeirin 1988; Steffens 1992). All these indices can be mapped as a continuous surface throughout the species range.

Generally, discrete models are built associating the presence of a species to polygons of land unit types (e.g., vegetation categories), most often with a deductive approach; in fact, transferring into the realm of GIS, the traditional way of producing distribution maps is based on a similar but more arbitrary partitioning of the study area (e.g., administrative boundaries, regular grids; see also "Habitat Definitions and Use"). There are also some examples of binary classifications of continuous environmental variables (e.g., slope, aspect, elevation) using statistical techniques such as logistic regression (Pereira and Itami 1991) or discriminant analysis (Corsi et al. 1999). Categorical–discrete models do not account for species mobility and tend to give a static description of species distribution. Nevertheless, this approach can be used to address the problem of defining areas of occupancy (Gaston 1991) and thus can be used successfully for problems of land management and administration. On the other hand, probabilistic models can describe part of the stochasticity typical of locating an individual of a species and can be used to address problems of corridor design and metapopulation modeling (Akçakaya 1993), introducing the geographic dimension in the analysis of species viability.

## LITERATURE REVIEW

Table 11.2 indicates the results of our bibliographic review. Papers are classified according to the categories described in the previous paragraph.

We have considered GIS and RS as two different views of the same tool, the former being more devoted to spatial correlation analysis and the later more concerned with basic data production. In fact, the two families of software tools share many basic functions and are evolving toward integration into a single system. It should be noted that the review includes not only papers that use GIS or RS but also some that deal with HSI, HEP and general assessment of species' ecological requirements. The papers in this last group do not generally represent examples of spatial models (Scott et al. 1993), in the sense that their products are not distribution maps, but they have been included because they are considered to be just a few steps away from a real distribution model. In fact, they describe the ecological requirements of the species in terms of mappable environmental conditions.

Most of the papers that use the deductive approach consider the a priori knowledge sufficient to define the ecological requirements of the species under investigation. This is especially true of papers that model distribution on the basis of interpretation of remotely sensed data; in fact, 15 out of 16 papers pertaining to the deductive group that used remotely sensed data to model species

Table 11.2    Classification of Reviewed Papers

|  |  | Deductive |  |  |  |  | Inductive |  |  |  |  |
|---|---|---|---|---|---|---|---|---|---|---|---|
| Descriptive | GIS | GIS and RS | RS | Non-spatial |  | GIS | GIS and RS | RS | Non-spatial |  |  |
|  | 9 | 8 | 7 | 8 | 32 | 3 | 1 | 0 | 0 | 4 | 36 |
| Analytical | GIS | GIS and RS | RS | Non-spatial |  | GIS | GIS and RS | RS | Non-spatial |  |  |
|  | 3 | 0 | 1 | 4 | 8 | 14 | 4 | 4 | 16 | 38 | 46 |
|  |  |  |  |  | 40 |  |  |  |  | 42 |  |

Papers are classified according to the approach used to define the species–environment relationship and whether their approach was descriptive or analytical. Further subtopics indicate whether the author considers the research to pertain to the domain of RS, GIS, or both. *Nonspatial* is used for papers that do not contain an explicit distribution model but define species–environment relationship in terms of mappable variables.

distributions fall within the descriptive group. In these papers, image classification techniques tend to receive more emphasis, whereas the ecological application is most often seen as an excuse to apply a specific classification algorithm.

The time trend of the papers published shows rather stable use of RS technology and increasing use of GIS. Up to 1986, no paper makes explicit reference to the term *GIS,* even though some of the papers dealing with the use of RS do use raster GIS-style overlay procedures to define their distribution models (e.g., Lyon 1983) and others do use a spatial approach but do not mention *GIS* (e.g., Mead et al. 1981).

Little is generally said about model assumptions. Of the 82 papers reviewed, only 21 discuss their assumptions. Those that do generally limit their discussion to the statistical assumptions of the technique used to perform the analysis. Very few deal with the biological and ecological assumptions and tend to take them for granted. When dealing with ecological modeling, we need to take into account both biological and methodological assumptions, along with some general assumptions that may limit the applicability of the results produced (Starfield 1997).

Validation, a step that is evidenced at different levels in the data flow diagram (figure 11.2), is generally limited to the accuracy of the result of the analysis (e.g., distribution map); nothing is said about the accuracy of the original data sets (e.g., GIS data layers, observation locations) and no consideration is given to issues such as error propagation in GIS overlay (Burrough 1986).

Only 15 papers validate of the accuracy of their results based on an independent estimate of the distribution (either through comparison with an independent set of observations or through comparison with the known distribution of the species); interestingly, 50 percent of these papers are based on the deductive approach. In fact, it should be noted that because observation data sets are the most expensive data to be collected within the general framework of setting up a GIS species distribution model, the deductive approach is the most cost-effective if seen from the validation point of view. In fact, to avoid bias, a model developed with an inductive approach cannot be validated using the same data set used to derive the species–environment relationship. Thus validation can be performed either with a second, independent data set or by dividing the original data set into two subsets, one of which is used to derive species–environment relationships and the other to validate the resulting model.

Finally, it is interesting to note that the multidimensional power of GIS is still not backed up by adequate quantity and quality of geographic data sets (Stoms et al. 1992). This is reflected in the number of environmental variables used in analysis. In the papers reviewed, the average is just below 4.8, and only 9 out of 82 analyze more than 9 environmental variables, whereas 23 papers base their distribution models on only one environmental variable, generally vegetation.

## ■ Modeling Issues

Based on the results of the literature review, we have identified five major issues that must be addressed to allow a sound GIS modeling of species distributions. These range from uncertainties in the objectives of the research to the lack of adequate support for the assumptions underlying the implementation of GIS models. A problem that is gaining awareness is that of scale, in both time and space, but it still suffers from inadequate tools.

Slightly different is the issue of data availability, which is rarely addressable by the biologist concerned with species distribution modeling but limits the type of models that can be developed.

Finally, a review of sources of errors and ways of estimating the accuracy of a GIS model addresses the problem of validation.

### CLEAR OBJECTIVES

When setting up an ecological model, the very first step to be considered is clear statement of the model's objective (Starfield 1997). There is great confusion about the objectives of many published papers. This may caused by overqualification of the tool, in the sense that use of the tool becomes the objective of the paper, or by uncertainty in defining the model's goals, along with coexisting purposes of predicting or understanding (Bunnell 1989). For instance, most of the papers based on the inductive approach deal with the definition of a species–environment relationship without specifying whether they intend to analyze the relationship of cause and effect or just use the relationship as a functional description of the effect. In the first case, the goal would be to evidence the limiting factors that are related to the species' biological needs and that drive the distribution process; in the second, it would be the simple use of correlated variables whose distribution is functional to the description of the species' distribution.

Basically, we can summarize species needs as food, shelter, and adequate reproduction sites (Flather et al. 1992; Pausas et al. 1995). When using the distribution of an environmental variable to describe the species' distribution we implicitly assume that there is a correlation between these basic needs and the environmental variables used. This correlation can be causal; that is, it describes the species' basic needs. In such cases we can identify a function that within a reasonable range of values associates each value of the environmental variable to a measure of the fulfillment of the species' basic needs (e.g., reproductive success). But it can also be a functional description; that is, we don't really know why some ranges of values of the environmental variable are preferred by the species but we observe that the species tends to occur more frequently within those ranges. The variable might influence all the species' basic needs simultaneously or be correlated to another variable that describes one of the species' needs.

Generally speaking, the quantity and quality of the locational data and the GIS layers used in analyses are not sufficient to assess cause–effect relationships that determine the species' distribution. Furthermore, cause–effect relationships spring from the interactions of biophysical factors that range through different time and space scales (Walters 1992); few papers take scale dependency into account in their analysis. Moreover in this kind of analysis causal effects can be hidden by independent interfering variables (Piersma et al. 1993) or by the unaccounted stochasticity of natural events such as weather fluctuations, disturbance, and population dynamics (Stoms et al. 1992) and should be assessed in controlled environments.

We believe such uncertainties could be addressed by defining the overall goal as the assessment of the relationship that best describe the species distribution. In other words, even if the causal understanding of a relationship is not clear, whenever the species–environment relationship is able to describe the distribution of a species satisfactorily, the overall goal is achieved (Twery et al. 1991).

Obviously the approach just described has some drawbacks. Without an adequate description of the cause–effect relationship between the species and environmental variables, models lose in transferability, in both space and time, and this limits their predictive capabilities (Levin 1992).

## ASSUMPTIONS

All models analyzed extrapolate their results to an entire study area on the assumption of space independence of the phenomenon observed at a given place. That is, in the case of both a deductive and an inductive approach, the species–environment relationship is built on evidence that a certain species occurs somewhere and that we know the values of the environmental variables at those locations. Obviously we know only that a species occurs at locations where it has been observed, only part of these locations have measurements of the environmental variables, and usually these measurements are collected only for the limited time range during which the investigation was carried out. Thus, when building distribution models, evidence collected in a portion of the range is extrapolated to the entire range of occurrence of a species. In order to do so, it is assumed that the species–environment relationship used to build the model is invariant in space and time. Most of the time this is not the case, especially for species with a wide range and for generalist species. In fact, the higher the variance of the species–environment relationship, the higher the number of locations required to provide an adequate ecological profile for the species.

Second, it is generally implicitly assumed that variables that are not included in the analysis have a neutral effect on the results of the model. That is, we need to assume either that the species' ecological response to these environmental variable is constant or that the response is highly correlated with the other variables included.

Even though both of these general assumptions are very difficult to test, we believe that they should be discussed on a case-by-case basis because the result of their violation is species-specific. Errors may be negligible in certain cases but can introduce major interpretation problems in other cases.

### Biological assumptions

Biological assumptions are direct consequences of the general assumptions discussed in the previous paragraph. We nevertheless believe that they are probably the most critical, but have received minimal attention in the literature.

The first assumption, which follows from the general assumption of space and time independence, states that observations reflect distribution. In other words, information on absence can be derived from observation data (Rexstad et al. 1988; Clark et al. 1993), which is obviously seldom the case. In fact, any time we have a record for a species we can be sure that the species (at least occasionally) occurs at that location. In contrast, if there is no observation for a species, we can only assume that we have a record of absence if there is no bias in our sampling scheme and that we have conducted our observations over a sufficiently long period. Even then we have no way of evaluating the random effects that are intrinsic in observing animals.

These assumptions can have statistical relevance in dealing with inductive–analytical approaches, but must hold true also for the deductive models. If there is a constant bias in the visibility of a species' individuals, for instance because part of their range is less accessible than others to researchers and thus cannot be as carefully investigated, the species–environment relationship reflects this bias. For instance, observation data are often gathered through sightings carried out by volunteers (Stoms et al. 1992; Hausser 1995), which do not follow a predefined (e.g., random) sampling scheme. Habitat cover may limit observations to areas where the species is visible (Agee et al. 1989). This may create an artificial response curve that associates a positive relationship to the values of the environmental variables measured in the locations where the species is more visible and a negative one in the ones measured in areas were the species has been less investigated. In such cases, we would end up mapping the areas where the species and the observers are most likely to meet, not the true distribution of the species.

This example is tailored to inductive–analytical models but can easily be extended to deductive ones, both descriptive and analytical, considering that the deductive approach is based on the a priori knowledge of specialists who rely on series of observations to gain experience and define the species–environment relationship. Again, these observations can suffer from accessibility or visibility biases.

A further assumption is that observations reflect the environmental selection of the species. Obviously this is not always true; for example, occurrences of migrant or vagrant individuals whose presence in a given location is occasional may be considered among observations. An extreme case is represented by locust swarms blown into the middle of the desert by strong winds. Clearly, their presence does not reflect any ecological preference. Nevertheless, if we consider only the observation per se, we would conclude that high densities of locusts are found in the desert and that locusts do prefer (with all the limitations that this term carries along in such an analysis) desert environments.

Obviously the strong wind of the example should be regarded as a stochastic event and thus be treated as an outlier in the definition of a possible GIS distribution model. In other words, observations should be analyzed for their content of unconstrained selection by the species.

We will see, when dealing with the issues of scale, that GIS distribution models tend to describe only the deterministic components that drive a species' distribution pattern, so stochastic events must be either averaged on the long term or eliminated as outliers. When observations are carried out for a limited time and the biology of the species under investigation is scarcely known, this problem can become increasingly important because the identification of outliers will be virtually impossible.

## Statistical assumptions

Most of the statistical techniques used to define species–environment relationships rely on the identification of two observation sets: one that identifies locations in which the species is present and one in which it is absent. Even though this cannot be identified properly as a statistical assumption, it is probably the most important factor limiting the applicability of the statistical techniques that rely on the two groups of observations.

The most common way to define the two subsets is to compare locations of known presence with a random sample of locations not pertaining to the previous set. Obviously some of the random locations can represent a suitable environment for the species, thus introducing, for that particular environment, a bias that underestimates the species–environment association.

To overcome this problem, data sets can be screened for outliers (Jongman et al. 1995), using for instance a scatter plot of the variables taken two by two. Once an outlier is identified, it can be checked to identify possible reasons for the absence of the species and, if necessary, removed from the analysis. Similar results can be achieved through analyses such as decision trees, where additional rules can be introduced to predict outliers (Walker 1990; Skidmore et al. 1996).

Another way to get around the problem is to eliminate the absence subgroup. Skidmore et al. (1996), for example, used both the BIOCLIM approach and the supervised nonparametric classifier, which use only observation sites to derive distribution patterns. The same result can also be achieved by using distance (or similarity) measures from the environmental characteristics of locations in which the species has been observed. A measure of distance that seems particularly promising for this application is the Mahalanobis distance

(Clark et al. 1993; Knick and Dyer 1997). It has many interesting properties as compared to other measures of similarity and dissimilarity, the most appealing of which is that it takes into account not only the mean values of the environmental variables measured at observation sites, but also their variance and covariance. Thus the Mahalanobis distance reflects the fact that variables with identical means may have a different range of acceptability and eliminates the problem that the use of correlated variables can have in the analysis.

Along with the identification of presence–absence data sets, each statistical method has some specific assumption that must be satisfied for correct application of the technique. For example, nonparametric statistical tests may assume that a distribution is symmetric, whereas a parametric test may assume that the test data are normally distributed. We will not discuss further the assumptions of the different statistical methods because they are beyond the scope of this chapter; we refer the reader to more specific books and journal articles on statistical methods.

### SPATIAL AND TEMPORAL SCALE

Scale is a central concept in developing species distribution models with GIS. As mentioned earlier in this chapter, this concept is common to both geography and ecology, the two main disciplines involved in the development of GIS species distribution models. The concept of scale evolves from the representation of the earth surface on maps and is the ratio of map distance to ground distance. Scale determines the following characteristics of a map (Butler et al. 1986): the amount of data or detail that can be shown, the extent of the information shown, and the degree and nature of the generalization carried out.

This group of characteristics determines the quality of the layers derived, that is, the quality of the environmental variables stored in the GIS database and the type of species–environment relationship that can be investigated (Bailey 1988; Levin 1992; Gaston 1994) using the capabilities of the GIS.

The scale of the analysis influences the type of assumptions that need to hold true for sound modeling. To clarify this concept, we need to consider that species distribution is the result of both deterministic and stochastic events. The former tend to be described in terms of the coexistence of a series of environmental factors related to the biological requirements of the species, whereas stochastic processes are regarded as disturbances caused by unpredictable or unaccountable events (Stoms et al. 1992). Generally distribution models are built on deterministic events and are averaged over wide spatial and temporal ranges to minimize the error related to the unaccounted stochasticity.

As we have seen, GIS distribution models rely on species–environment relationships to extrapolate distribution patterns based on the known distribution of the environmental variables. We have also seen that the relationships reflect the biological needs of the species. The extent to which we need to coarsen our temporal and spatial scales depends on the stochastic events that must be minimized, which in turn depend essentially on the dynamics of the species under investigation. To this extent, it is important to note that major population dynamics events happen on different scales in both time and space. In figure 11.3 (modified from Wallin et al. 1992) the two axes indicate the increasing temporal and spatial scale at which population dynamics events happen. In accordance with the hypothesis formulated by other authors (O'Neill et al. 1986; Noss 1992), the figure shows a positive correlation between space and time scales; that is, events that happen on a broader spatial scale are slower and thus take more time.

As a tool for distribution modeling this graph can be of great help in defining scale thresholds toward both a minimum and a maximum scale for an analysis. For instance, when considering cause–effect species–environment relationships the processes involved (e.g., feeding behavior) must be analyzed at an adequate scale (e.g., in our example, very detailed scale both in time and space). On the other hand, if we need to overcome the stochasticity introduced in our observation scheme by, for instance, individual foraging behavior we must average our results on a coarser scale in both time and space.

Thus, in GIS distribution models, both temporal and spatial scales are generally broadened so that stochastic events can average to a null component and thus be ignored. For instance, the stochasticity associated with the individual selection of a particular site, which greatly influences the distribution at a local scale, is overcome when dealing with distributions at regional scale averaging the selection of different individuals. In a similar way, stochastic events such as local fires, which influence regional distributions when measured over a short time interval (e.g., 5–10 years), are considered outliers in an analysis that takes into account the average vegetation cover over a longer time or a wider spatial span. Similarly, we know that in short time intervals the population dynamics status of a population is highly unpredictable, whereas it may be more easily averaged on longer time scales (Levin 1992) to become scarcely predictable again at even longer intervals.

A similar consideration is intrinsic in the minimum mappable unit (MMU), a concept used largely to address spatial scale issues in GIS species distribution models (Stoms 1992; Scott et al. 1993) that can be readily extended to the time scale. MMU can be seen from two points of view. On one hand, it is a

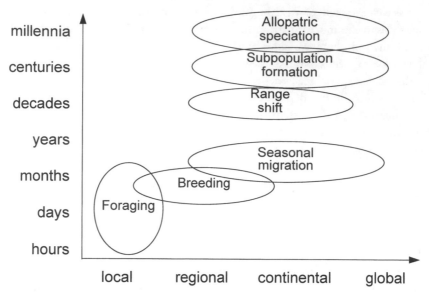

**Figure 11.3**  Population dynamics event in relation to time and space scales (modified from Wallin et al. 1992).

property of the data set that is being analyzed, that is, the minimum dimension of an element (e.g., a polygon representing vegetation types of a given category, the time span between successive manifestations of a given ecological event) that can be displayed and analyzed. On the other, it indicates the kind of averaging that must be carried out to smooth noise introduced by stochasticity. In fact, in the case of local fires, if the MMU is defined as larger than the extent of the fire in both time and space, the fire is automatically excluded from the analysis.

When dealing with scales on a practical basis, it should be noted that the structural complexity of distribution modeling can be simplified according to the hierarchical hypothesis (O'Neill et al. 1986) that states that at any given scale particular environmental variables drive the ecological processes. Thus weather becomes important at very broad spatial scales (e.g., continental scale). This is the basis of approaches behind models such as BIOCLIM (Busby 1991), that of Walker (1990), and that of Skidmore et al. (1996); all of them describe species distribution at a continental scale in terms of their direct relationship to climatic data. At successively finer scales such as regional landscapes, land form and topography play an important part (Haworth and

Thompson 1990; Aspinall 1992; Flather et al. 1992; Aspinall and Veitch 1993), whereas at the most local scales, indigenous land use structures become increasingly significant (Thomasma et al. 1991; Picozzi et al. 1992; Herr and Queen 1993) to the extent that even an individual stand of timber (Pausas et al. 1995) or a single pond (Genard and Lescourret 1992) can play a role. Generally speaking, the factors that are important vary according to scale, meaning that factors that are important at one scale level can lose their importance (Noss 1992), or at least much of it, at others.

As with any type of classification, the relationship between scale and environmental variables that drive ecological processes should not be taken too rigidly, and although most authors tend to agree that for broader scales climate is the most important factor, the same cannot be said when trying to identify the driving forces at finer scales. For instance, variables considered useful at coarser scales are used in detailed studies, as in the cases of Pereira and Itami (1991) and Ross et al. (1993), which use topography to explain species distribution at a much finer scale than the regional one. The same consideration applies to the studies of Aspinall and Matthews (1994), which use climatic data on a regional scale. On the other hand, land use is often used in distribution models developed at regional scale (Livingston et al. 1990; Flather and King 1992).

Finally, we must consider that distribution is the result of the interaction of many different biological events and that an ecological event cannot be described exhaustively on any single specific scale, but is the result of complex interactions of phenomena happening at different scales (Levin 1992; Noss 1992). Thus the limit of the applicability of a given environmental variable to describe distribution on any given scale may not be so sharp and the challenge is toward the integration of different scales in the description of the species' distributions. Buckland and Elston (1993) gave an example of the integration of environmental variables stored at different resolutions within the same distribution model.

It is important to note that the concept of scale not only determines the biological extent to which a distribution model can be applied but also affects the use that can be made of such a model for conservation. Also, conservation actions can be seen as having a hierarchical approach (Kolasa 1989). For instance, Scott et al. (1987) identified six different levels of intervention: landscape, ecosystem, community, species, population, and individual. Not surprisingly, conservation actions tend to become more effective and less expensive when the assessment moves toward broader scales, that is, when one moves from the individual to the landscape approach (Scott et al. 1987). Obviously this relates only to the extent of the analysis, not to its resolution. Nevertheless, on a cost–benefit basis, it is generally more efficient to address conservation-related issues at a coarser scale, which enables a landscape approach, than to concentrate on a more detailed scale (e.g., individual or population level), which requires high-resolution data to be analyzed that are either too precise or simply too abundant in terms of storage requirements to be analyzed profitably with a landscape approach.

What economics suggests is that conservation science needs to have a broader view of phenomena. A broad-scale approach and the possibility of predicting the potential dynamics of spatial patterns are needed to manage fragmentation of suitable environments and the inevitable metapopulation structure of the resulting population (Noss 1992). May (1994) indicates that when multiple levels of biological organization are concerned, as in a typical conservation action, the best management approach can be achieved on the regional landscape scale ($10^3$ to $10^5$ km$^2$). This scale level has suffered historically from limitations in the tools available for consistent analysis and is the one that has gained the most from the evolution of GIS; in fact, most of the distribution models based on GIS address problems at regional landscape level.

## DATA AVAILABILITY

Data availability and quality are two of the three limiting factors in the development of GIS-based species distribution models (the other being reliability of the models themselves [Stoms et al. 1992], which is discussed later in this chapter). The problem of developing extensive data sets of environmental variables is limited by economic and political rather than technical constraints. Estes and Mooneyhan (1994) list a number of different attitudes of governments throughout the world that limit the availability of high-resolution, "science-quality"[5] environmental data sets. These range from military classification of the data, thereby precluding the use of the data to the scientific community, to the low political priority that certain governments give to environmental issues. Moreover, even when policy is not an obstacle to the production and availability of data sets, entire nationwide data sets are sometimes lost during revolutions, wars, and civil disturbances. To this it should be added that some governments (e.g., the European Union countries) ask high prices for data sets, which are generally acquired with tax money, actually preventing their broad use in any type of activity and more specifically in environmental research.

In many cases, high-quality site-specific data sets are generated for a particular research project but are compiled with nonstandard techniques, rendering them unsuitable for combination and the achievement of more extensive knowledge of an area.

In the past few years there has been an increasing effort to develop meta-databases of available data sets throughout the world, and the problem is being addressed by national and international organizations (e.g., United Nations Environmental Programme, World Bank, U.S. Geological Survey [USGS], European Environmental Agency). These initiatives still do not address the problem of producing high-quality data sets, but at least they are a start in collating existing data sets. An important example is given by the joint efforts of the USGS, the University of Nebraska–Lincoln, and the European Commission's Directorate General Joint Research Centre, which are generating a 1-km-resolution Global Land Cover Characterisation (GLCC) database suitable for use in a wide range of environmental research and modeling applications from regional up to continental scale. All data used or generated during the course of the project (source, interpretations, attributes, and derived data), unless protected by copyrights or trade secret agreements, are distributed through the Internet. This effort goes in the direction of producing and distributing homogeneous medium-resolution high-quality data sets with known standards of accuracy.

Further aspects of raw data sets are discussed in the next section, where the quality of the data used to build models is discussed. We do not discuss this issue further here because we do not believe it to be a problem that can be addressed directly by conservation biologists or ecologists, although they can contribute to developing awareness of the need for standardization of data sets and for their production and dissemination.

## VALIDATION AND ACCURACY ASSESSMENT

Generally, the main function of a GIS-based species distribution model is to produce a map or its digital analogue for assessment of management and conservation actions. Possibly the most important question to be asked by a user is 'how accurate is the distribution map that has been produced?'

Many articles have been written on the sources of error in the data layers that may be included in a GIS. Nevertheless, few authors of papers dealing with animal distribution include an assessment of the accuracy of their model and a validation of the product. Because we believe this issue to be central to the entire process of species distribution modeling, the aim of this section is to review sources of error in GISs, to discuss methods of assessing mapping accuracy, and to evaluate the accumulation of thematic map errors in GISs, thus providing a framework for assessment of the accuracy of distribution models developed with GIS.

*Source of errors*

GIS data layers are traditionally classified according to their data structure, either raster or vector. To a certain extent, both error sources and accuracy evaluation methods have been investigated following this traditional classification.

Raster images may be obtained from remote sensing instruments carried by aircraft or spacecraft platforms, or by converting an existing line map (vector data structure) to a raster data structure. Two types of error are inherent in remotely sensed images: geometric and radiometric. These error sources are addressed in detail in numerous monographs and papers, including Colwell (1983) and Richards (1986).

A raster image is usually made up of a regular grid of adjacent rectangular cells or pixels (i.e., a rectangular tessellation). Geometric error in a remotely sensed image is caused by movement in the remote sensing platform; distortion caused by the earth's curvature and terrain; different centrifugal forces from earth affecting spacecraft movement; the earth's rotational skew; distortions introduced by the remote sensing device itself, including systematic distortions caused by sampling sequentially from each detector and nonlinear scanning (Adomeit et al. 1981); and errors introduced by the georeferencing process. Geometric error causes a point on the remotely sensed image to occur in the wrong position relative to other points in the image.

Correction of geometric errors in remotely sensed data is now a routine aspect of their preprocessing. The map or image is usually rubber-sheeted to fit it to an appropriate map projection. Corrected images with geometric errors of less than 0.5 pixel are now obtainable and acceptable[6] (Ford and Zanelli 1985; Ehlers and Welch 1987; Skidmore 1989b). However, the base maps from which control point information is derived may be of poor quality. Bell (1986) reported that maps used to geometrically correct images of the Great Barrier Reef contained errors of up to 1 km. The accurate selection of control points is crucial in obtaining acceptable results.

Points within a rubber-sheeted image are no longer on a regular grid because they have been warped to fit into the projection defined by the ground control points (GCPS). To obtain a regular grid, an interpolation method is used to nominate a value for a regular grid point that falls between the points in the rubber-sheeted image. Lam (1983) provides an excellent review of other interpolation methods, including splines, finite difference, and kriging.

Radiometric errors occur as a result of differential scattering of electromagnetic radiation at varying wavelengths, sensors that have poorly calibrated multiple detectors within a band, sensor calibration error, signal digitization error, and scene-specific error such as off-nadir viewing, irradiance variation, and terrain topography (Richards 1986). Correction of band-to-band distortion is performed using image histograms (shifted to the origin to remove atmospheric scattering effects), whereas line striping effects are reduced by calibration of detectors or by matching detector statistics during computer processing (Teillet 1986).

A final type of error may be caused by a time lag between ground truthing and image collection. In this case, pixels may be noted as incorrect in the error matrix (described later in this chapter) when they may be actually correct at the time of image acquisition.

Vector images have been traditionally recorded and stored as maps. Maps are subject to many errors. Some errors are introduced during the creation of the map, such as the original line smoothing, which may not follow the true isolines on the ground (Chrisman 1987). Other errors may be associated with the physical medium used to store the map (e.g., paper stretch and distortion).

Maps may be represented in computer GISs by a variation of the vector data structure (Peuquet 1984) or converted to a raster data structure. In its simplest form, the vector data structure has map lines approximated to a set of points (nodes), which are linked by lines (or arcs). Vector data may be obtained by digitization.

Digitization introduces a number of errors. Varying line thickness on the original map requires automatically scanned vector lines to be thinned. During manual digitization the center of the map line must be followed carefully if the map lines vary in thickness (Peuquet and Boyle 1984). This requires very careful hand digitizing or high-accuracy automatic scanners. The number of vertices (points) used to approximate a curve is also critical (Aldred 1972). Too few vertices will result in the line appearing stepped, and too many vertices create large data volumes. Thus, even with extreme care, error is introduced during digitization.

As for raster images, the main method of correcting geometric error in vector images is by using ground control points from a cartographically correct map to transform the vector image to a known projection.

## Data layer error quantification

Methods for quantifying error in a raster data layer are based on the error matrix (also called a contingency table or confusion matrix) concept, first expounded for remotely sensed data in the 1970s (Hoffer 1975).

The aim of the error matrix is to estimate the mapping accuracy (i.e., the proportion of correctly mapped pixels) within an image. An error matrix is constructed from points sampled from the image. The reference (or verification) data are normally represented along the columns of the matrix, and are compared with the classified (or image) data represented along the rows. The major diagonal of the matrix represents the agreement between the two data sets (table 11.3).

To check every cell for correctness would be impossible except in the smallest map area, so various sampling schemes have been proposed to select pixels to test. The design of the sampling strategy, the number of samples required, and the area of the samples have been debated by the remote sensing and GIS community.

As with any sampling problem, one is obviously trying to select the sampling design that gives the smallest variance and highest precision for a given cost (Cochran 1977). A number of alternative designs have been proposed for sampling the pixels to be used in constructing the error matrix. Berry and Baker (1968) recommended the use of a stratified systematic sample.[7] The advantage of systematic sampling over random sampling is that sample units are distributed equitably over the area. The disadvantage is that the resulting sample is weighted in favor of the class covering the largest area and that classes with a small area may not be sampled at all.

Simple random sampling in land evaluation surveys emphasizes larger areas and undersamples smaller areas (Zonneveld 1974). Zonneveld (1974) suggested that a stratified random sample was preferable, and Van Genderen et al. (1978:1135) agreed that a stratified random sample is "the most appropriate method of sampling in resource studies using remotely sensed data."

Rosenfield et al. (1982; see Berry and Baker 1968) suggested a stratified systematic unaligned sampling procedure (i.e., an area weighted procedure) as a first-stage sample to assist in identifying categories occupying a small area, followed by further stratified random sampling for classes with fewer than the desired minimum number of points. Todd et al. (1980) argued that single-stage cluster sampling is the cheapest sampling method because multiple observations can be checked at each sample unit on the ground.

Congalton (1988) simulated five sampling strategies (simple random sampling, stratified random sampling, cluster sampling, systematic sampling, and stratified systematic unaligned sampling) using a different number of samples over remotely sensed images of forest, rangeland, and grassland. The aim of the study was to ascertain the effect of different sampling schemes on estimating map accuracies using error matrices. He concluded that great care should be

**Table 11.3  Typical Error Matrix**

| | Class | \multicolumn{9}{c}{Classification of Pixels (ground truth)} | |
| --- | --- | --- | --- | --- | --- | --- | --- | --- | --- | --- | --- |

| | Class | I | II | III | VI | V | VI | VII | VIII | IX | Total |
| --- | --- | --- | --- | --- | --- | --- | --- | --- | --- | --- | --- |
| Classification | I | 14 | 3 | 7 | 2 | | 5 | | 7 | | 38 |
| of pixels | IV | 4 | | 8 | 3 | | 2 | 1 | 3 | | 21 |
| (model) | V | 2 | | | | 16 | | | | 1 | 19 |
| | VI | | | 1 | | 1 | | | | | 2 |
| | VII | | 1 | 3 | | | | 3 | | | 7 |
| | VIII | | | | | | | | 2 | | 2 |
| | IX | | | | | | | | | | |
| Total no. of pixels | | 20 | 19 | 38 | 6 | 17 | 8 | 8 | 17 | 2 | 135 |

Overall classification accuracy 50.4%.[a]

Example drawn from the classification of vegetation types from a satellite image. I = yertchuk, II = gum/stringybark, III = silvertop ash, IV = blue-leaf stringybark, V = clearcut/road, VI = tea tree, VII = gum/silvertop ash, VIII = black oak, IX = unclassified.
[a]Ratio of the sum of correctly classified pixels in all classes to the total number of pixels tested.

taken in using systematic sampling and stratified systematic unaligned sampling because these methods could overestimate population parameters. Congalton (1988) also stated that cluster sampling may be used, provided a small number of pixels per cluster are selected (he suggested a maximum of 10 sample pixels per cluster). Stratified random sampling worked well and may be used where small but important areas must be included in the sample. However, simple random sampling may be used in all situations.

The number of samples may be related to two factors in map accuracy assessment: the number of samples that must be taken in order to reject a map as being inaccurate and the number of samples required to determine the true accuracy within some error bounds for a map.

Van Genderen et al. (1978) pointed out that we want to know, for a given number of sample pixels, the probability of accepting an incorrect map. In other words, when high mapping accuracy is obtained with a small sample (e.g., 10 items), there is a chance that no pixels that are in error may be sampled (i.e., a type II error[8] is committed). The corollary, as stated by Ginevan (1979), is also important: The probability of rejecting a correct map (i.e., committing a type I error) must also be determined.

Van Genderen and Lock (1977) and Van Genderen et al. (1978) argued that only maps with 95 percent confidence intervals (i.e., $b = 0.05$) should be accepted and proposed a sample size of 30. Ginevan (1979) pointed out that Van Genderen et al. (1978) made no allowance for incorrectly rejecting an accurate map. The tradeoff one makes using Ginevan's more conservative approach is to take a larger sample, but in so doing reduce the chance of rejecting an acceptable map. Hay (1979)[9] concluded that minimum sample size should be 50, greater than that of Van Genderen and Lock (1977).

The number of samples must be traded off against the area covered by a sample unit, given a certain quantity of money to perform a sampling operation. Curran and Williamson (1986) asked whether many small-area samples or a few large-area samples should be taken. The answer is that it depends on the cover type being mapped; a highly variable cover type such as rainforest is better suited to many small-area samples, whereas for more homogeneous cover types it is more efficient to take fewer large-area samples.

Generally, mapping of heterogeneous classes such as forest and residential land is more accurate at 80-m resolution than at finer resolutions such as 30 m (Toll 1984); however, more homogeneous classes such as agricultural land and rangeland are more accurately mapped at 30 m than at 80 m (Toll 1984). The reason for this is the tradeoff between ground element size and image pixel resolution.

Based on the error matrix, different measures of accuracy can be derived. A commonly cited measure of mapping accuracy is the overall accuracy, which is the number of correctly classified pixels (i.e., the sum of the major diagonal cells in the error matrix) divided by the total number of pixels checked (table 11.3). Anderson et al. (1976) suggested that the minimum level of interpretation accuracy in the identification of land use and land cover categories should be 85 percent.

Overall classification accuracy is the ratio of the total number of correctly classified pixels to the total number of pixels in each class (Kalensky and Scherk 1975).

Cohen (1960) and Bishop et al. (1975) defined a measure of overall agreement between image data and the reference (ground truth) data called Kappa or $K$. $K$ ranges in value from 0 (no association, that is, any agreement between the two images equals chance agreement) through 1 (full association, or perfect agreement between the two images). $K$ can also be negative, which signifies a less than chance agreement.

The methods just discussed for quantifying error in raster images are equally applicable to quantifying error in vector polygons. Instead of checking whether an image pixel is correctly classified, a point within the polygon is ver-ified against the ground truth information. A specific problem encountered with vector images is ground truth samples that occur across boundary lines; in this case the class with the largest area within the sample area may be selected to represent the vector map image.

A method of assessing map accuracy based on line intersect sampling was described by Skidmore and Turner (1992). Line intersect sampling is used to estimate the length of cover class boundaries on a map[10] that coincide with the true boundaries of the cover classes on the ground. A ratio of coincident boundary to total boundary is proposed as a measure of map accuracy and this ratio is called the boundary error. Although this technique was developed for vector maps, it is equally applicable to raster maps.

Skidmore and Turner (1992) found that the true boundary lengths were not significantly different from the estimated boundary lengths sampled using line intersect sampling, with $a' = 0.05$. The estimated boundary accuracy (64 percent) was extremely close to the true boundary accuracy (65.1 percent), and there was no significant difference between the true and estimated boundary accuracy.

### Reliability of the output

Apart from assessing the accuracy of the original data sets used to produce GIS models, the final products of the modeling effort must be validated.

As previously evidenced, the process of model building takes its move from a number of data maps of the same region (e.g. elevation, soils, ground cover), which are digitized and geographically rectified to a common projection in a GIS. Such maps may be stored as a series of layers in a GIS. Each point within a polygon, or each cell in a raster layer (where each raster cell is assigned one value), takes the values of the layers directly above the point. The model is then built defining specific questions that may be asked about specific points or cells, such as the slope or aspect at the point. Eventually the biologist is inter-ested in mapping areas that satisfy the known ecological requirements of a species (e.g., slope greater than 30°, with erodible soils, occurring on a southerly aspect). By overlaying the map layers, such simple queries may be answered.

Nevertheless, the simple process of overlaying layers propagates the errors of the original data and sometimes amplifies them. If the sources of errors are identified and their accuracy is known, an estimate of the accuracy of the final output can be achieved through error propagation analysis.

Sources of errors range from those evidenced in the previous paragraphs to those inherent to nonspatial or text data that are also part of a GIS model. For instance, nonspatial data include knowledge or rules used by expert systems (Skidmore 1989b). All these errors contribute to error accumulation when overlaying GIS data layers.

Composite overlaying is the simplest overlaying technique. Two or more layers are combined, and the raster locations (or polygons formed) describe the union of the classes on the layers. The composite overlay is in effect a universal Boolean *and* operation over the whole map. That is, for a two-layer data set comprising layer X and layer Y, we note $X_{i=1,n} \cap Y_{j=1,m}$ (i.e., the intersection of X and Y for the *n* classes in map X and the *m* classes in map Y) at all points over the map.

Arithmetic and mathematical operators that may be applied to two or more layers include addition, subtraction, multiplication, division, maximum, minimum, average, and exponent.

The method by which error is accumulated during the overlaying process is important for modeling error in the final map products. The first necessity for modeling map error accumulation is to quantify the error in the individual layers being overlaid. As discussed earlier, a lot of work has been done on this problem, with some tangible results.

Newcomer and Szajgin (1984) used probability theory to calculate error accumulation through two map layers. They assumed that the two map layers were dependent; that is, if we select a cell that is in error in layer 1 then that act reduces the probability of selecting an erroneous cell from layer 2. If the data layers are independent, then an erroneous cell selected from layer 1 does not reduce the probability of selecting an incorrect cell from layer 2.

Using the statistics of Parratt (1961) with empirical data, Burrough (1986) concluded that with two layers of continuous data, the addition operation is unimportant in terms of error accumulation. The amount of error accumulated by the division and multiplication operations is much larger. The largest error accumulation occurs during subtraction operations. Correlated variables may have higher error accumulation rates than noncorrelated data because erroneous regions tend to coincide and concentrate error rates there.

The use of Bayesian logic for GIS overlaying is explained in Skidmore (1989b). As with Boolean, arithmetic, and composite overlaying, there is inherent error in the individual data layers when overlaying using Bayesian logic. In addition, Bayesian overlaying uses rules to link the evidence to the hypotheses; the rules have an associated uncertainty and are an additional source of error.

A number of possible solutions are suggested for modeling accumulation of error in a GIS. These include reliability diagrams and a probabilistic approach.

Wright (1942) suggested that reliability diagrams should accompany all maps. He also emphasized that the sources used to generate different regions of the map have varying accuracy and these sources should be stated clearly on the map. For example, one region may have been mapped using low-altitude aerial photography and controlled ground survey, and would therefore be more accurate than another region mapped using high-altitude photography and only reconnaissance survey. This theme was taken up by Chrisman (1987) and MacEachren (1985), who suggested that such a reliability diagram showing map pedigree should be included as an additional layer accompanying each map layer in a GIS. However, for the purposes of error accumulation modeling, reliability diagrams do not provide a quantitative statement about the accuracy (or error) of the map.

The supervised nonparametric classifier described by Skidmore and Turner (1988) and Skidmore et al. (1996) classified remotely sensed and GIS digital data. The classifier gives for all cells the empirical probability of correct classification for each class according to the training area data and thereby gives an indication of map accuracy.

The few methods proposed for modeling error accumulation are limited in their application. Working with ideal data, these methods do allow some conclusions to be drawn about error accumulation during GIS overlay operations. However, the methods break down when used with map layers created under different conditions than assumed by the methods.

Newcomer and Sjazgin (1984) used probability theory to model error accumulation, whereas Heuvelink and Burrough (1993) modeled the accumulation of error in Boolean models, using surfaces interpolated by kriging as the estimated error source.

*Sensitivity analysis*

When no information is available on the extent of the errors of the original data sets or on the type of error propagation function applicable to the model, a way of defining levels of reliability of the output is to analyze its variability subject to changes in the input parameters. In the this chapter we have seen possible sources of variability and uncertainties that can arise in the deductive and inductive approaches in species distribution modeling. They range from subjective errors introduced by the specialist who defines the species–environment relationship to locational errors of species observation caused by possible biases in the sampling scheme or inaccuracy of the instrument used to locate the species (e.g., radiotelemetry, global positioning system). A through discussion of the sources of uncertainties in species distribution modeling can be found in Stoms et al. (1992).

Once the sources and ranges of variability are identified, different input data sets can be systematically produced, selecting the variates (sensu Sokal and Rohlf 1995) from the variability range of the original input variables. These alternative data sets are used to build alternative models that can be compared with the original one, identifying the variability induced in the output by the uncertainty of the input variables. The variability induced in the output is a measure of the overall performance of the model and can be compared with a predetermined acceptable significance level. As a general rule, when dealing with great uncertainties in the measures of the input variables, a greater inertia (less subject to changes in the results) of the model is generally preferable.

Sensitivity analysis does not replace validation but can be used at any stage of the model-building process to identify the parameters that should be monitored more carefully to maximize the reliability and the accuracy of the results.

## ■ Discussion

The use of GIS in species distribution modeling should follow precise steps in which each of the issues discussed in this chapter is accounted for. First of all, we recommend more unambiguous use of some key terms such as *scale* and *habitat*. The latter seems to be a particular problem not limited to GIS applications but spanning the entire field of ecological studies (Hall et al. 1997). We believe that GIS can be a valid tool to overcome the current ambiguity between the species-related and land-related concepts of the term *habitat*. As a matter of fact, the latter was introduced as a way of dealing with problems related to environmental mapping using traditional tools, and the enhancements introduced by GIS do overcome those problems. In the meantime, however, we suggest replacing the word *habitat* with more unproblematic terms such as *environment*.

The ambiguity of the term *habitat* and most of the works on habitat use and habitat selection have also given rise to the question of whether GIS-based models can be used to explain the causal event of a species–environment relationship. In our opinion, use of GIS is not central to a better understanding of causal effects in a species–environment relationship, especially if the quality of data does not support both high-resolution and large-extension analyses. Currently our analytical capabilities are limited by the lack of high-resolution, global coverage data sets. Nevertheless, the use of GIS's spatial analysis tools in the framework of a controlled environment in which all the key variables are monitored at an adequate resolution can increase our ability to assess causal effects in species–environment relationships.

Apart from further generic considerations, we think that a few important issues have been overlooked in these first years of application of GIS to the field of ecological modeling and especially in the field of species distribution modeling. There has been inadequate discussion and consideration of the assumptions underlying the model-building process and the related issues of spatial and temporal scale, which are of paramount importance for sound scientific[11] use of GIS. Adequately discussed assumptions can justify the development of a model. Whenever a hypothesis is stated and a model is built to test its congruence, it should be regarded as a problem-solving tool. For instance, we will never really know what will be the outcome of alternative management options, but we can state different hypotheses, state the assumptions that must be met to make each hypothesis hold true, and try to model the result of the different options. In such cases we don't have direct control over the results of the management action; we can only ensure that the assumptions are met. This means that the output of the model will hold true if the assumptions are met and if the model is built on the logical consequences of these assumptions. In such cases validation, meant as an independent estimate of the truth, can to a certain extent be neglected (Starfield 1997). Nevertheless, most of the time assumptions are not adequately discussed and this is particularly evident when dealing with the constraints of scale dependency of biological events. Probably the issues of scale still suffer from inadequate support from the available tools. For instance, we still lack convenient ways of handling spatiotemporal data in GIS software packages, not to speak of analyzing the two components together.

If validation can be neglected somewhat when dealing with hypothesis-testing models, it becomes a fundamental issue when building analytical models, which are built to assess species–environment relationships and ecological processes. In such cases, validation steps must be included from the beginning of the model building process, first assessing the quality and reliability of the raw data used, then evaluating the limits of the relationships that drive the process and finally analyzing the correspondence of the output with the truth.

Validation can be a costly exercise in model building, and efforts are being made to find a cost-effective approach to this issue. Because the issue of validation is general to GIS modeling and especially GIS ecological applications, it can benefit from the experience derived from other fields of science (e.g., RS). Methods for summarizing the accuracy of raster and vector maps using point samples and error matrices are now widely used in GIS and are beginning to find their way into ecological applications as well. However, standard techniques have not had universal acceptance for a number of reasons. For example, a number of alternative sampling designs have been proposed for analyzing the accuracy of imagery. The choice of sampling design is often subject to the particular problems associated with the area to be ground-truthed. However, a number of general trends are obvious from the GIS and RS literature. Random and stratified random sampling are acknowledged to maximize precision and accuracy (though at a higher cost than cluster sampling or systematic sampling). It should be noted that in highly heterogeneous landscapes (e.g., native forests, especially tropical forests), stratification is often too costly to consider. Cluster sampling offers reduced sampling costs, but in order to be effective it depends on low intracluster variance. Systematic sampling schemes may lead to a bias in parameter estimation if periodic errors align with the sampling frame (e.g., as a result of image banding or linear topographic features, as in the Allegheny Mountains of Pennsylvania).

GIS data layers contain numerous errors. These pose a number of problems as errors accumulate during the process of analysis and model building. Although modeling the accumulation of error during GIS overlay analysis is still in its infancy, some methods for measuring error accumulation during GIS analysis have been discussed.

Any procedure to reduce mapping error in individual layers in a GIS will improve the mapping accuracy of an overlay generated from the GIS. Until better error-modeling techniques are developed for GISs, descriptive statistics should ideally be calculated for each layer in a GIS, as well as for each layer produced by GIS modeling. The descriptive statistics should include overall mapping accuracies as well as class mapping accuracies. An alternative way of defining the performance of a GIS model, thus assigning a level of reliability to its results, is sensitivity analysis that identifies crucial parameters. These parameters are those that, within their range of variation, determine the highest variation in the model output.

## ■ Conclusions

A common opinion among epistemologists is that we are facing a break between the development of advanced technologies and our needs and abilities to use them. As a result of this break, sophisticated systems are often used for simple operations and their use becomes a goal in itself. Toraldo di Francia (1978) names this paradox "the law of inversion between the goal and the instrument." This has also been the case with GIS: the emphasis was on the use of the tool rather than on solving the problems to which it was applied. During the infancy of the tool's development its enthusiastic and acritical application was a clear evidence of the low awareness of the tool's limitations. Although we don't believe that the infancy of GIS is over, in recent years we have seen a growth in the capabilities of the tool and the awareness of its users. We think that we have enough case studies to define a logical framework in which the process of GIS modeling and more specifically species distribution modeling should be kept. We have tried to accomplish this by identifying the issues that must be addressed during the entire process.

In an era in which the need to acquire and analyze data at wider scales is increasing and globalization in environmental assessment applications is becoming urgent, we should not waste the opportunity GIS offers to wildlife biologists to cope with these needs. This does not mean that GIS modeling can substitute for fieldwork and direct observation; especially in the case of rare or endangered species, all management action should be based on direct site-specific studies. Nevertheless, sound GIS distribution models, which can be achieved by addressing the different issues we have tried to address in this chapter, can help to identify areas that require more detailed investigation.

### Acknowledgments

We thank Luigi Boitani and Todd Fuller for giving us the opportunity to organize these ideas. The chapter is also the result of suggestions and comments from Luigi Boitani, Alessandra Pellegrini, Iacopo Sinibaldi, and two anonymous reviewers.

### Notes

1. *Area of occupancy* is defined as the area within the species' extent of occurrence that is occupied by a taxon, excluding cases of vagrancy. The measure reflects the fact that a taxon will not usually occur throughout the area of its extent of occurrence, which may contain unsuitable habitats, for example. The area of occupancy is the smallest area essential at any stage to the survival of existing populations of a taxon (e.g., colonial nesting sites, feeding sites for migratory taxa) (IUCN 1995).

2. Bibliographic database of literature in earth sciences, ecology, and geography published by Elsevier.

3. *Habitat* is also used in connection to humankind. In this chapter the term *habitat* refers to plant and animal species excluding human beings.

4. Here *habitat* relates to species, whereas it refers to land in the next sentence.

5. "'Science-quality' means that, in so far as both practical and possible, the errors inherent in the overall production of [these] maps have been documented" (Estes and Mooneyhan 1994).

6. Note that the accuracy of the geometric correction is sometimes expressed as root mean square (RMS) error, which is the standard error (of the difference between the transformed GCPs and the original GCPs) multiplied by the pixel size.

7. Each stratum has an unaligned systematic sample.

8. Type I errors have been called "consumer's risk" and type II errors "producer's risk" by Fung and LeDrew (1988) and others. These terms are taken from a branch of statistics called acceptance sampling. For the sake of consistency and in order to use conventional statistical terms, *type I* and *type II errors* are used here.

9. Hay (1979) noted that he developed the ideas expounded by Van Genderen and Lock (1977).

10. The map may be generated from remotely sensed imagery or by traditional cartographic methods such as aerial photograph interpretation.

11. Here *scientific* is used in the sense of scientific method: the recognition and formulation of a problem, the collection of data through observation and experiment, and the formulation and testing of the hypotheses.

## Literature Cited

Adomeit, E. M., D. L. B. Jupp, C. I. Margules, and K. K. Mayo. 1981. The separation of traditionally mapped land cover classes by Landsat data. In A. N. Gillison and D. J. Anderson, eds., *Vegetation classification in Australia,* 150–165. Canberra, Australia: ANU Press.

Agee J. K., S. C. F. Stitt, M. Nyquist, and R. Root. 1989. A geographic analysis of historical grizzly bear sightings in the North Cascades. *Photographic Engineering and Remote Sensing* 55: 1637–1642.

Akçakaya, H. R. 1993. *RAMAS/GIS: Linking landscape data with population viability analysis.* New York: Applied Biomathematics.

Akçakaya, H. R., M. A. McCarthy, and J. L. Pearce. 1995. Linking landscape data with population viability analysis: Management options for the helmeted honeyeater *Lichenostomus melanops cassidix. Biological Conservation* 73: 169–176.

Aldred, B. K. 1972. *Point-in-polygon algorithms.* Peterlee, U.K.: IBM.

Allen, A. W., P. A. Jordan, and J. W. Terrell. 1987. *Habitat suitability index models: Moose, Lake Superior region.* U.S. Fish and Wildlife Service Biological Report 82.

Anderson, J. R., E. E. Hardy, J. T. Roach, and R. E. Witmer. 1976. *A land use and land cover classification system for use with remote sensor data.* Washington, D.C.: U.S. Geological Survey.

Aspinall, R. 1992. An inductive modeling procedure based on Bayes' theorem for analysis of pattern in spatial data. *International Journal of Geographic Information Systems* 6: 105–121.

Aspinall, R. and K. Matthews. 1994. Climate change impact on distribution and abundance of wildlife species: An analytical approach using GIS. *Environmental Pollution* 86: 217–223.

Aspinall, R. A. and N. Veitch. 1993. Habitat mapping from satellite imagery and wildlife survey data using a Bayesian modelling procedure in a GIS. *Photographic Engineering and Remote Sensing* 59(4): 537–543.

Atkinson, S. F. 1985. Habitat based methods for biological impact assessment. *Environmental Professional* 7: 265–282.

Bailey, R. G. 1988. Problems with using overlay mapping for planning and their implications for geographic information systems. *Environmental Management* 12(1): 11–17.

Begon, M., J. L. Harper, and C. R. Townsend. 1990. *Ecology: individuals, populations and communities*. Cambridge, U.K.: Blackwell.

Bell, A. 1986. Satellite mapping of the Great Barrier Reef. *Ecos* 47:12–15.

Berry, B. J. L. and A. M. Baker. 1968. Geographic sampling. In B. J. L. Berry and D. F. Marble, eds., *Spatial analysis: A reader in statistical geography*, 230–248. Englewood Cliffs, N.J.: Prentice Hall.

Bishop, Y. M., S. E. Fienburg, and P. W. Holland. 1975. *Discrete multivariate analysis: Theory and practice*. Cambridge, Mass.: MIT Press.

Bozek, M. A. and F. J. Rahel. 1992. Generality of microhabitat suitability models for young Colorado River trout (*Oncorhynchus clarki pleuriticus*) across site and among years in Wyoming streams. *Canadian Journal of Aquatic Science* 49: 552–564.

Breininger, D. R., M. J. Provancha, and R. B. Smith. 1991. Mapping Florida scrub jay habitat for purposes of land-use management. *Photographic Engineering and Remote Sensing* 57: 1467–1474.

Buckland, S. T. and D. A. Elston. 1993. Empirical models for the spatial distribution of wildlife. *Journal of Applied Ecology* 30: 478–495.

Bunnell, F. L. 1989. *Alchemy and uncertainty: What good are models?* General Technical Report PNW-GTR-232. Portland, Oreg.: U.S. Department of Agriculture, Forest Service, Pacific Northwest Research Station.

Burrough, G. H. 1986. *Principles of geographical information systems for land resources assessment*. London: Clarendon Press.

Busby, J. R. 1991. BIOCLIM: A bioclimatic analysis and prediction system. In C. R. Margules and M. P. Austin, eds., *Nature conservation: Cost effective biological surveys and data analysis*, 64–68. Melbourne, Australia: CSIRO.

Butler, M. J. A., C. Le Blanc, J. A. Belbin, and J. L. MacNeill. 1986. Marine resources mapping: An introductory manual. FAO Fisheries Technical Paper 274.

Chrisman, N. 1987. Efficient digitizing through the combination of appropriate hardware and software for error detection and editing. *International Journal of Geographic Information Systems* 1(3): 265–278.

Clark, J. D., J. E. Dunn, and K. G. Smith. 1993. A multivariate model of female black bear habitat use for geographic information systems. *Journal of Wildlife Management* 57 (3): 519–526.

Cochran, W. G. 1977. *Sampling techniques.* New York: Wiley.

Cohen, J. 1960. A coefficient of agreement for nominal scales. *Educational and Psychological Measurement* 20(1): 37–46.

Collin, P. H. 1988. *Dictionary of ecology and the environment.* Teddington, U.K.: P. Collin Publishing.

Colwell, R. N. 1983. *Manual of remote sensing.* Falls Church, Va.: ASPRS.

Congalton, R. G. 1988. A comparison of sampling schemes used in generating error matrices for assessing the accuracy of maps generated from remotely sensed data. *Photographic Engineering and Remote Sensing* 54: 593–600.

Congalton, R. G., J. M. Stenback, and R. H. Barrett. 1993. Mapping deer habitat suitability using remote sensing and geographic information systems. *Geocarto International* 3: 23–33.

Corsi, F., E. Duprè, and L. Boitani. 1999. A large-scale model of wolf distribution in Italy for conservation planning. *Conservation Biology* 13(1): 1–11.

Crance, J. H. 1987. Guidelines for using the Delphi technique to develop habitat suitability index curves. U.S. Fish and Wildlife Service Biological Reports 82 (10.134).

Curran, P. J. and H. D. Williamson. 1986. Sample size for ground and remotely sensed data. *Remote Sensing of Environment* 20: 31–41.

Daubenmire, R. 1976. The use of vegetation in assessing the productivity of forest lands. *Botanical Review* 42: 115–143.

Donovan M. L., D. L. Rabe, and C. E. Olson. 1987. Use of geographic information systems to develop habitat suitability models. *Wildlife Society Bulletin* 15: 574–579.

Dubuc, L. J., W. B. Krohn, and R. B. Owen Jr. 1990. Predicting occurrence of river otters by habitat on Mount Desert Island, Maine. *Journal of Wildlife Management* 54(4): 594–599.

Duncan, B. W., D. R. Breininger, P. A. Schmalze, and V. L. Larson. 1995. Validating a Florida scrub jay habitat suitability model, using demography data on the Kennedy Space Center. *Photographic Engineering and Remote Sensing* 61: 1361–1370.

Edwards, P. J., R. M. May, and N. R. Webb, eds. 1994. *Large-scale ecology and conservation biology.* Oxford, U.K.: Blackwell Scientific.

Ehlers, M. and R. Welch. 1987. Stereocorrelation of Landsat TM images. *Photographic Engineering and Remote Sensing* 53: 1231–1237.

Encyclopaedia Britannica. 1994. *The new encyclopaedia Britannica.* Chicago: Encyclopaedia Britannica.

Estes, J. E. and D. W. Mooneyhan. 1994. Of maps and myths. *Photographic Engineering and Remote Sensing* 60(5): 517–524.

Flather, C. H., S. J. Brady, and D. B. Inkley. 1992. Regional habitat appraisal of wildlife communities: A landscape-level evaluation of a resource planning model using avian distribution data. *Landscape Ecology* 7(2): 137–147.

Flather, C. H. and R. M. King. 1992. Evaluating performance of regional wildlife habitat models: Implication to resource planning. *Journal of Environmental Management* 34: 31–46.

Ford, G. E. and C. I. Zanelli. 1985. Analysis and quantification of errors in the geometric correction of satellite images. *Photographic Engineering and Remote Sensing* 51: 1725–1734.

Fung, T. and E. LeDrew, 1988. The determination of optimal threshold levels for change detection using various accuracy indices. *Photographic Engineering and Remote Sensing* 54(10): 1449–1454.

Gaston, K. L. 1991. How large is a species geographic range? *Oikos* 61(3): 434–438.

Gaston, K. L. 1994. Measuring geographic range size. *Ecography* 17(2): 198–205.

Genard, M. and F. Lescourret. 1992. Modelling wetlands habitats for species management: The case of teal (*Anas crecca crecca*) in the Bassin d'Arachon (French Atlantic Coast). *Journal of Environmental Management* 34: 179–195.

Ginevan, M. E. 1979. Testing land-use map accuracy: Another look. *Photographic Engineering and Remote Sensing* 45: 1371–1377.

Hall, L. S., P. R. Krausman, and M. L. Morrison. 1997. The habitat concept and a plea for standard terminology. *Wildlife Society Bulletin* 25(1): 173–182.

Haslett, J. R. 1990. Geographic information systems: A new approach to habitat definition and study of distributions. *Trends in Ecology and Evolution* 5(7): 214–218.

Hausser, J. 1995. *Säugetiere der Schweiz/Mammifères de la Suisse/Mammiferi della Svizzera*. Geneva: Commission des Mémoires de l'Académie Suisse des Sciences Naturelles.

Haworth, P. F. and D. B. A. Thompson. 1990. Factors associated with the breeding distribution of upland birds in the South Pennines, England. *Journal of Applied Ecology* 27: 562–577.

Hay, A. M. 1979. Sampling design to test land use accuracy. *Photographic Engineering and Remote Sensing* 45: 529–533.

Herr, A. M. and L. P. Queen. 1993. Crane habitat evaluation using GIS and remote sensing. *Photographic Engineering and Remote Sensing* 59: 1531–1538.

Heuvelink, G. B. M. and P. A. Burrough. 1993. Error propagation in cartographic modelling using Boolean logic and continuous classification. *International Journal of Geographic Information Systems* 7: 231–246.

Hodgson, M. E., J. R. Jensen, H. E. Mackey Jr., and M. C. Coulter. 1987. Remote sensing of wetland habitat: A wood stork example. *Photographic Engineering and Remote Sensing* 53: 1075–1080.

Hodgson, M. E., J. R. Jensen, H. E. Mackey Jr., and M. C. Coulter. 1988. Monitoring wood stork habitat using remote sensing and geographic information systems. *Photographic Engineering and Remote Sensing* 54: 1601–1607.

Hoffer, R. M. 1975. *Computer-aided analysis of* KYLAB multispectral scanner data in mountainous terrain for land use, forestry, water resource, and geological applications. West Lafayette, Indiana: Laboratory for Applications of Remote Sensing, Purdue University.

IUCN: World Conservation Union, Species Survival Commission. 1995. *IUCN red list categories.* Gland, Switzerland: IUNC/World Conservation Union.

Jensen, J. R., S. Narumalani, O. Weatherbee, and K. S. Morris Jr. 1992. Predictive modelling of cattail and waterlily distribution in a South Carolina reservoir using GIS. *Photographic Engineering and Remote Sensing* 58(11): 1561–1568.

Jones, K. B. 1986. Data types. In A. Y. Cooperrider, J. B. Raymond, and H. R. Stuart, eds., *Inventory and monitoring of wildlife habitat,* 11–28. Denver, Colo.: U.S. Department of the Interior, Bureau of Lands.

Jongman, R. H. G., C. J. F. ter Braak, and O. F. R. van Tongeren, eds. 1995. *Data analysis in community and landscape ecology.* Cambridge, U.K.: Cambridge University Press.

Kalensky, Z. D. and L. R. Scherk. 1975. Accuracy of forest mapping from Landsat CCTs. Proceedings of the 10th International Symposium on Remote Sensing of Environment, 2: 1159–1163. Ann Arbor, Mich.: ERIM.

Kerr, R. M. 1986. Habitat mapping. In A. Y. Cooperrider, J. B. Raymond, and H. R. Stuart, eds., *Inventory and monitoring of wildlife habitat,* 49–72. Denver, Colo.: U.S. Department of the Interior, Bureau of Lands.

Klopatek, J. M., J. R. Krummel, J. B. Mankin, and R. V. O'Neil. 1983. A theoretical approach to regional environmental conflicts. *Journal of Environmental Management* 16: 1–15.

Knick, S. T. and D. L. Dyer. 1997. Distribution of black-tailed jackrabbit habitat determined by GIS in southwestern Idaho. *Journal of Wildlife Management* 61(1): 75–85.

Kolasa, J. 1989. Ecological systems in hierarchical perspective: Breaks in community structure and other consequences. *Ecology* 70: 36–47.

Krebs, C. J. 1985. *Ecology, the experimental analysis of distribution and abundance.* New York: Harper & Row.

Lam, N. S. 1983. Spatial interpolation methods: A review. *American Cartographer* 10: 129–149.

Lehmkuhl, J. F. and M. G. Raphael. 1993. Habitat pattern around northern spotted owl locations on the Olympic Peninsula, Washington. *Journal of Wildlife Management* 57: 302–315.

Levin, S. A. 1992. The problem of pattern and scale in Ecology. *Ecology* 73: 1943–1967.

Livingston, S. A., C. S. Todd, W. B. Krohn, and R. B. Owen Jr. 1990. Habitat models for nesting bald eagles in Maine. *Journal of Wildlife Management* 54(4): 644–653.

Lyon, J. G. 1983. Landsat-derived land-cover classification for locating potential kestrel nesting habitat. *Photographic Engineering and Remote Sensing* 49(2): 245–250.

Mace, G. M. 1994. An investigation into methods for categorizing the conservation status of species. In P. J. Edwards, R. M. May, and N. R. Webb, eds., *Large scale ecology and conservation biology,* 293–312. Oxford, U.K.: Blackwell Scientific.

MacEachren, A. M. 1985. Accuracy of thematic maps: Implications of choropleth symbolization. *Cartographica* 22: 38–58.

Maurer, B. A. 1994. *Geographical population analysis: Tools for the analysis of biodiversity. Methods in ecology.* Oxford, U.K.: Blackwell Scientific.

May, R. M. 1994. The effects of spatial scale on ecological questions and answers. In P. J. Edwards, R. M. May, and N. R. Webb, eds., *Large-scale ecology and conservation biology,* 1–17. Oxford, U.K.: Blackwell Scientific.

Mayhew, S. and A. Penny. 1992. *The concise Oxford dictionary of geography.* Oxford, U.K.: Oxford University Press.

McCullagh, P. and J. A. Nelder. 1988. *Generalized linear models.* London: Chapman & Hall.

Mead, R. A., T. L. Sharik, S. P. Prisley, and J. T. Heinen. 1981. A computerized spatial analysis system for assessing wildlife habitat from vegetation maps. *Canadian Journal of Remote Sensing* 7: 34–40.

Merriam-Webster. 1981. *Webster's third new international dictionary.* Chicago: Merriam-Webster.

Miller, R. I., ed. 1994. *Mapping the diversity of nature.* London: Chapman & Hall.

Moore, W. G. 1967. *A dictionary of geography.* London: A. & C. Black.

Morrison, M. L., B. G. Marcot, and R. W. Mannan. 1992. *Wildlife–habitat relationships: Concepts and applications.* Madison: University of Wisconsin Press.

Newcomer, J. A. and J. Szajgin. 1984. Accumulation of thematic map errors in digital overlay analysis. *American Cartographer* 11: 58–62.

Norton, T. W. and H. P. Possingham. 1993. Wildlife modelling for biodiversity conservation. In A. J. Jakeman, M. B. Beck, and M. J. McAleer, eds., *Modelling change in environmental systems,* 243–266. New York: Wiley.

Noss, R. S. 1992. Issues of scale in conservation biology. In P. L. Fiedler and S. K. Jain, eds., *Conservation biology: The theory and practice of nature conservation, preservation and management,* 240–250. New York: Chapman & Hall.

Odum, E. P. 1971. *Fundamentals of ecology.* Philadelphia: WB Saunders.

O'Neill, R. V., D. L. DeAngelis, J. B. Waide, and T. F. H. Allen. 1986. *A hierarchical concept of ecosystems.* Princeton, N.J.: Princeton University Press.

Palmeirin, N. 1988. Automatic mapping of avian species habitat using satellite imagery. *Oikos* 52: 59–68.

Parratt, L. G. 1961. *Probability and experimental errors in science.* New York: Wiley.

Pausas, J. G., L. W. Braithwaite, and M. P. Austin. 1995. Modelling habitat quality for arboreal marsupials in the south coastal forest of New South Wales, Australia. *Forest Ecology and Management* 78: 39–49.

Pearce, J. L., M. A. Burgman, and D. C. Franklin. 1994. Habitat selection by helmeted honeyeaters. *Wildlife Research* 21: 53–63.

Pereira, J. M. C. and L. Duckstein. 1993. A multiple-criteria decision-making approach to GIS-based land suitability evaluation. *International Journal of Geographic Information Systems* 7(5): 407–424.

Pereira, J. M. C. and R. M. Itami. 1991. GIS-based habitat modeling using logistic multiple regression: A study of the Mt. Graham red squirrel. *Photographic Engineering and Remote Sensing* 57: 1475–1486.

Peuquet, D. J. 1984. A conceptual framework and comparison of spatial data models. In D. F. Marble, H. W. Calkins, and D. J. Peuquet, eds., *Basic readings in geographic information systems,* 55–70. Williamsville, N.Y.: Spad Systems.

Peuquet, D. J. and A. R. Boyle. 1984. *Raster scanning, processing and plotting of cartographic documents.* Williamsville, N.Y.: Spad Systems.

Picozzi, N., D. C. Catt, and R. Moss. 1992. Evaluation of capercaillie habitat. *Journal of Applied Ecology* 29: 751–762.

Piersma, T., R. Hoekstra, A. Dekinga, A. Koolhaas, P. Wolf, P. Battley, and P. Wiersma. 1993. Scale and intensity of intertidal habitat use by knots *Calidirs canutus* in the western Wadden Sea in relation to food, friends and foes. *Netherlands Journal of Sea Research* 31(4): 331–357.

Rexstad, E. A., D. D. Miller, C. H. Flather, E. M. Anderson, J. W. Hupp, and D. R. Anderson. 1988. Questionable multivariate statistical inference in wildlife habitat and community studies. *Journal of Wildlife Management* 52(4): 794–798.

Richards, J. A. 1986. *Remote sensing: Digital analysis.* Berlin: Springer-Verlag.

Rosenfield, G. H., K. Fitzpatrick-Lins, and H. S. Ling. 1982. Sampling for thematic map accuracy testing. *Photographic Engineering and Remote Sensing* 48: 131–137.

Ross, R. M., R. M. Bennet, and T. W. H. Backman. 1993. Habitat use by spawning adult, egg, and larval American shad in the Delaware River. *Rivers* 4(3): 227–238.

Sanderson, G. C., E. D. Ables, R. D. Sparrowe, J. R. Grieb, L. D. Harris, and A. N. Moen. 1979. Research needs in wildlife. *Transactions of the North American Wildlife Natural Resource Conference* 44: 166–175.

Scott, J. M., B. Csuti, J. D. Jacobi, and J. E. Estes. 1987. Species richness. *BioScience* 37(11): 782–788.

Scott, J. M., F. Davis, B. Csuti, R. Noss, B. Butterfield, C. Groves, H. Anderson, S. Caicco, F. D'Erchia, C. E. Thomas Jr., J. Ulliman, and R. G. Wright. 1993. Gap analysis: A geographic approach to protection of biological diversity. *Wildlife Monographs* 123: 1–41.

Shaw, D. M. and S. F. Atkinson. 1990. An introduction to the use of geographic information systems for ornithological research. *Condor* 92: 564–570.

Skidmore, A. K. 1989a. An expert system classifies eucalypt forest types using Landsat Thematic Mapper data and a digital terrain model. *Photographic Engineering and Remote Sensing* 55: 1449–1464.

Skidmore, A. K. 1989b. Unsupervised training area selection in forests using a nonparametric distance measure and spatial information. *International Journal of Remote Sensing* 10: 133–146.

Skidmore, A. K., A. Gauld, and P. W. Walker. 1996. A comparison of GIS predictive models for mapping kangaroo habitat. *International Journal of Geographic Information Systems* 10: 441–454.

Skidmore A. K. and B. J. Turner. 1988. Forest mapping accuracies are improved using a supervised nonparametric classifier with SPOT data. *Photographic Engineering and Remote Sensing* 54: 1415–1421.

Skidmore, A. K. and B. J. Turner. 1992. Assessing map accuracy using line intersect sampling. *Photographic Engineering and Remote Sensing* 58: 1453–1457.

Sokal, R. R. and F. J. Rohlf. 1995 (3d ed.). *Biometry.* New York: W. H. Freeman.

Starfield, A. M. 1997. A pragmatic approach to modeling for wildlife management. *Journal of Wildlife Management* 61: 261–270.

Steffens, F. E. 1992. Geostatistical estimation of animal abundance in the Kruger National Park, South Africa. In A. Soares, ed., *Geostatistics Troia Vol. 2: Quantitative Geology and Geostatistics,* 887–897. Pretoria, South Africa: Klumer.

Stelfox, H. A. and G. R. Ironside. 1982. Land/wildlife integration no. 2. Proceedings of a technical workshop to discuss the incorporation of wildlife information into ecological land surveys. Ecological Land Classification Series no. 17, Lands Directorate, Environment, Canada.

Stoms, D. M. 1992. Effects of habitat map generalization in biodiversity assessment. *Photographic Engineering and Remote Sensing* 58: 1587–1591.

Stoms, D. M., F. W. Davis, and C. B. Cogan. 1992. Sensitivity of wildlife habitat models to uncertainties in GIS data. *Photographic Engineering and Remote Sensing* 58: 843–850.

Stoms, D. M. and J. E. Estes. 1993. A remote sensing research agenda for mapping and monitoring biodiversity. *International Journal of Remote Sensing* 14: 1839–1860.

Teillet, P. M. 1986. Image correction for radiometric effects in remote sensing. *International Journal of Remote Sensing* 7: 1637–1651.

Thomasma, L. E., T. D. Drummer, and R. O. Peterson. 1991. Testing habitat suitability index model for the Fisher. *Wildlife Society Bulletin* 19: 291–297.

Todd, W. J., D. G. Gehring, and J. F. Haman. 1980. Landsat wildland mapping accuracy. *Photographic Engineering and Remote Sensing* 46: 509–520.

Toll, D. L. 1984. An evaluation of simulated Thematic Mapper data and Landsat MSS data for discriminating suburban and regional land use and land cover. *Photographic Engineering and Remote Sensing* 50: 1713–1724.

Toraldo di Francia, G. 1978. *Il rifiuto.* Torino, Italy: Einaudi, Nuovo Politecnico 99.

Twery, M. J., G. A. Elmes, and C. B. Yuill. 1991. Scientific exploration with an intelligent GIS: Predicting species composition from topography. *AI Applications* 5(2): 45–53.

U.S. Fish and Wildlife Service. 1980a. Habitats as a basis for environmental assessment (HEP.). Ecological Service Manual 101. Washington, D.C.: U.S. Department of Interior, Fish and Wildlife Service, Division of Ecological Services.

U.S. Fish and Wildlife Service. 1980b. Habitat evaluation procedures (HEP). Ecological Service Manual 102. Washington, D.C.: U.S. Department of Interior, Fish and Wildlife Service, Division of Ecological Services.

U.S. Fish and Wildlife Service. 1981. Standards for the development of habitat suitability index models. Ecological Service Manual 103. Washington, D.C.: U.S. Department of Interior, Fish and Wildlife Service, Division of Ecological Services.

Van Apeldoorn, R. C., C. Celada, and W. Nieuwenhuizen. 1994. Distribution and dynamics of the red squirrel (*Sciurus vulgaris* L.) in a landscape with fragmented habitat. *Landscape Ecology* 9(3): 227–235.

Van Genderen, J. L. and B. F. Lock. 1977. Testing land-use map accuracy. *Photographic Engineering and Remote Sensing* 43: 1135–1137.

Van Genderen, J. L., B. F. Lock, and P. A. Vass. 1978. Remote sensing: Statistical testing of thematic map accuracy. *Remote Sensing of Environment* 7: 3–14.

Verbyla, D. L. and J. A. Litvaitis. 1989. Resampling methods for evaluating classification accuracy of wildlife habitat models. *Environmental Management* 13: 783–787.

Walker, J., D. L. B. Jupp, L. K. Penridge, and G. Tian. 1986. Interpretation of vegetation structure in Landsat MSS imagery: A case study in disturbed semi-arid eucalypt woodlands. Part 1. Field data analysis. *Environmental Management* 23: 19–33.

Walker, P. A. 1990. Modelling wildlife distributions using geographic information system: Kangaroos in relation to climate. *Journal of Biogeography* 17: 279–289.

Walker, P. A. and D. M. Moore. 1988. SIMPLE: An inductive modelling and mapping tool for spatially oriented data. *International Journal of Geographic Information Systems* 2(4): 347–363.

Wallin, D. O., C. C. H. Elliott, H. H. Shugart, C. J. Tucker, and F. Wilhelmi. 1992. Satellite remote sensing of breeding habitat for an African weaver-bird. *Landscape Ecology* 7(2): 87–99.

Walters, C. 1992. Trends in applied ecological modelling. In D. R. McCullough and R. H. Barrett, eds., *Wildlife 2001: Populations,* 117–122. Barking, U.K.: Elsevier.

Whittaker, R. H., S. A. Levin, and R. B. Root. 1973. Niche, habitat and ecotone. *American Naturalist* 107: 321–338.

Williams, G. L. 1988. An assessment of HEP (habitat evaluation procedures) applications to Bureau of Reclamation Projects. *Wildlife Society Bulletin* 16: 437–447.

Wright, J. K. 1942. Map makers are human: Comments on the subjective in maps. *Geographical Review* 32: 527–544.

Yapp, R. H. 1922. The concept of habitat. *Journal of Ecology* 10: 1–17.

Zonneveld, I. S. 1974. Aerial photography, remote sensing and ecology. *ITC Journal* 4: 553–560.

Zonneveld, I. S. 1989. The land unit: A fundamental concept in landscape ecology, and its applications. *Landscape Ecology* 3: 67–86.

Zonneveld, I. S. 1995. *Land ecology: An introduction to landscape ecology as a base for land evaluation, land management and conservation.* Amsterdam: SPB.